HISTORY OF TWENTIETH-CENTURY PHILOSOPHY OF SCIENCE

ERNST MACH

PIERRE DUHEM

RUDOLPH CARNAP

WILLARD VAN QUINE

WERNER HEISENBERG

KARL POPPER

THOMAS KUHN

PAUL FEYERABEND

RUSSELL HANSON

HERBERT SIMON

HISTORY OF TWENTIETH CENTURY PHILOSOPHY OF SCIENCE

THOMAS J. HICKEY

Manufactured in the United States of America.

Library of Congress Catalog Card Number: 94-96838
Dewey Decimal Classification: 190.904
Published in the United States of America by:
Thomas J. Hickey, Forest Park, Illinois.

Key Words/Subjects:

1- Pragmatism, Positivism, Romanticism;
2- Language, Semantics, Ontology, Theory;
3- Discovery, Criticism, Explanation, Aim of Science;
4- Artificial Intelligence Discovery Systems;
5- Philosophy of Physics incl. Quantum Theory;
6- Philosophy of Economics incl. Econometrics;
7- Philosophy of Sociology.

Key Words/Names:

1- Mach, 2- Duhem, 3- Carnap, 4- Quine,
5- Heisenberg, 6- Popper, 7- Kuhn, 8- Feyerabend,
9- Hanson, 10- Simon.

ISBN 0-9644665-0-3
The current mailing address for the purchase of this book can be obtained at any time by writing to:

ISBN U.S. Agency,
R.R. Bowker,
121 Chanlon Road,
New Providence, NJ 07974.

Do *not* send purchase orders to the ISBN Agency.

Agency telephone is: 908-665-6770.
Agency fax is: 908-464-3553.

PREFACE

This *History of Twentieth-Century Philosophy of Science* follows my *Introduction to Metascience: An Information Science Approach to Methodology of Scientific Research* (1976). The earlier book was originally intended to be the thesis for a Ph.D. dissertation in philosophy of science at the University of Notre Dame, Indiana. It set forth my metatheory of semantical description and my Pragmatic definition of "theory", and it also described my working computerized **METAMODEL** discovery system. Each of the several discovery systems described in the last chapter of this *History* was written by a graduate student for his Ph.D. dissertation - every one, that is, but mine. The Notre Dame philosophy faculty was more interested in my criticizing my metaphysical views than my developing my methodological views. Rather than accept their demand for my metaphysical reform, I discontinued my enrollment in their doctoral program and persisted in my unsanctioned interest.

I developed my philosophical ideas independently, and I have copyrighted my discovery system; I did not receive any of these ideas from the Notre Dame philosophy faculty, and I had never heard these ideas mentioned by any faculty member or by any graduate student, while I was enrolled there. In 1976 I sent copies of *Metascience* to members of the Notre Dame philosophy faculty, hoping that the book would have some influence, even if not acknowledged by the faculty, if only to improve their philosophy of science graduates' academic employment opportunities, especially outside the Roman Catholic parochial school system. I have never seen any book or paper by any member of the Notre Dame philosophy faculty setting forth my ideas either prior to my 1976 *Metascience* or in the decade that preceded my "Pragmatic Turn..." (1985), a copy of which I had also sent to members of the Notre Dame philosophy faculty.

Since I am not an academician, I do not have a subsidized university press to produce and distribute my publications. As a self-publishing author I was thus pleasantly surprised when my 1976 book went out of print in the mid-1980's. Not surprisingly my views have continued to evolve since 1976. Consequently I chose not to issue a reprint of *Metascience*, but instead decided to write a new book. Since some readers found the brief historical notes in the appendix of the 1976 monograph to be particularly interesting, the new book was to be in three sections with the first part consisting of a short survey history of twentieth-century philosophy of science concluding with the appearance of metascience, which I now call "Logical Pragmatism". But with the entry of Nobel laureate Herbert Simon and various other contributors into the field in the 1980's the history took on a life of its own and became this book, while the second part of the original plan, which is a shorter work setting forth a statement of principles of metascience, is becoming another book now slowly gestating in my word processor and destined to be christened with the title, *Logical Pragmatism.*

Let the reader note that this history is written with a prospective vision that controls and animates its retrospective viewpoint. I view myself as a participating contributor to twentieth-century philosophy of science rather than merely a reporting observer. I am not a professional historian; I am a philosopher of science with a philosophy and an agenda. I advocate the contemporary Pragmatist philosophy of science, and I attack its less adequate nineteenth-century predecessors, notably Positivism and Romanticism. My agenda is Logical Pragmatism, the combination of computer systems analysis and the Pragmatist philosophy of language. Contemporary Pragmatism relativizes semantics and ontology to the syntactical structures of belief systems, thereby enabling the development of computerized discovery systems, which are generative grammars that produce new empirically warranted belief systems or theories. Bertrand Russell included reference to himself and Whitehead and description of their symbolic logic in his *Wisdom of the West*, a survey history of Western philosophy from the pre-Socratics to modern times in three-hundred twenty pages. I have followed Russell's precedent. I include herein reference to myself and description of my discovery system and Logical Pragmatist agenda. And I furthermore affirm that the discovery systems and their implementation in Logical Pragmatism will be far more consequential both for philosophy of science and for the empirical sciences, than Russell's symbolic logic and the Logical Positivists' implementation of it could ever have been. Both computers and Pragmatism are here to stay. This book is for those opportunistic philosophers of science and empirical scientists, who would explore and exploit the possibilities occasioned by the propitious combination of computer technology and Pragmatism.

July 26, 1994
Oak Park, IL

CONTENTS

INTRODUCTION

The Evolution of Contemporary Pragmatism

This history of twentieth-century philosophy of science is the story of the evolution of the contemporary Pragmatist philosophy of science. The evolution, however, was not a development internal to the earlier versions of Pragmatism. The twentieth-century philosophers who developed the contemporary Pragmatism seldom referenced earlier Pragmatists or set forth their own views as a gloss on the views of Peirce, James or Dewey. They simply drew upon the common property of the Pragmatist tradition including such themes as realism, the centrality of belief, the importance of empiricism for scientific criticism.

The contemporary philosophy of science is distinguished by a new philosophy of language, in which the relation among ontology, semantics and truth is reversed relative to the relation found in earlier philosophies including earlier Pragmatisms. Underlying this reversal is the replacement of the naturalistic view of the semantics of language with the artifactual view. On the artifactual thesis the semantics of language is not uniquely determined by nature, but is determined as a linguistic artifact by the context of a set of beliefs that are accepted as empirically warranted, and so are given a definitional role. In earlier philosophies of science some chosen ontology is assumed as the true constitution of reality, and then certain types of statements are accepted as descriptive of this ontology, because the meanings of terms in this descriptive language are viewed as representations determined both by nature, where "nature" is nothing other than the ontology itself, and by the natural causal cognitive processes thought to produce the mental representations of this ontology that constitute "true" knowledge.

In the Positivist philosophy the assumed ontology has historically been called "phenomenalist", "sense datum", or "sensationalist" ontology, and the type of language having the semantics describing this ontology is called "observation language". In the sciences dominated by the Positivist philosophy such as turn-of-the-century physics and contemporary behavioristic psychology, language that does not conform to an observational ontology is rejected as "theoretical" and "meaningless", unless a reductionist connection is established to relate the theory to its observational reduction basis, thereby supposedly giving the theory a valid semantics. In the Romanticist philosophy the assumed ontology is the mental experiences of the subjects or social actors whose behavior is explained in terms of those experiences, and the type of language that has the semantics that describes the mentalistic ontology and explains the social behavior in terms of it, is called "theory". In the Romantic sciences such as sociology language that does not conform to the mentalistic ontology is rejected out of hand as "atheoretical" and "meaningless", even if empirical criteria are satisfied. The Romanticist and the Positivist philosophies share the same philosophy of language; their doctrines differ only about which type of ontology is required *a priori* to empirical testing. In the Pragmatist philosophy the difference between observation and theory is merely the distinction between what is presumed for testing, namely the test-design statements, and what is proposed for testing, namely the theory. It is a difference that depends on the use to which a statement is put in a research strategy in basic science, and is therefore a pragmatic distinction.

The scientist's freedom to choose which statements he will presume true and which he will propose for testing, is a central part of his research strategy. This freedom offers room for judgments of strategy about which scientists may differ, and to which they may be prejudicially yet productively faithful. But the freedom is not completely arbitrary. It is regulated within limits by the empirical testing constraint, the expression of which may be included in a statement of a rationality postulate setting forth the institutionalized aim of basic empirical science. Today "theory" may be used conventionally either in an almanac sense, that prevails in the Positivist and Romantic philosophies, such that we still speak of "Newton's theory" or say "observation is theory-laden", or in its contemporary Pragmatist context, as when we reference its strategic function in empirical testing in research science. In what N.R. Hanson called "research science" where the distinctively Pragmatic sense applies, there is no theory that is not current theory and no natural observation language.

On Archetypes and Architectonics

In his *Types of Economic Theory* the Columbia University institutionalist economist, Wesley C. Mitchell, says that the process that constitutes the development of the social sciences is an incessant interaction between logically arranged ideas and chronologically arranged events. This memorable institutionalist refrain may be paraphrased for describing the development of philosophies of science as well: the social process which constitutes the development of philosophy of science is an episodic interaction between logically arranged ideas in philosophy of science and chronologically arranged events in the history of the empirical sciences. And in the present context the most important chronologically arranged events in twentieth-century history of science are the two great scientific revolutions in physics: relativity theory and quantum theory with the latter having the greatest impact on the development of the contemporary Pragmatist philosophy of science. No more recent development in any empirical science has occasioned any revision to the contemporary Pragmatist philosophy of science, although future science will eventually surprise the Pragmatists.

It might also be noted that the principal theses of the contemporary Pragmatist philosophy of science were initially formulated by two of the physicists who perpetrated these two scientific revolutions, namely Albert Einstein and (even more so) Werner Heisenberg. But the situation is complex. The evolution of the contemporary Pragmatist philosophy has not been a simple linear development originating in events in physics, and it certainly has not been a development limited to academic philosophy. There are several parallel lines of development both within and without academic philosophy. These lines have occurred in philosophy of science, in history of science, in physics, and in the social sciences, and there have been several within academic philosophy. Consequently the organization of this book could not be a simple chronology. While chronological order has not been ignored, the chapter organization has instead displayed an architectonic reflecting the orientations that gave rise to the various developmental lines.

Each chapter is built around one or two central figures, each of whose architectonic is archetypal for the line of development that the chapter presents, and each contains a section summarizing the central figure's architectonic in terms of the four basic topics in philosophy of science.

Those topics are: (1) scientific discovery, (2) scientific explanation, (3) scientific criticism, and (4) the aim of science. A Pragmatist's statement of the aim of science would today be given in terms of the other three topics, but there have been other statements of the aim of science in the history of philosophy of science. Each chapter also contains an exposition and discussion of several other contributors to the orientation it describes, and a concluding section with comments. Consider next an overview of the chapters describing these several lines of development.

Overview of A Century of Progress

There are seven chapters following this introductory chapter. The second chapter of the book sets the stage, as it were, by exhibiting two archetypal Positivist philosophies from the age of Newtonian physics, those of Ernst Mach and Pierre Duhem. Both were research physicists, and both believed that their respective Positivist architectonics are sufficiently adequate, that each wrote a history of his science from the viewpoint of his philosophical interpretation. The third chapter describes the philosophies of Rudolf Carnap and Willard Van Quine, and both views may be taken as archetypal of the "analytic" developmental line. It might be said that Carnap took Mach as his point of departure, and that Quine took Duhem as his point of departure, but the principal interaction was Quine's Pragmatist criticism of Carnap's Logical Positivism, and specifically of Carnap's metatheory of semantical systems. Einstein's theory of relativity, and particularly the idea of applying a pre-existing semantically uninterpreted or "analytic" formal calculus consisting of the non-Euclidian geometry for the mathematical development of the relativity theory, was the inspiration for Carnap's metatheory of semantical systems, the central thesis of his philosophy of science. Quine criticized Carnap and took up a strategy of elevating Duhem's philosophy of physical theory into a general philosophy of language, which is based on the artifactual view of the semantics of language, by discarding the Positivist and metaphysical dimensions in Duhem's philosophy, which are based on the naturalistic view of the semantics of language. Quine attacked Carnap's Positivism at its foundations by attacking reductionism and by attacking the thesis of analytic truth, which Carnap used as the basis for a semantically independent, formal calculus in his metatheory of semantical systems. Quine also developed several of the central theses of the contemporary

Pragmatist philosophy including the theses of contextual definition, empirical underdetermination, semantical indeterminacy, and most notably his thesis of ontological relativity, the basis for his scientific realism.

The fourth chapter's archetypal view is that of Werner Heisenberg. No development in twentieth-century science has been more inspiring (or disturbing) to philosophers than the modern quantum theory. Heisenberg was one of the principal contributors to the modern quantum theory, and developed a philosophy of science for the "Copenhagen interpretation" of the quantum theory. Most of the early literature setting forth a philosophy of science for the Copenhagen interpretation was written by Bohr, and even today some old philosophers conceive the quantum theory in Bohr's terms instead of Heisenberg's. Bohr formulated an ersatz naturalistic philosophy of the semantics of terms involving "forms of perception", which philosophers have variously construed as neo-Kantian, Positivist, or operationalist. This naturalism led Bohr to construe the mathematical expressions of the quantum theory instrumentally instead of descriptively. Heisenberg was influenced by Bohr, and in his own doctrine of closed-off theories he maintained that the mathematically expressed quantum theory cannot be used as an observation language.

However, Heisenberg was also influenced by Einstein, and consequently he developed a descriptive and realistic variation of the Copenhagen interpretation, that anticipates the three fundamental theses of the contemporary Pragmatist philosophy of science. Firstly Heisenberg accepted Einstein's statement that it is the theory that decides what the physicist can observe, and then used this idea in his development of his indeterminacy principle, the discovery for which he received the Nobel prize. It is noteworthy that Heisenberg's acceptance of Einstein's view of observation is contrary to his doctrine of closed-off theories, and that Heisenberg never resolved this inconsistency in his lifetime. Secondly he referenced Einstein's realistic interpretation of the relativity theory, and then contrary to Einstein invoked this precedent for his own realistic interpretation of the quantum theory duality thesis. The duality thesis is common to both Heisenberg's and Bohr's versions of the Copenhagen interpretation of quantum theory, but the complementarity thesis is an instrumentalist interpretation that is distinctive to Bohr. Thirdly, contrary to Einstein's programmatic aim of physics Heisenberg affirmed ontological pluralism.

The fifth chapter's archetypal view is that of Karl Popper. Popper is one of the earlier post-Positivist philosophers of physics, and no history of twentieth-century philosophy of science could exclude him. Because of Einstein's influence, Popper anticipated several of the central thesis of the contemporary Pragmatism, even though he had rejected Pragmatism. Like Einstein, Heisenberg, and other post-Positivists, Popper recognizes the dependence of observation on theory and maintains that all observation is "theory-impregnated". This recognition in turn led Popper to reject explicitly the naturalistic view of the semantics of language, and he criticized Carnap's Positivism from this standpoint. He also redefined the meaning of "theory", such that it is not language based on so-called "theoretical terms"; he held that all terms are a type of theoretical term that Positivists called "disposition terms", and thus cannot distinguish theory langauge. He said that "theory" language consists of statements that are logically quantified universally, and that function to realize the aim of science, which in turn he defines in terms of testability and nonfalsifying test outcomes.

Furthermore, like Einstein, Popper was not sympathetic to the Copenhagen quantum theory duality thesis, and he developed his own interpretation of the formalism, which he called the "propensity-particle" interpretation. Popper's interpretation is consistent with Einstein's programmatic aim of physics, with the noteworthy exception that Popper rejects Einstein's determinism. Popper views indeterminism realistically, and he influenced the physicist Alfred Lande, who had previously been sympathetic to the Copenhagen interpretation, to develop his own indeterministic particle interpretation of the quantum theory. In his philosophy of quantum theory Popper does not follow the Pragmatists' philosophy of scientific realism, which subordinates ontology to *a posteriori* empirical adequacy, but instead accepts an *a priori* ontological commitment that he calls "commonsense realism". Nor does he accept the contemporary Pragmatists' thesis of ontological pluralism, but instead rejected what he calls the "schism" in physics and spent most of his career trying to reconcile this schism in Einstein's favor. His propensity-particle interpretation resulted from his effort to reconcile this schism.

The sixth chapter is an exposition of what may be called the historical orientation, and the archetypal views are those of Thomas Kuhn and Paul Feyerabend. This orientation has its origins in twentieth-century history of science in the writings of James Conant. Conant was struck by the fact that in the history of science many contributing scientists maintained strong prejudices, which led them to ignore falsifying empirical evidence for their theories until a new conceptual scheme or theory was advanced, and especially by the fact that these prejudices had in some cases been later justified. For Conant this phenomenon was nonetheless merely incidental to the development of science, but for his former assistant at Harvard, Thomas Kuhn, it assumed a strategic and functional role in the development of science. What Conant thought was fortuitous prejudice, Kuhn viewed as functional prejudice which he called consensus paradigm. The function of the consensus paradigm, which produces what Kuhn calls "normal science", is to provide a necessary precondition for later revolutionary developments in a science. Furthermore, these revolutionary developments involve a radical semantical change producing what Kuhn described as incommensurable discontinuities between successive paradigms.

Conant's observations of prejudice in the history of science and Kuhn's hypothesis of consensus paradigm and normal science are interesting illustrations of the element of voluntaristic belief that is found in the contemporary Pragmatist philosophy of language. But these two historians of science offered little linguistic philosophical analysis. The language analysis of such phenomena was left to Paul Feyerabend, the only one of the three considered in this chapter who is a professional philosopher of science. In the same year that Kuhn published his book on scientific revolutions, Feyerabend published an article in which he explicitly rejected the naturalistic causal theory of the semantics of language, and advocated his contextual theory, which is an artifactual theory of language. Using this theory of semantics and affirming the dependence of observation on theory, he attacked the Logical Positivist thesis of the reduction of theories. He maintained that semantical change or "meaning variance" occasioned by the development of a new theory affects even the observation language common to the old and new theories, thus creating equivocations that make logical reduction impossible.

Feyerabend proposed that there are cases of such developments of theory which result in semantical change in which the semantical discontinuity is so complete, that the semantics of the old and new theories are "incommensurable", a term also used by Kuhn to describe scientific revolutions. Later Feyerabend invoked the linguistic theory of Benjamin Lee Whorf to give further explication to the idea of incommensurability. And for Feyerabend this incommensurability implies not Kuhn's intermittent normal-science consensus but rather continuing pluralism in science. Thus what was seen as fortuitous incidental prejudice for Conant, and then functional consensus prejudice for Kuhn, became functional pluralistic prejudice for Feyerabend in the latter's "principle of tenacity". Prejudice thus assumes a function that is the opposite of what Kuhn had postulated. Tenacity is justified while consensus is unjustified, because the dependency of observation on theory requires the stubborn continuing existence of a multiplicity of theories for scientific criticism to be possible, and therefore for science to develop.

Feyerabend is also an advocate of realism, but in his philosophical career realism has had a uniquely problematic status. Feyerabend had long aspired to extend the realism which he calls the Galileo-Einstein tradition to microphysics. His problem was that the contemporary quantum physics, by which is meant the Copenhagen interpretation with its complementarity principle expounded by Bohr, is in Feyerabend's view a Positivist philosophy. For many years Feyerabend had placed his hopes in the development of the "hidden-variable" interpretation of the physicist, David Bohm, but no empirical evidence of such a subquantum level phenomenon has been found. Eventually Feyerabend was persuaded by Russell Hanson to accept the Copenhagen interpretation, although for Feyerabend, unlike Hanson, the Copenhagen interpretation always meant the Bohr's interpretation. Then he found a way to construe quantum theory realistically: he viewed Bohr's complementarity as a unique case of incommensurability that occurs within a theory instead of between successive theories. It is ironic that Feyerabend had never made a careful study of Heisenberg's realistic Copenhagen interpretation. Except for the doctrine of closed-off theories, Einstein's influence on Heisenberg's philosophy is nothing other than an extension of the Galileo-Einstein tradition, so that there is a Galileo-Einstein-Heisenberg tradition.

Both of the two final chapters pertain to the topic of scientific discovery. The central figure in the seventh chapter is Russell Hanson, whose principal interest was the semantics of scientific discovery. Just as Einstein told Heisenberg that it is the theory decides what the physicist can observe, so too Hanson was led by Wittgenstein to maintain that all observation is "theory-laden". But unlike Heisenberg, Hanson was not influenced by what he called the naive epistemology of Bohr including notably Bohr's complementarity principle, which requires that the duality thesis be expressed in terms of the semantics of Newtonian or "classical" physics. Because of his personal acquaintance with the physicist, P.A.M.

Dirac at Cambridge, Hanson was willing to pursue the contextual semantical thesis of observation with a consistency and conviction that had eluded Heisenberg, who retained his doctrine of closed-off theories.

Hanson was more sophisticated. He distinguished the Copenhagen interpretation from the Bohr interpretation, and defined the former as the semantical interpretation of the duality thesis of the quantum theory that is determined wholly by the theory's formalism, on the grounds that it is the formalism that supplies the intelligible pattern and conceptual organization for the observations for microphysics. This is a variant of the contextual determination of semantics. Hanson gave Bohm's hidden-variable interpretation serious but critical consideration. He defends the Copenhagen interpretation by denying that it is supplied by any philosophical or metaphysical ideas appended to its formalism, such as the supplementary language constituting the interpretations of Bohr and Bohm. The correct Copenhagen interpretation is the one that is determined contextually by the structure of the equations of the theory. This semantical thesis makes Hanson one of the most consistently Pragmatic of the twentieth-century philosophers of science. But Hanson's contribution to the topic of scientific discovery consists almost exclusively in exemplifying in the history of science how semantics operates as a constraint on discovery because of the semantical change required for the development of new theories. Notwithstanding his references to Peirce's thesis of "abduction", Hanson offered no explanation of how semantics might be used to effect scientific discovery, or of how it operates in the development of new theories in any manner except as an impediment. This chapter also considers two other contributors to the topic of the semantics of scientific discovery, the Cambrian philosopher Mary Hesse and the physicist David Bohm, who explore the use of semantics in figures of speech, notably analogy and metaphor, in attempts to develop explanations of theory development as semantical processes.

The archetypal view in the eighth and last chapter is that of Herbert Simon, and since he is one of the founding fathers of artificial intelligence, this chapter may be said to describe the artificial-intelligence developmental line and more specifically his discovery-systems orientation toward the topic of scientific discovery. This chapter is not likely to be found in a typically academic history of twentieth-century philosophy of science. One reason is that curricula in academic philosophy do not include a systems orientation. Graduate-level philosophy students still routinely take course work in the symbolic logic even though rejection of the Logical Positivism has long ago deprived symbolic logic of its relevance for philosophy of science, if not also for everything else as well. But it would be a rare philosophy student whose academic curriculum in philosophy included course work in computer systems analysis.

A second reason is that much of the development of the discovery systems has been in the social sciences. Academic philosophers of science have tended justifiably to disregard the social sciences. Neoclassical economists, neo-Parsonian sociologists, and sociologists who naively seek to imitate the neoclassical econometricians even now at the close of the twentieth century, dogmatically follow the Romanticist philosophy of science much as the physicists in the opening of the century dogmatically followed the Positivist philosophy. In other words the social sciences are a kind backwater, in which the philosophy of science that is practiced and enforced is anachronistic by about a century. Both the social science profession and the national macrosocieties that have optimistically applied Romantic social theory have paid dearly for this philosophical retardation.

At the same time the principal arguments against the Romantics are essentially the same as those that the contemporary Pragmatists had leveled against the Positivists. Consequently, while Romantic social scientists would undoubtedly benefit greatly from elementary instruction in the contemporary Pragmatist philosophy of science, professional philosophy on the other hand would seem to have little occasion to develop further as a result of any consideration of the backward social sciences. However, the development of the discovery systems in the social sciences introduces a new perspective both to philosophy and to the social sciences. This novelty may be called "Logical Pragmatism". Simon started his career with a Ph.D. in political science, and then turned to economics and years later received the Nobel prize for economics. But his principal interest since the 1950's has been artificial intelligence and specifically the topic of human problem solving. His interest in this area did not extend to the topic of scientific discovery until the late 1970's. Simon's colleagues and students at Carnegie-Mellon University applied his theory of human problem solving by heuristic search, in order to design and implement computer systems that mechanically simulate the discoveries of numerous theories in the history of physics, chemistry, and biology. Simon's work in economics resulted in his thesis of bounded rationality, which he proposes as an alternative to the global optimization thesis set forth in the neoclassical rationality postulate that has prevailed in economics since Adam Smith. His basis for his thesis of bounded rationality is the computational limitations of the human mind, a limitation that he found to be strategic in his theory of human problem solving, and that serves as the basis for his theory of heuristic search in both intuitive and mechanized problem solving. He also recognizes other reasons for the bounds imposed on rationality, and these reasons are the same as those of the Pragmatist Institutionalist school of economics, which is the principal alternative in the United States to the prevailing Romantic neoclassical school of economics.

The principal contemporary alternative to Simon's thesis of bounded rationality is the rational expectations hypothesis, which extends the global optimization in the neoclassical rationality postulate to include expectations by the economic actors whose behavior is described by economic theory. As it happens, the rational expectations advocates have also developed a new type of econometric model known as the Bayesian vector autoregression or "**BVAR**" type of model, and this chapter considers both this type of model and the computer procedures used for its development. The computerized procedures are viewed as a kind of discovery system, even though its inventor, Robert Litterman formerly of the Federal Reserve Bank of Minneapolis, has not referred to it in such terms. One noteworthy aspect of the new kind of model is that unlike Haavelmo's structural-equation type of econometric model which is still used in neoclassical economics, the new type relies almost completely on data analysis for specifying the model's structure. The result is a philosophically ambiguity. On the one hand a social scientist subscribing to the Romantic philosophy, such as either the neoclassical or the rational expectations economist, would be expected to require that the selection of variables for his model be justified by economic theory with its rationality postulate. On the other hand, the rational expectations advocate maintains that expectations are formed by consideration of all economic factors, such that he needs no theoretical justification for the specification of his models, and he therefore relies almost exclusively on data analysis for the choice of variables for the model. Ironically the outcome is that reliance on data analysis has in practice made Romanticism an irrelevant rhetoric, while the **BVAR** modeling has made economics as Pragmatic as any other contemporary empirical science, whether natural or social science.

Two other contributing figures are also considered. Both have developed discovery system designs that are based on combinatorial procedures. The more recent of these is Hickey's **METAMODEL** system, and the earlier one is the **AID** (Automated Interaction Detection) system developed in the 1960's by the sociologist, John Sonquist. Combinatorial or factorial design systems are very extravagant in their consumption of computer resources, although they are very thorough search strategies. Simon's theory of heuristic search is a strategy for mechanized discovery, that is more efficient than combinatorial procedures. However, as the advance of computer technologies such as parallelism and the RISC technology have increased the processing speeds of the machines, the combinatorial system design has become more plausible, and the tradeoff between efficiency and thoroughness in solution search has been moved in the latter direction. Both the **METAMODEL** and the **AID** systems have been applied in sociology as well as in economics, and both have seen successful applications in government and/or commerce over the years. However, since these systems are data-analysis techniques, they have not been welcomed by the academic sociologists, whose philosophy of science has been dominated by Romantic dogmatism. In this chapter there is a review of Hickey's correspondence with the editors of four sociological journals, all of which rejected his submitted paper describing an application of his **METAMODEL** system to macrosociology. The reasons for rejection stated by the referees of three of the journals contain no contrary empirical evidence, and except for the critics' technical incompetence in modeling, the reasons are exclusively philosophical objections based in the Romanticist philosophy of science. The editor of the fourth sociological journal refused to communicate any reasons for rejection, except to report that the one and only critic he consulted "threw up his hands". Sonquist, like Simon, consciously attempts to simulate the decisions that the research analyst makes in the discovery process. Unlike Simon, however, Sonquist considers the discovery process in the social sciences rather than in the natural sciences, and furthermore he does not replicate historical discoveries, but instead applies his **AID** system to contemporary social science. The **AID** system is designed for analysis of survey research data, which is usually compatible with the Romanticist mentalistic ontology. But Sonquist's concept of scientific discovery does not contain the *verstehen* criterion for scientific criticism; he identifies the concept of causality with the occurrence of invariance found in data by the discovery system.

ERNST MACH AND PIERRE DUHEM ON PHYSICAL THEORY

Ernst Mach (1838-1916) is the representative figure of the early Positivist philosophy of science in physics during the turn of the century. He earned a doctorate in physics from the University of Vienna in 1860. He taught experimental physics for most of his career at the University of Prague (1867-1895), and then held the chair of Inductive Philosophy at the University of Vienna (1895-1901). He set himself the philosophical task of implementing the phenomenalist philosophy of David Hume in physics while Newtonian mechanics still prevailed, and he proposed contributions to physics, psychology, and education, as well as to philosophy.

Pierre Duhem (1861-1916) studied physics at the Ecole Normale in Paris. He received a doctorate in physics, and for most of his career was a professor of physics at the University of Bordeaux. His most significant contributions to physics were in physical chemistry, where he aspired to recast the theoretical foundations of chemical processes on the basis of a generalized thermodynamics. Contrary to Mach, he accepted the Aristotelian metaphysics, which he viewed as separate from Positivist physics, and he believed that progress in physical theory asymptotically approaches a "natural classification", which he viewed as analogous to the cosmology of Aristotle. Duhem departed from Mach by accepting physical theory in physics, and he developed a semantical metatheory to locate theory in Positivist physics. It is his semantical metatheory in his variation of Positivism that retrospection reveals as his more lasting philosophical contribution.

Mach's Phenomenalism

Prior to the contemporary Pragmatist philosophy philosophers of science developed their philosophies of science from one or another metaphysical point of departure. Though the Positivists philosophers including Mach were explicitly "antimetaphysical" (Mach even denied that he was a philosopher), they were actually advocating their own metaphysics while calling that which they opposed "metaphysical" using the term in a pejorative sense. Positivism is a philosophy that evolved in reaction against the various Romantic philosophies, and what the Positivists meant by the "metaphysics" they opposed was the metaphysics of the Romantics. Just as the views of the Romantics evolved out of the tradition of the Rationalists, so too those of the Positivists evolved out of the tradition of the Empiricists. Thus Mach's epistemology is very similar to the views of the Empiricists Berkeley and Hume, and he expressed indebtedness to them in his works. Mach's principal work setting forth his phenomenalist philosophy is his *Analysis of Sensations* (1885), which went through five editions in both German and English, although Mach also discussed his epistemological views in many of his other works. His epistemology postulates "elements" such as individual sounds, temperatures, pressures, spaces, times and colors. When these elements are considered in relation to one another, they are studied by the physical sciences, and when they are considered in relation to the human mind or rather the nervous system of the human body, they are called "sensations" and are studied by psychology. One of the central theses of Mach's *Analysis of Sensations* is that the only difference between elements and sensations is the aspect under which they are viewed, and that physics and psychology therefore have the same subject matter. The distinction between the physical and the psychical is entirely a matter of convenience or practicality, because everything is merely a function of these elements, everything is a mental construct consisting of complexes of sensations. All material things including our own bodies and even the ego, are nothing but complexes of elements that have been constructed by the human mind, because these complexes have some fixedness or constancy in sense experience. The world is not such that bodies produce sensations, but rather that complexes of sensations are associated together by the human mind to produce material bodies. Ultimately all that is valuable in science is the discovery of functional relations of dependency of sensations upon one another. The constancies that enable our mental construction of physical bodies have no privileged reality status. This is even more so with such mental constructs as the physicists' molecules and atoms; atoms and molecules are mental constructs, which unlike those of physical bodies are not found in experience. The Positivist philosophy is conventionally viewed as a nonrealist metaphysics, and if it is inaccurately said to contain an ontology, the ontology consists merely of the phenomenal elements.

Mach's Philosophy of Science

Mach addresses all the four basic topics conventionally considered in a philosophy of science: explanation, criticism, discovery, and the aim of science. He offers several statements of the aim of science. One sets forth the "biological task of science", which is to give the fully developed human individual with as perfect a means of orienting himself as possible. In another statement he says that the aim of all science is the representation of facts in thought either for practical purposes or for removing intellectual discomfort, since every practical and intellectual need is satisfied when our thoughts can represent the facts of the senses completely. He adds that our knowledge of a phenomenon of nature is as complete as possible, when thoughts are set before the mind's eye such that all the relevant sensible facts can be regarded as a substitute for the phenomenon itself. Then the facts appear as familiar and are not able to occasion any surprise. In a third statement he says that the goal of science is the simplest and most economical abstract expression of facts. The noted economy of science involves uncompleted facts, judgments or laws. The last two statements of the aim of science are essentially the same as Mach's theory of scientific explanation, and do not represent different aims for science. But unlike the first, the second and third statements give an aim that is intrinsic to science. Mach set forth his theory of scientific explanation in many places including his *Analysis of Sensations* and in his "The Economical Nature of Physical Inquiry" (1882) and "On the Principle of Comparison in Physics" (1894) reprinted in his *Popular Scientific Lectures* (1898). Explanation is the economical description of experience in terms of elements. Mach says that when we examine facts for the first time they appear confusing. In time we discover simple stable elements out of which we can mentally construct the entire factual domain, and when we have reached the point where everywhere we can discuss the same facts with other persons, then we no longer feel lost and the phenomenon is explained. The explanation offers a survey of a given domain of facts with the least expenditure of thought. The representation all the facts of a domain by some one single mental process is economical. The greatest perfection in mental economy occurs when science uses mathematics.

Not all descriptions are explanations. Mach distinguishes direct and indirect descriptions, not to be confused with Mach's distinction between complete and incomplete descriptions. Incomplete description refers to what is presently unobserved but observable, and what is associated by a law, as for example the movement of a comet that is presently unobserved or the body of a man who disappears behind a pillar. The description is completed by the human mind by means of the associations made by a scientific law. Direct descriptions are truly explanations, while theories on the other hand are indirect descriptions and are not explanations. A direct description is one in which a single feature of resemblance among facts is called from memory, while a theory such as the description of light as a wave motion is an appeal to another description that had previously been made elsewhere. A theoretical idea offers more than what we actually observe in a new fact. It can be used to extend a fact and enrich it with features, which we are firstly induced to seek from its suggestions and, which are often actually are found. But the adoption of a theory always carries a danger: the theory may lead to discoveries, but even the most fruitful theory may be an obstacle to inquiry; e.g. the theory that light is an undifferentiated straight line of particles impeded the discovery of the periodicity of light. The ideal of a given domain of facts is direct description; such description accomplishes all that the scientific investigator could wish.

In the *Analysis of Sensations* Mach states that he has taken Hume as his starting point, and this starting point is reflected in his views on scientific criticism. The scientist like everyone else knows the elements with complete certainty as sensations. But scientists and other persons also make judgments that are laws or generalizations; as the aim of science is the adaptation of thoughts to facts, a new fact may require a new adaptation, which finds its expression in the operation of judgment. A judgment is the supplementing of a sensational presentation in order to represent more completely a sensational fact. In the adaptation of thoughts to facts the adaptation can be made only to what is constant in the facts; it is only the mental construction of constant elements that can yield economy. But our confidence in the constancy in our judgments or generalizations rest entirely on the supposition, which in a given case has been substantiated by numerous trials, that our mental adaptation is sufficient. And we must be prepared to find this supposition contradicted at any moment. Therefore empirical laws as well as theories are provisional in Mach's view, but for different reasons. The empirical generalizations are provisional, because they impute constancies to an infinite number of individual occurrences of sensations while only a limited number have actually been experienced. On the other hand theories postulate things that have never been experienced; no one for example has ever (in Mach's time) actually seen atoms or molecules nor has anyone ever experienced Newtonian absolute space or absolute time. Mach did not seem to find the provisional status of empirical laws to be very disturbing and in fact he considered them to be necessary for science to have its economy. But he considered the provisional status of theories to be an unsatisfactory expediency for science. His theory of scientific criticism includes an phenomenalist criterion that rejects theories. Initially the Logical Positivists who followed Mach were reluctant to accept Hume's skeptical views on scientific criticism, and instead accepted the idea of "verification", the view the scientific laws or empirical generalizations can be established in some permanent sense, an idea that historically had been definitive of truly scientific knowledge. Then as Carnap and the Logical Positivists moved away from Mach's phenomenalism, they also moved toward his acceptance of scientific laws as provisionally true instead of permanently true.

Mach's philosophy of scientific discovery is set forth in his "The Part Played by Accident in Invention and Discovery" (1895) in his *Popular Scientific Lectures*. In this paper Mach notes the importance of accident in invention and discovery, but maintains that the inventor is not passive. In fact Mach compares the discoverer to the artist. No man should consider attempting to solve a great problem unless he has thoroughly saturated his mind with the subject and everything else recedes into relative insignificance. Then the discoverer can detect the uncommon features in an accidental occurrence and their determining conditions. Mach believed that it is the idea that dominates the thinking of the inquirer and not vice versa. The movement of thought obeys the laws of association, and in a mind rich with experience every sensation is connected with so many others that the course of thought is easily influenced by apparently insignificant circumstances, the accidental occurrence of which turn out to be decisive.

Therefore there is a process of discovery, and Mach considered how this process can be guided. He rejected a combinatorial approach as too laborious and extensive. The man of genius in Mach's view consciously or unconsciously pursues systematic methods, and in his deliberate presentiment he omits many alternatives and abandons others after hasty trial, alternatives on which less endowed minds would squander their energies. From the abundance of fancies that a free and active imagination produces, there emerges one particular configuration which fits perfectly with a basic design or idea. Mach does not elaborate further upon this process; although he believes that it may be guided, he does not propose any procedure. Perhaps he could go no further in this investigation, because he also believed in gestalt qualities and accepted a wholistic view of complexes of sense impressions. In any event his belief that the process can be guided leads him to conclude that genius may be regarded as only a small deviation from the average mental endowment. He states that the way to discovery must be prepared long beforehand, and that in due course the truth will make it appearance as if by divine necessity, apparently therefore rejecting the heroic theory of invention.

Mach's History of Mechanics

Mach's most popular work was his *Science of Mechanics: A Critical and Historical Account of Its Development* (1883), also known as *The History of Mechanics*. This book went through nine editions both in German and in English; seven of the editions were published in Mach's lifetime. The physicists whose works Mach examined were not phenomenalists, and he set out to write a critical history of mechanics from the perspective of his own philosophy of science. As he stated in the introduction to the first edition, the book's purpose is to clarify ideas, reveal the real significance of the matter, and to purge it of its metaphysics. With this aim in mind Mach criticized the contributors of the past as he salvaged and reconstructed what he found to be of lasting value. Even the achievements of the great Isaac Newton did not escape his phenomenalist criticism unscathed; Mach criticized Newton's principle of reaction, his concept of mass, and his concepts of absolute space and absolute time. Starting from his view that all phenomena are related, Mach concluded contrary to Newton that all masses, all velocities, and all forces are relative, a thesis known as Mach's phenomenalistic relativity; and he proposes his own set of definitions and empirical propositions to replace Newton's. The outcome of this effort was to have a large impact on the history of physics, that would be very different from what Mach expected or wanted.

There are two lines of criticism of Mach's rejection of theory in physics. Duhem's was one, and it was basically philosophical in nature. It involves a new philosophy of language, and was eventually taken up into

Pragmatic philosophy of language of Willard Van Quine, whose philosophy is taken up in the next chapter. The second line of criticism evolved within physics, and evolved because of the two great scientific revolutions in physics, the relativity and quantum theories. It was eventually taken up into the Pragmatic philosophy of science of the philosopher Russell Hanson. This line of development is examined in greater detail in later chapters, but since both Einstein and Heisenberg were initially Positivists and were led to reject Positivism by reflection on their own work in physics, Mach's influence on them is also considered briefly in this chapter. Consider firstly Duhem's philosophy of science and his distinctive semantical metatheory of physical theory.

Duhem on Physical Theory and Metaphysics

Duhem was influenced by Mach, and he called his own philosophy of science Positivist. But there were other intellectual influences in his thought, and as a result Duhem differed from Mach in at lease two important respects: firstly Duhem accepted scientific theory as a valid part of science, and secondly he reserved a place in human knowledge for metaphysics. Mach's philosophy is often called "scientistic", by which is meant the view that only science offers valid knowledge and that no nonphenomenalist discourse, which is summarily called "metaphysical", is valid. While Mach was a physicist, philosopher, historian of science, and atheist, Duhem was a physicist, philosopher, historian of science and believing Roman Catholic. Like Mach, Duhem rejected the mechanistic, atomistic physics although for different reasons than Mach; but unlike Mach, Duhem wished to retain the natural philosophy and cosmology of the Aristotelian and Scholastic philosophies upon which had been built the theology of his religion since Thomas Aquinas.

The outcome of these differences between Mach and Duhem is a complex philosophy of science, that affirms and protects the autonomy of physics from any encroachment by metaphysics, while conversely affirming and protecting the autonomy of metaphysics from any encroachment by physics. This isolation of physics and metaphysics from each other is due to Duhem's view that metaphysics and natural philosophy or cosmology on the one hand pertain to realities that are hidden and that underlie the phenomenal appearances accessible by the senses, while physics on the other hand pertains to observed phenomena. Furthermore and contrary to

Mach, Duhem maintained that theories are integral to physics. Unlike the phenomenal descriptions in physics, the only criterion that may operate in the scientific criticism of theories is their ability to make predictions that are correct with a sufficient degree of approximation, that is, correct within the range of indeterminacy produced by some degree of measurement error that always exists in experimental data. Thus when Duhem rejected mechanism, one reason that he gave is that no mechanical atomic theory has been found to be sufficiently accurate, when judged by this purely scientific criterion for the criticism of theories. But his more fundamental reason for this thesis is that the autonomy of physical theory is protected from the metaphysical thesis that physics must be mechanistic, is that theory in physics has a special semantics that forbids interpreting the hypothetical postulates realistically, even if a mechanistic hypothesis were proposed that were scientifically adequate. Physical theory in Duhem's view can never be given a realistic semantics. No metaphysical or cosmological philosophy may be called upon to supply theoretical physics with its axioms. For this reason Duhem denies that physical theory has any explanatory function in science; only metaphysics is able to explain, and metaphysics has no place in physics. The distinctive semantics of physical theory is a very strategic part of Duhem's philosophy of science. His religious and other intellectual influences may have operated in his developing this distinctive philosophy of science, but his stratifying the semantics of the language of science into the realistic and the nonrealist, has as its basis reasons that are entirely integral to his concept of empirical science itself. These reasons must be examined before attempting any exposition of his philosophy of science.

Duhem's Stratified Semantics for Physics

As mentioned above, the second respect in which Duhem differs from Mach is the former's views on physical theory, and the difference is the most distinctive aspect of Duhem's philosophy of science. Mach had rejected theory as "metaphysical", meaning nonphenomenalist, and he maintained that ultimately in the ideal state of science all theory would be eliminated from science. Duhem's alternative view is set forth in his *Aim and Structure of Physical Theory* (1906). In this work as well in other works he not only recognized a valid metaphysics distinct from science, but also considered theory to be characteristic of science in its highest state of development. Over and above the economy that Mach saw in the empirical laws of science, Duhem furthermore saw an additional economy offered by theory. Theory is a hypothetical axiomatized system of equations, that orders the multiplicity of experimental laws by means of a symbolic structure, which is not identical with the empirical laws but which "represents" them in a parallel language.

This symbolic structure, the axiomatized mathematical system which constitutes the theory, is a distinctive language in science relative to all the other language of science including the realistic semantics of common discourse and the nonmathematical generalizations of descriptive sciences such as physiology, and even to the phenomenalist semantics of mathematically expressed empirical laws of science such as Kepler's laws. This language, which constitutes theory, is distinctive from all nontheory language, because the nontheory language has a semantics which describes either the phenomenal or real world, while the language of theory does not have these semantics. The semantics of theory language is called "symbolic", which means that its meaning is a sign of the meanings of the nontheory language. Thus the semantics of science in Duhem's philosophy is stratified into two levels, in which one represents to the other.

The basis for Duhem's distinguishing the semantics of theory language from that of all other language, is the existence of a numerical indeterminacy caused by the fact that measurements, which may occur in the equations of theory, are always approximate. There are two reasons for the indeterminacy between the equations of theory and the nontheoretical language. The first reason is simply the approximate character of all measurements. When measurements are made, a "translation" must also be made from what Duhem called a practical fact to a theoretical fact. The practical fact describes the observed phenomena and circumstances of the experiment; the theoretical fact is the set of mathematical data that replaces the practical fact in the equations of the theory. Duhem calls the method of measurement the dictionary that enables the physicist to make this translation, and he thus seems to have anticipated the later Positivists' idea of operationalist definitions.

For any practical fact there is always an infinity of potential theoretical facts, even though the degree of indeterminacy is reduced with improved instruments and measurement procedures. So long as the one or several equations of a theory are correct, the numbers that are the solution set for the equations will fall within the range of measurement indeterminacy. Duhem illustrates the semantical duality caused by this numeric indeterminacy in his discussion of the different meanings of the phrase "free fall". There is one meaning that is contained in a phenomenal description given by any person who knows nothing about physical theory. And there is a second meaning that occurs in the physical theory that includes the idea of uniform acceleration. These are two distinct meanings; the former may be either a realist or phenomenalist meaning, while the latter is called the symbolic meaning. The latter is a sign of the former, so long as the theory is accurate enough to be accepted as true.

However, the numerical indeterminacy that occasions the semantical distinction between practical facts and theoretical facts is not unique to the variables occurring in the equations of theories, the equations that are the conclusions drawn from the hypotheses which are the postulates of the theory; it also occurs in the variables occurring in the equations of empirical laws, the equations that are developed by experimental or other observational judgments. This creates another occasion for numerical indeterminacy, one which exists between the values of the variables in the equations of theory and the values of the corresponding variables in the equations of the empirical laws that a theory orders. Duhem discusses this numerical indeterminacy and the semantical duality to which it gives rise, when he criticizes Newton's claim that his theory of gravitation is not based on hypotheses. The basic question is whether or not Newton's theory was or could be developed empirically by generalizing from Kepler's laws. Duhem argues that Newton had actually created hypotheses, because the mathematical deduction from these hypotheses produces conclusions that formally contradict Kepler's observational laws. In other words the solution set for the empirical law and that for the theory are not the same. But Kepler's laws are approximate, and therefore admit to an infinity of small deviations. The measurements by Tycho Brahe permit the theorist to choose a variation of Kepler's laws which is also produced by deduction from Newton's theory. Just as there must be a translation from practical facts to theoretical facts resolving the indeterminacy in measurements, so too there must be a translation from empirical laws such as Kepler's laws to "symbolic" laws such as Newton's dynamics. Here again the numeric indeterminacy causes a semantic duality, and a translation is made in which the new symbolic formulas that are derived from Newton's hypotheses are substituted for the old realistic formulas which are Kepler's observational laws.

Having shown that there is are different semantics for theory and nontheory language in science, Duhem then gives two ways in which the meanings of the symbols in theory language differ from the meanings in all the other language of science. The first way, which is most important to him, is that the semantics of theory language is neither realistic nor phenomenalist; it does not describe the world of phenomena as does the semantics of empirical laws, nor the real world as does the semantics of common sense discourse. When Duhem states, therefore, that theories represent laws, he means to be taken literally; he means that theories do not represent the world but instead represent the empirical statements which in turn represent the phenomenal world. He is not a conventionalist or an instrumentalist in the sense that he denies that theory language has any semantics. He has stratified the semantics of science such that theory has its own higher level semantics.

He also states that when a theory agrees with experimental laws to the degree of approximation corresponding to the measuring procedures employed, and furthermore when the theory predicts the outcome of an experiment before the outcome has occurred, then there is reason to believe that the theory is not merely an economical representation of the experimental laws, but is also a natural classification of these laws in which the logical order in which the theory organizes the experimental laws is a reflection of the ontological order considered by metaphysicians, and that underlies the phenomenal order considered by physicists. However, professionally the physicist cannot pass judgment on this analogical apprehension of the underlying ontological order that is the proper subject only of metaphysics or natural philosophy.

The second way in which the meanings of the symbols in theory language differ from those in the other language of science, is that the meanings of theory are determined by their context, by the statements that constitute the theory itself. Therefore, according to whether the physicist adopts one or another theory, the very words that occur in the symbolic law change their meaning, so that the law may be accepted by one physicist who admits one theory while it may be rejected by another physicist who admits an alternative theory. Duhem illustrates this contextual determination of meaning in theory language in his discussion of Kepler's observational laws and the symbolic laws of Newton's theory. The formulas that constitute Kepler's laws refer to orbits, but when they are replaced by the symbolic formulas that are deduced from Newton's dynamics, the symbolic law contains variables referring to forces and masses also. The translation from

Kepler's laws into symbolic laws presupposes the physicist's prior adherence to the hypotheses of the theory. The contextual determination of the meanings of theories is Duhem's wholistic concept of theory, a concept that is strategic to his views about scientific criticism of theories. In his wholistic view theoretical physics is not like a machine but is more like an organism.

Finally it should be noted that although the higher level semantics of theory language is relatively remote from the phenomena described by the semantics of the nontheory language, nevertheless theory is not remote from the experimental situation. He states that an experiment in physics is not simply the observation of a phenomenon, but is furthermore the theoretical interpretation of it. And this theoretical interpretation is not just a technical language, but one that makes possible the use of instruments. He illustrates this distinction between observation and interpretation in physical experiment by offering two descriptions of some experimental apparatus in a laboratory, one description given in the vocabulary of the physicist who understands the theory of electricity, and the other description given in the observational language of the observer innocent of such theoretical understanding. The experimenting physicist actually has two distinct representations of the instrument in his mind. One is the phenomenal image of the concrete instrument that he manipulates in reality. The other is a schematic model of the same instrument constructed mentally with the aid of the symbols from the theories that the physicist accepts. Without knowing the theories which the physicist regards as established and which he uses for interpreting the facts he observes, it is impossible for anyone to understand the meaning he gives to his statements. And when a physicist discusses his experiments with another physicist, who accepts an alternative theory, it is necessary for the two physicists to seek to establish a correspondence between their ideas of the two physicists, and then to reinterpret the experiment. Twenty years before the development of the quantum theory Duhem cited as an example the two alternative theories of light: Newton's emission theory and Frensel's wave theory. He maintained that the observations and experiments interpreted in the concepts of one theory can be translated into the concepts of the other theory. In his philosophy this is possible not because he anticipated quantum theory, but because the two theories share a theory-neutral phenomenal domain.

Duhem's stratification of the semantics of the language of theoretical science is central and strategic to his philosophy of science. It is not surprising that he stated that the approximate fit between measurements and theory creates a semantical difference, although it might seem more correct were he to have said that the resolution of the indeterminacy in measurement by the calculated value for a variable in a theory actually resolves a semantic vagueness instead of saying, as he does, that it creates two distinct meanings. But it is surprising to find him concluding that the distinct meaning of the symbol in the theory is a "sign" of the phenomenal meaning defined by the experimental measurement method. It is this latter position that stratifies the semantics of science, so that theory cannot be given a realistic or phenomenalistic interpretation. Nonetheless Duhem has a reason for taking this position. In his "The Physics of a Believer", an appendix to *Aim and Structure of Physical Theory*, he reports that earlier in his career after attempting unsuccessfully to conform to Newton's methods set forth in the "General Scholium", he concluded that physical theory is neither a metaphysical explanation nor a set of general laws, whose validity is established, but rather that theory is an artificial construction manufactured with the aid of mathematical magnitudes, and that the relations of the magnitudes to the abstract notion emergent from experiment, is that of sign to thing signified. The key concept seems to be the idea of artificial construction. The artificial nature of theory gives it an artificial semantics, if it has any semantics at all, and this artificial semantics is of a different kind than the natural semantics of language that describes the phenomenal world which is referenced to establish the law's validity.

Throughout most of the history of philosophy, philosophers believed that while the multiplicity of languages argues for the existence of a conventional aspect in human language, still, as Aristotle said, while men speak different languages, they have the same cognitive experiences. This is the thesis of a naturalistic semantics; all men have the same cognitive experience when in the presence of the same reality, because there is a natural relation between knowledge and reality. A variation of this thesis was also maintained by Mach with his theory of sensations and their identification with elements of the phenomenal world. But Duhem could not fit this thesis to the language of physical theory, even while he like Mach maintained it for the language of observation. He viewed physical theory as so artifactual, that its meaning could not be natural but had to be artificial. Therefore the language of theory does not describe either the real or the phenomenal world, the world of nature. At the same time he was not led to conclude that theory is meaningless either in the sense of purely conventional or in the sense of metaphysical. Thus his reconciliation strategy was to make the artificial semantics of theory language describe or represent the language of science, which is not a phenomenon of nature but rather is an artifact.

Duhem's Philosophy of Science

Duhem's statement of the aim of science is similar to Mach`s: the aim of science is economy of thought. Like Mach, Duhem believes that experimental laws contribute an intellectual economy, because they summarize a large number of concrete facts. But unlike Mach, Duhem furthermore says that theories also contribute to the realization of the aim of science. The economy achieved by the substitution of a law for concrete facts is redoubled for the mind, when the mind substitutes theories for the numerous experimental laws. A theory is a system of mathematical propositions deduced from a small number of principles, which aim to represent as simply, as completely, and as exactly as possible a set of experimental laws; its aim in other words is economy of thought by schematically representing and logically organizing experimental laws. But theories do not explain the laws nor do the laws explain the facts. Explanation is proper only to metaphysics and not to science. Duhem developed a sophisticated theory of scientific criticism, and it is central to his philosophy of science. He is very emphatic in defending the autonomy of empirical science from any encroachment by metaphysics or natural philosophy. Metaphysics pertains to realities that underlie the phenomenal appearances that are hidden by the phenomena, while science pertains to these appearances. Consequently whatever may be the criteria and procedures for criticizing a metaphysical thesis, they are not relevant to empirical science. In empirical sciences that are nonmathematical, the generalizations such as "all men are mortal" may be accepted or rejected as simply true or false. But in mathematical physics the equations both of the empirical laws and of the hypothetical theories are not simply regarded as true or false, but are approximate. The amount of indeterminacy due to the approximate nature of the values of the variables in these equations, will be reduced as experimental and measurement techniques improve. These improvements occur because instruments are improved; and because instruments depend on physical theory, the improvement of instruments occurs because of the improvement in theory. As the range of this indeterminacy becomes smaller, the equations of either the empirical laws or the hypothetical theories that represent the laws may no longer be able to predict values for their variables which fall within the smaller range. When this happens the equations are no longer satisfactory. Duhem is very emphatic in his thesis that the only criterion that may validly operate in scientific criticism is the ability of the law or theory to make accurate predictions. This exclusion of all ontological or metaphysical criteria from scientific criticism has been carried forward into the contemporary Pragmatist philosophy of science. In his theory of scientific criticism Duhem rejected the use of so-called crucial experiments as a means of establishing the validity of a theory. His thesis is that if the physicist is confronted with several alternative theories, the rejection of all but one cannot imply the establishment of the one remaining. As an example he cites the two alternative theories of light: one theory is the hypothesis that light is a stream of high speed projectiles, and the other is the hypothesis that light consists of vibrations whose waves are propagated in an ether. This is not an anticipation of the Copenhagen duality thesis; Duhem is thinking of the wave and particle theories as alternative theories. His position is that the choice is not mutually exclusive, because no one can ever enumerate completely all of the various hypotheses, which may pertain to a determinate group of phenomena. He thus maintains that several alternative theories may fall within the range of indeterminacy of the measurement data and experimental laws, so that more than one theory may be satisfactory. This represents a pluralistic thesis about science, and in the crucial experiment discussion, it means that even if all hypotheses could somehow be enumerated, elimination could not leave but one to be considered as established. More generally his pluralism means that the indeterminacy in measurement, laws, and theories produces an indeterminacy in scientific criticism. This pluralism is another aspect of his philosophy of physical theory that has been carried forward into the contemporary Pragmatist philosophy of science.

His theory of scientific criticism also reflects his wholistic or organic view of theories. This wholistic view not only makes the meanings of the mathematical symbols determined by the context consisting of the equations of the theory, but also makes it necessary to test the theory as a whole. Furthermore the test involves all the theories used in the experiment and in the measuring instruments. Thus if the prediction in the test is wrong, not only may the proposition being tested be at fault, but so may be the whole theoretical scaffolding used by the physicist. The only thing that the experiment reveals is that among all the theoretical propositions used to predict the phenomenon, there is at least one error. But the failure of the prediction does not inform the physicist where the error lies. The physicist can never subject an isolated hypothesis to experimental test, but only a whole group of hypotheses. And a failure does not reveal which hypothesis should be modified. In Duhem's view physics is not like a machine which lets itself be disassembled; the physicist cannot test each piece in isolation and then make adjustments to the isolated part found wanting. He compares physics to an organism in which one part cannot be made to function except when the parts that are most remote from it are called into play. If there is a malfunction felt in the organism, the physicist must ferret out through its effects on the entire system, which organ needs to be remedied or modified without the possibility of isolating the organ and examining it apart. Duhem says that the physicist confronted with a failed prediction is more like a physician than a watchmaker.

Duhem also has a philosophy of scientific discovery. Unlike Mach's view on discovery and invention in science, Duhem's is not principally a theory of perception; he anticipates later philosophers including the Logical Positivists with his emphasis on the language of science. For him scientific discovery is not reduced to noticing what had previously been overlooked in perception; for him discovery is also the construction of theoretical hypotheses. The construction of a theory involves four successive operations: Firstly certain physical properties are taken as simple, so that other things are combinations of these simple properties. These properties are not simple in any absolute sense like Mach's elements, but are taken as simple for purposes of the theory only. The simple properties are measured, and the magnitudes are assigned to symbolic variables. Secondly the magnitudes are connected by propositions that are hypotheses, and that serve as postulates of the deductive system. Thirdly the postulates are not realistic or phenomenalist, but are freely created; using them requires only that the logic of algebra be correctly applied for making deductions. Fourthly the conclusions drawn from the postulates are compared with the experimental laws that the theory is intended to represent.

If the conclusions agree with the laws within the degree of approximation corresponding to the measurements taken in the experiments, then the theory is said to be a good theory. Such acceptable theory may in turn be used for the further development of measuring instruments used in experiments, as well as constituting the final product of the scientific endeavor with its maximum economy. Improved theory produces improved instruments, which in turn produce better measurements. These better measurements reduce the range of the indeterminacy in the numerical data, which may cause the theories to fail in their predictions. This failure will occasion two types of responses. The initial response is to modify the theory with corrections, which will enable the predictions made with the theory to fall within the smaller range of indeterminacy produced with improved measurements. But these corrections also complicate the theory, and in due course "good sense" may lead some physicists to decide to refrain from adding more complicating corrections, and instead to attempt to revise the hypothetical postulates of the symbolic schema, the theory itself. The accomplishment of such a revision is the work of the genius. But Duhem does not subscribe to the heroic concept of invention; history creates the genius as much as the genius creates history. The physicist does not choose the hypotheses on which he will build a new theory; the theory germinates within him. This germination is not sufficiently explained by the contemplation of the experimental laws that the theory must represent. It is a larger cultural development. In due course when the cultural process that he calls universal science has prepared minds sufficiently to receive a theory, it arises in a nearly inevitable manner. Often physicists who do not know one another and who are working great distances from one another, generate the same theory at the same time. In the course of his studies the historian of science according to Duhem often observes this simultaneous emergence of the same theory in countries far from one another.

In the opening sentence of the introduction to his *Aim and Structure of Physical Theory*, Duhem says that he offers a simple logical analysis of the method by which physical science makes progress. While affirming the autonomy of physics with his thesis that agreement with experiment is the sole criterion of truth for a physical theory, Duhem has a distinctive concept of scientific progress, which he elaborates in the appendices to the book. He says that there are two types of development in physics, that are occurring simultaneously. One is what today would be called the revolutionary type of development consisting of a succession of alternative theories, in which one theory arises, dominates the scene for the moment, and then collapses to be replaced by another theory. The other is an evolutionary progress in which constantly more ample and more precise mathematical representation of the phenomenal world is disclosed by experiment. When the progress of experimental science goes counter to a theory and compels it to be modified or transformed, the purely representative part enters nearly whole into the new theory, bringing to it the inheritance of all the valuable possessions of the old theory, while the hypothetical part falls out in order to give way to another theory. The first type is identified with the mechanistic physical systems including Newtonian physics as well as Cartesian and atomic physics. The second type is identified with general thermodynamics, which Duhem believes will lead physical theory toward its goal. He envisioned this goal as the convergence toward an analogy with Aristotle's physics. He concludes in his discussion of the value of theory, that the physicist is compelled to recognize that it would be unreasonable to work for the progress of physical theory, if theory were not the increasingly better defined and more precise reflection of a metaphysics. He concludes his book with the statement that belief in an order transcending physics is the sole justification of physical theory.

Duhem's History of Physics

Just as Mach had written a history of physics viewed through the lenses of his philosophy of science, so too did Duhem. However, Duhem's effort was relatively monumental; it is a work originally intended to be twelve volumes of which ten were actually written before its author's death in September 1916. This magnum opus was his *System of the World: A History of Cosmological Doctrines from Plato to Copernicus.* The central thesis of this work is summarized in a much smaller book begun earlier, *To Save the Phenomena: An Essay on the Idea of Physical Theory from Plato to Galileo* (1908). The thesis is that the hypotheses of physics and especially the heliocentric hypothesis in astronomy, are mere mathematical contrivances for the purpose of saving the phenomena.

Galileo was condemned by Pope Urban VIII in 1633 for maintaining that Copernicus' heliocentric theory is not merely a mathematical contrivance, but is rather a description of the real world. The Pope, formerly Cardinal Bellarmine, had admonished Galileo saying that regardless of how numerous and exact may be the confirmations of a theory by experience, these confirmations can never transform a hypothesis into a certain truth that can be taken realistically, since this transformation would require that the experimental facts should contradict any other hypotheses that might be conceived, a requirement that can not logically be satisfied. Galileo maintained that because Copernicus's theory saved the phenomena more adequately than any alternative hypothesis, the Copernican theory had to be a realistic one. Contemporary Pragmatists agree with Duhem's rejection of any ontological criteria for the criticism of scientific theory, but contrary to Duhem they furthermore agree with Galileo's realism. Contemporary Pragmatists are realists, who let the most empirically adequate theory decide the ontology. Galileo's argument for realism is the same as Quine's doctrine of ontological relativity, and Feyerabend calls it the Galileo-Einstein tradition of realism. And Heisenberg invoked this tradition, when he referenced Einstein's realistic interpretation of relativistic time in the relativity theory, and then used it as a precedent for his own realistic interpretation of the quantum theory's duality thesis, notwithstanding Bohr's instrumentalist complementarity principle. Duhem, however, denied that science is realistic, and he construed Galileo's argument as a case of the fallacy of the crucial experiment. He argued that it is impossible to enunciate all the possible hypotheses, and establish the truth of one by elimination of all others. The accomplishment that Duhem credits to Kepler and Galileo is the rejection of Aristotle's view that celestial and terrestrial physics are fundamentally different, and that hypotheses of physics must save all the phenomena of the inanimate world.

Mach, Einstein, and Positivism vs Relativity Theory

In his "Autobiographical Notes" in Schilpp's *Albert Einstein* (1949) Einstein stated that Mach's *History of Mechanics* had exercised a profound influence on him when he was a student. He related that all physicists of the last century saw in classical mechanics a firm and final foundation not only for all physics but also for all natural science, and that it was Ernst Mach who in his book shook this dogmatic faith. At sixty-seven years of age, when he was writing these autobiographical notes, Einstein saw Mach's greatness in the latter's incorruptible skepticism and independence, even though Einstein himself had since rejected Mach's epistemology. Einstein was specifically influenced by Mach's critique of the Newtonian concept of absolute space, time and motion, ideas that are also rejected in Einstein's relativity theory. Initially Mach seemed to support Einstein's views. But

Mach and Einstein were fundamentally working at cross purposes: Mach attacked the Newtonian concepts of absolute space, time and motion as part of his attack on all theoretical physics, while Einstein discarded these ideas as a means for developing a new theoretical physics.

Another influence on Einstein was a thought experiment that Einstein reports he imagined, when he was sixteen years of age. In this thought experiment Einstein wondered what would happen if an observer traveled at the speed of light, riding on a beam of light. The light would then be at rest relative to the rider, but since light is defined in terms of its frequency of motion, Einstein concluded that the idea of a light beam at rest is a self contradiction. This thought experiment was imagined many years before Einstein was introduced to Mach's book by his friend Besso while they were students at Zurich, and it contributed to his forming the idea that the velocity of light in a vacuum is constant in all reference systems. However, from the Positivist view the constancy of light is no less objectionably absolute than the concepts of absolute space or time; Mach's phenomenalist relativity states that all sensations are dependent on all other sensations, while Einstein's relativity theory states that the velocity of light in a vacuum is independent of other phenomena.

Throughout Mach's lifetime Einstein continued to view his relativity theory to be a continuation of Mach's philosophy, and in his obituary on Mach in 1916 Einstein expressed the opinion that Mach would have come across the theory of relativity, if when he was younger the constancy of the velocity of light had been accepted by physicists. In 1921 Mach's son published his father's *Principles of Physical Optics*. The preface of the book is dated July 1913, and in it Mach opposes Einstein's relativity theory and rejects the idea that he was a forerunner of relativity theory. As it happens, in June of 1913 Einstein had sent Mach a preliminary draft of the general theory of relativity, which uses non-Euclidian geometry. As it happens, in the 1912 edition of his *Science of Mechanics* Mach introduced a lengthy footnote (Ch. IV, Sec IV, 9) opposing Minkowski's use of four-dimensional geometry in physics, and stated that the space of sight and touch is three dimensional. It is unlikely that Mach was pleased when he received Einstein's 1913 correspondence, and it may have provoked Mach's comments in the 1913 preface to the book on optics. Eventually Einstein accepted the existence of basic differences between his relativity theory and the philosophy of Mach, and he ultimately rejected Mach's epistemology.

Einstein's general theory of relativity departed even further from

Mach's philosophy than did the special theory of relativity, because in the general theory it is not possible to restrict the equations to relations among observable magnitudes. But as the theory became accepted among physicists, the Positivists who followed Mach did not want to reject it, and instead they modified their philosophy. These later or "Logical" Positivists, as the Positivists of the Vienna Circle came to be known, replaced Mach's rejection of theories with the less restrictive idea, that the language of science may contain theoretical terms referring to nonobservable entities and magnitudes, on condition that statements referring only to observables could logically be related to those that contain these theoretical terms referring to the nonobservable magnitudes or entities. This later Positivist program is considered below in the discussion of the Logical Positivists and particularly of Rudolf Carnap, who accepted Einstein's relativity theory, and who persuaded Moritz Schlick, founder of the Vienna Circle and successor to the chair of inductive philosophy held by Mach at Vienna, to accept Einstein's theory also. With this acceptance of Einstein's relativity theory one of the basic theses of the Positivist philosophy was changed.

Einstein, Heisenberg, and Positivism vs Quantum Theory

Positivism was not without some influence on the contributors to the new quantum physics, whose views became known as the "Copenhagen interpretation". Adherents to this interpretation included Niels Bohr, Werner Heisenberg, and Wolfgang Pauli. Its opponents included Albert Einstein, Erwin Schroedinger, Max Planck, Louis de Broglie and David Bohm. The member of the University Institute for Theoretical Physics in Copenhagen, Denmark, who was initially influenced by the Positivist philosophy was Heisenberg. In his *Physics and Beyond* (1971) he relates how Mach's philosophy operated in his own thinking. In the chapter titled "Understanding in Modern Physics (1920-1922)" he described his Positivist views during the years that preceded his development of his matrix mechanics. At that time he believed that true understanding in physics consists in using only language that refers to direct sense perceptions, and that while the ability to make correct predictions is often a consequence of this Positivist kind of understanding, nonetheless making correct predictions is not the same as having true understanding. Because he accepted this philosophy of science, Heisenberg rejected the hypothesis of electron orbits, since the orbits are not observable, but unlike Mach he admitted the existence of the electron itself because of the observable vapor trails in the Wilson cloud chamber experiments.

In the chapter titled "Quantum Mechanics and a Talk with Einstein (1925-1926)" Heisenberg relates that on the day that he presented his matrix mechanics to the Physics Colloquium at the University of Berlin, Einstein, who was present in the assembly, expressed interest and invited Heisenberg to talk with him at his home that evening. The matrix mechanics does not postulate the existence of electron orbits around the nucleus of the atom, and when Einstein questioned Heisenberg about his Positivistic views that evening, Heisenberg replied that he did not believe that postulates about orbits are appropriate, because the orbits are not observable. Heisenberg affirmed the view that the physicist should consider only observable magnitudes, and for that reason he developed the matrix mechanics, which treats only of the frequencies and amplitudes associated with the lines in the spectrum of the atom. Heisenberg also stated that he was using the same philosophy that Einstein had used, when the latter had rejected the concept of absolute space and time in developing relativity theory. Einstein then replied that he no longer accepted the Positivist view, because it is the physical theory that describes what the physicist can observe. This idea that theory determines what is observed, is philosophically very strategic, because it contradicts the underlying Positivist assumption that there is a dichotomous distinction between the descriptive language about what is observable on the one hand, and the theoretical language about what is not observable on the other hand. When this dichotomy is denied, the Positivist program of building science on firm foundations of observation is rendered untenable.

In the chapter titled "Fresh Fields (1926-1927)" Heisenberg describes the arguments between Niels Bohr and Erwin Schroedinger concerning the issue of the wave verses the particle views in microphysics, and the statistical approach taken by Max Born in 1927. Born maintained that Schroedinger's wave function can be construed as the measure the probability of finding an electron at a given point in space and time. Heisenberg accepted Born's probability interpretation, but there still remained a problem in Heisenberg's mind: Born's interpretation did not explain how the trajectory of an electron in the cloud chamber could be reconciled with the wave mechanics. Trajectories did not figure in the quantum mechanics, and wave mechanics could only be reconciled with the existence of a densely packed beam of matter if the beam spread over areas much larger than the diameter of an electron. With this problem in mind Heisenberg remembered his conversation with Einstein the previous year, specifically Einstein's statement that it is the theory that determines what the physicist can observe. Einstein's discussion with Heisenberg on the day that Heisenberg had first presented his matrix mechanics in 1926 in Berlin led Heisenberg to recognize in 1927 that it was the classical theory that led him to think that the vapor trails in the Wilson cloud chamber represent the movement of a particle as having a definite position and velocity that defined its trajectory. Recognition of the interpenetration of theory and observation led Heisenberg to reconsider what is observed in the cloud chamber. He then rephrased his question about trajectories in terms of the quantum theory instead of the classical theory; he asked: Can the quantum mechanics represent the fact that an electron finds itself approximately in a given place and that it moves approximately at a given velocity? In answer to this new question he found that these approximations could be represented mathematically, and he called this mathematical representation the "indeterminacy principle". On this principle the limit of accuracy with which both position and momentum can be known is defined in terms of Planck's constant, and in the view of Heisenberg and those who advocate the "Copenhagen interpretation" this necessary degree of approximation is imposed by the nature of the universal quantum of action. Einstein's semantical principle, that theory decides what the physicist can observe, became one of the corner stones of the postPositivist philosophy of science as articulated both by Karl Popper and by the Pragmatists; it led the contemporary philosophers to reject the Positivist separation of theory and observation.

Heisenberg also describes his thought processes in this discovery experience in his chapter on the history of quantum theory in his *Physics and Philosophy* (1958), where he says that he turned around a question; instead of asking how the known formalism of Newtonian physics could be used to express a given experimental situation, he instead asked whether or not only such experimental situations can arise in nature as can be expressed in the mathematical formalism of the matrix mechanics. This recounting of his thinking gives greater emphasis to the ontological commitment that characterizes the "indeterminacy principle", according to which there does not simultaneously exist in reality both a determinate position and a determinate momentum for the electron. The Pragmatist philosophers have accepted this criterion, which Quine has called "ontological relativity", in which a carte blanche realism is presumed while the ontology describing reality is supplied by empirical science, a realism which is a blank check for which scientific theory specifies its cash value, and for which scientific criticism backs its negotiability. As it happens, Einstein was never willing to accept the ontology of the Copenhagen interpretation, even though in principle Heisenberg did the same thing with his matrix mechanics that Einstein did with the Lorentz transformation, when the latter interpreted the Lorentz equation in terms of actual time instead of apparent time and redefined the concept of simultaneity. Nevertheless Einstein maintained that a more "complete" microphysical theory is needed, which would satisfy his own deterministic criterion for physical reality, and from which Heisenberg's formalism should be mathematically derived. Such a more "complete" microphysical theory has yet to be developed by physicists. The recent attempt to reconcile general relativity theory with quantum theory by means of the "superstring" theory, and thereby to decide between the Einstein and Bohr ontologies, may or may not succeed. But whatever the outcome with superstrings, when resolution is achieved, it will have been done with theoretical physics and not with any prior ontology set forth in any autonomous metaphysics. Heisenberg did not completely escape the influence of Positivism, even though he had departed from it in a very fundamental way to develop the indeterminacy principle. Another influence on his thinking in addition to Einstein's was that of Bohr. Bohr did not explicitly embrace Positivism, but in his view classical physics is permanently valid and must serve as the language of observation in which all accounts of evidence must be expressed in physical science. Heisenberg's attempt to reconcile the influences of Einstein and Bohr resulted in his developing his semantical theory of "closed-off theories", which is different from the Logical Positivist philosophy, but which is more similar to Positivism than it is to contemporary Pragmatism. It was left to the Pragmatists to explore Einstein's semantical thesis to its limits, but not until after Positivism had run its course over a period of at least thirty years following Heisenberg's development of the indeterminacy principle.

Comment and Conclusion

Twentieth-century philosophy of science saw the eventual ascendency of the Pragmatism, but this development did not result by any exegesis on the writings of Charles S. Peirce, the founder of Pragmatism, or of William James, his colleague at Harvard and expositor. Twentieth-century philosophy of science evolved in response to developments in twentieth-century physics and proceeded by way of criticism of the prevailing Positivist philosophy of science, of which Ernst Mach was the principal advocate at the opening of the century. Mach is important to twentieth-century philosophy of science as its point of departure. But there is a lengthy time lag between developments in science and developments in the philosophy of science. Even before the beginning of this century the development of the atomic theory of matter had made Mach's antitheoretical Positivism anachronistic in his own time. In retrospect it is not surprising that a contemporary of Mach, Pierre Duhem, should attempt to modify the Positivist philosophy of physics, in order to locate a place for physical theory. His philosophy of physical theory anticipated many characteristics of the contemporary Pragmatist philosophy of science. It was left to Harvard's Willard Van Quine fifty years later to collapse these three types into one metatheory that describes all language as both wholistic, pluralistic, and realistic.

RUDOLF CARNAP ON SEMANTICAL SYSTEMS AND W.V.O. QUINE'S PRAGMATIST CRITIQUE

Rudolf Carnap (1891-1970) was a leading member of a group of philosophers and scientists who gathered in Vienna during the interwar years, and who called themselves the Vienna Circle. A statement of the group`s manifesto, "The Scientific Conception of the World", written by Otto Neurath with Carnap's collaboration can be found in English translation in Neurath's *Empiricism and Sociology*. The group was scattered when the National Socialists came to power in Germany, and although Carnap was a native German citizen, he and several other members of the group migrated to the U.S. With the aid of Willard Van Quine of Harvard University Carnap received an appointment to the faculty of philosophy at the University of Chicago in 1935, which he retained until 1952, when he spent two years at the Institute for Advanced Study at Princeton. In 1954 he filled the vacancy created by the death of Hans Reichenbach at the University of California at Los Angeles, and he held the position until his retirement from teaching in 1961. However, he continued to write for the ten years of his intellectually active retirement. He died in 1970 and was memorialized in *Boston Studies in the Philosophy of Science* (1971).

Linguistic Constructionalism

In his "Intellectual Autobiography" published in *The Philosophy of Rudolf Carnap* (ed. Schilpp, 1963) Carnap reports that while he was studying at the University of Jena during the years just before the First World War, he was greatly influenced by one of his teachers, Gottlob Frege, who maintained that logic should be the foundation for mathematics. Shortly after the war Carnap read Bertrand Russell's *Principia*

Mathematica, and was greatly impressed by Russell's theory of relations. But Carnap was even more impressed by Russell's philosophical outlook expressed in *Our Knowledge of the External World.* This book states that the logical-analytical method gives a method of research in philosophy, just as mathematics gives the method of research in physics. Carnap reports that upon reading this text he felt that its words had been directed to him personally. As a result of these influences, the construction of logical systems would characterize all of Carnap's philosophical work during his long career. There would be many other influences, but they would only produce variations on the basic agenda of logical constructionalism.

Carnap's philosophy of science was Positivist, and he and the other members of the Vienna Circle were favorably disposed to the philosophies of Mach, Poincare, and Duhem. The antimetaphysical and scientistic character of Mach's philosophy was reinforced by the writings of Ludwig Wittgenstein, who maintained that all philosophical sentences including most notably all of metaphysics are pseudo sentences, and that in spite of their grammaticalness and common usage, these pseudo sentences are really devoid of any cognitive content. Later Wittgenstein departed from this view and moved away from the constructionalist approach in philosophy. But the earlier views of Wittgenstein expressed in his *Tractatus Logico-Philosophicus* had a lasting influence on the Vienna Circle Positivists. One of the central philosophical tasks that they set for themselves was the use of logical constructionalist methods to implement the Positivist philosophy, and for this reason they are known as the Logical Positivists.

Einstein and Mathematical vs Physical Geometry

Like many philosophers of his generation, Carnap was impressed by Einstein's revolutionary theory of relativity. But while other philosophers such as Popper found the significance of this successful overthrow of the three-hundred-year reign of Newtonian physics to lie in its philosophical implications for scientific criticism, Carnap found its significance in the distinction between mathematical and physical geometry, or more generally in the role of mathematics as the logic for the physical theory. The central role in the relationship between the formal and the empirical in the development of modern physics became the axis for Carnap's whole philosophical career. He made it the subject of a distinctive type of theory, which evolved into his theory of semantical systems.

Carnap had started out in experimental physics at the University of Jena before the First World War, and then later turned to philosophy after the war. In 1921 he wrote a Ph.D. dissertation titled *Der Raum*, in which he attempted to demonstrate that contradictory theories about the nature of space maintained by the mathematicians, philosophers and physicists, are entirely different subjects. He distinguished three meanings of the term "space" corresponding to the three disciplines that treat it. These are the formal meaning used by mathematicians, the intuitive meaning used by philosophers, and the physical meaning used by physicists. The intuitive meaning used by philosophers was based on the Kantian idea of "pure intuition"; Carnap later rejected this idea are retained only the formal and empirical meanings. A later development in Carnap's thinking on these matters occurred when he read Wittgenstein's *Tractatus*. Wittgenstein had defined formal meaning in terms of tautologies or logical truth. This was the origin of Carnap's thesis of analyticity, and he believed that the concept of logical truth supplied the key to the problem of formal systems such as mathematical geometry, which had enabled Einstein to make his revolutionary relativity physics. In his autobiography Carnap says that because of the doctrine of logical truth Wittgenstein had the greatest influence on his thinking besides Russell and Frege.

After many years of silence on the subject of geometry, Carnap returned to it in his *Philosophical Foundations of Physics*. There he says that he views the Euclidian, the Lobachevskian, and the Riemannian geometries as different languages in the sense of theories of logical structure, which as such are concerned only with the logical implications of axioms. In this work he references Einstein's *Sidelights on Relativity* (1921; English, 1923) where Einstein says that the theorems of mathematics are certain in so far as they are not about reality, and that in so far as they are about reality they are uncertain. Carnap states that the philosophical significance of Einstein's theory of relativity is that it made clear that if geometry is taken in an *a priori* or analytic sense, it tells us nothing about reality, like all logical truths, while physical geometry is *a posteriori* and empirical, and describes physical space and time.

Carnap notes that in relativity theory Einstein used the Riemannian mathematical geometry as the axiomatic system for his physical geometry, but the choice of which mathematical geometry to use for a physical theory is not obvious. Several years before Einstein developed his relativity theory the mathematician Poincare postulated a non-Euclidian physical space, and said that physicists have two choices. They can either accept non-Euclidian geometry as a description of physical space, or they can preserve Euclidian geometry for the description of physical space by adopting new physical laws stating that all solid bodies undergo certain contractions and expansions, and that light does not travel in straight lines. Poincare maintained that physicists would always choose to preserve the Euclidian description of physical space, and claim that any observed non-Euclidian deviations are due to the expansion or contraction of measurement rods, and to the deflection of light rays used for measurement. Einstein's choice of the Riemannian geometry and physical laws for measurement was based on the resulting simplicity of the total system of physics. Relativity theory using Riemannian geometry greatly simplifies physical laws by means of geodesics, such that gravitation as a force is replaced by a geometrical structure.

The *Aufbau* and "Rational Reconstruction"

In 1928 Carnap published his *Der Logische Aufbau der Welt,* translated in 1967 as *The Logical Construction of the World,* in which he attempted in this book to analyze concepts of ordinary language pertaining to things in our environment and to their observable properties and relations, and to construct definitions of these concepts with the help of symbolic logic. He also reports that his use of this method was influenced by Mach and the phenomenalist philosophers, but he maintains that he was the first philosopher who took the phenomenalist doctrine of these philosophers seriously. He was not content with their customary general statements that a material body is a complex of visual, tactile, and other sensations, but he tried actually to construct these complexes, in order to show their structure. Carnap described this constructionalist procedure as "rational reconstruction". He states that while his aim was guided by the psychological facts concerning the formulation of concepts of material things out of perceptions, his aim was not the description of the genetic process. Rather his aim was to produce a schematic description of an imaginary procedure consisting of rationally prescribed steps, which would lead to essentially the same results as the actual psychological processes. His idea of rational reconstruction is different from the views of some contemporary information scientists, who propose that their procedural reconstructions of historic scientific discoveries with computers are hypotheses in "cognitive psychology". However, artificial intelligence is a broader concept of proceduralizing discovery processes.

Logical Syntax of Language

When Carnap discovered *gestalt* psychology, he reconsidered the phenomenalist constructionalism that he had undertaken in his *Aufbau*, and he came to view that a physicalist language or thing language describing things in ordinary experience is more suitable as a basis of all scientific concepts. At about the same time he also learned of Hilbert's metamathematics programme, and while the influence of Russell had led the Vienna Circle to prefer the logistic program of the foundations of mathematics rather than Hilbert's formalist approach, Carnap was nonetheless attracted to the idea of a metalanguage, not just for mathematics but for the logic of all science. This was his idea of a metalogic, that he developed in his *Logical Syntax of Language* (1934). The metalogic is the logical syntax of language viewed as a purely analytic theory of the structure of its expressions. In his autobiography he reports that the whole theory of language structure and its possible applications in philosophy came to him like a vision during a sleepless night in January 1931, when he was ill, and that on the following day he wrote down the idea in a manuscript of forty pages titled *Attempt at a Metalogic*, which was the first draft of *Logical Syntax*.

In this book he distinguished the metalanguage from the object language. The former contains no reference to the meanings of linguistic signs occurring in the object language; it refers only to the logical structure of the expressions in the object language, which is to say that the syntax contains only logical constants. Carnap says that his chief motivation for developing this syntactical method was to formulate more precisely philosophical problems that have evaded resolution when expressed in traditional manner. In 1934 he published "On The Character of Philosophical Problems" in *Philosophy of Science*, which expounded his treatment of metaphysical issues in the German edition of *Logical Syntax* published in the same year. In this work he distinguishes the formal or syntactical perspective from the connotative or material perspective. He identifies logic as a set of metalinguistic transformation rules, and the logic of the language of science, which is the object language, is one in which logical entailment is a formal transformation rule. Carnap defines the "content" of a proposition in science as a class of entailments from a synthetic proposition in the science. Content is thus a purely formal concept, and the difference between the formal and material perspectives is merely a difference between modes of expression. Accordingly philosophical analysis consists of translating statements into the formal mode. Meaningful statements in science can be translated into the formal mode of speech, but the meaningless metaphysical statements cannot be translated into the formal mode. For this reason differences between Positivists and realists disappear, when their positions are translated into the formal mode. Similarly problems in the foundation of physics are problems in syntax. For example verification of physical laws is a matter concerning the syntactic deductive coherence between the general law-like propositions and singular propositions called protocol sentences. The problem of induction is a question of how transformation rules lead from protocol sentences to laws.

In 1937 Carnap published his English edition of *Logical Syntax*. This latter edition contains additional material not in the German edition, and its bibliography includes reference to Quine's "Truth by Convention" published in 1936. The book is divided into five parts. The first three set forth two artificial object languages. Language I is designed to be acceptable to philosophers persuaded of the intuitionist philosophy of mathematics, because it includes no infinities. Language II is adequate to all classical mathematics including what the intuitionists would not accept, and it includes Language I as a sublanguage. The fourth part sets forth the general procedures for constructing any artificial language, and is titled "General Syntax". Carnap defines general syntax as a system of definitions of syntactical terms. In general a language is any sort of calculus in the sense of a system of formation and transformation rules concerning expressions, which in turn are defined as finite, ordered series of elements called symbols. Formation rules determine concatenations of symbolic elements to form expressions, and transformation rules determine what transformations produce valid deductions and proofs. The interpretation of a language is the method of learning by explicit statements that are translations from an already interpreted language, and therefore can be formally represented and belong to syntax. A system of axioms in a calculus may firstly be given, and then interpreted in various ways by translations that establish correlations between the expressions of the language being interpreted and those already interpreted.

The fifth and concluding part of the book pertains to philosophy and syntax, where philosophy is identified with the logic of science. The material for the 1934 article in *Philosophy of Science* was taken from section A of this part. In section B Carnap considers the logic of science as syntax, stating that the logical analysis of physics is the syntax of the physical language. The language must have formation rules both for the protocol sentences by means of which the results of observation are expressed, and for the postulated or "P-primitive" laws which have the form of universal sentences of implication and equivalence. The transformation rules of the physical language consist either of purely logical or "L-rules", or of "L-rules" and empirical or "P-rules" together. A sentence in physics is tested by deducing consequences on the basis of transformation rules until finally sentences of the form of protocol sentences are reached. Then these are compared with the protocol sentences that have actually been stated as observation reports, and are either confirmed or refuted by the latter. If a sentence which is an L-consequence of certain P-primitive sentences, contradicts a sentence which has been stated as a protocol sentence, then some change must be made in the system. But there are no established rules for the kind of change that must be made, nor is it possible to lay down any sort of rules as to how new primitive laws are to be established on the basis of actually stated protocol sentences. There are no rules for induction because of the universality of laws; the laws are hypotheses in relation to protocol sentences. Furthermore not only general laws, but also singular sentences are formulated as hypotheses, i.e. as P-primitive sentences, as sentences about unobserved processes by which certain observed processes can be obtained.

With respect to scientific criticism Carnap maintains that there is no complete falsification or confirmation of any hypothesis. When an increasing number of L-consequences of the hypothesis agree with already acknowledged protocol sentences, then the hypothesis is increasingly confirmed, but it is never finally confirmed. He states that it is in general impossible to test even a single hypothetical sentence, because the test applies at bottom not to a single hypothesis but also to a whole system of physics as a system of hypotheses. In this context Carnap references Duhem and Poincare. He also says that both P-rules and L-rules including those of mathematics are laid down with the reservation that they may be altered as soon as it seems expedient to do so, and that in this respect P-rules and L-rules differ only in degree with some more difficult to renounce than others. The thesis that logical and descriptive language differs only in degree is maintained by Tarski. Carnap explains that if every new protocol sentence introduced into a language is synthetic, then L-valid (i.e. analytic) sentences differ from synthetic sentences, because such a new protocol sentence can be incompatible only with the P-valid synthetic sentence; it cannot be incompatible with the logical L-valid or analytic sentence. But then he further goes on to say that in spite of the above fact, it may come about that under the inducement of new protocol sentences the language may be altered to such an extent that the L-valid or analytic sentence is no longer analytic. He emphasizes in italics that the construction of the physical system is not effected in accordance with fixed rules, but is by means of convention. These conventions are not arbitrary; they must be tested. The choice of convention is influenced firstly by practical considerations such as simplicity, expediency, and fruitfulness, and secondly by their compatibility with the total system of hypotheses to which the already recognized protocol sentences belong. Thus in spite of the subordination of hypotheses to empirical control by means of protocol sentences, hypotheses contain a conventional element, because the system of hypotheses is never "univocally" determined by empirical material however rich it may be. Carnap never developed this thesis of the empirical underdetermination of a system of hypotheses, and the artifactual theory of language it implies, which was extensively developed by Quine in the 1950's and afterward. Later Carnap rejected Tarski's thesis that logic and descriptive language differ only in degree, but he always maintained that definitions of L-true sentences are relative to the specific language system under construction.

Semantical Systems: Definitions and Characteristics

Carnap's principal book in semantics is his *Introduction to Semantics* (1943). When he had written his *Logical Syntax* he had believed that metalogic should deal only with the form of expressions of the object language, and that no reference should be made to the meanings of the signs and expressions. In the preface to his *Introduction to Semantics* Carnap states that Tarski was the first to call his attention to the fact that the formal methods of syntax must be supplemented by semantical concepts, and also that these semantical concepts can be defined by means no less exact than those of syntax. He says that his *Introduction to Semantics* owes more to

Tarski than to any other single influence, although he also notes that he and Tarski are not in complete agreement on the distinction between syntax and semantics, and on the distinction between logical and descriptive signs. In this new semantical perspective semantical systems remained central to his philosophy for the remainder of his life. It is a concept that is fundamental to his views in philosophy of science, his philosophy of probability, and his philosophy of information theory.

Following the Pragmatist tradition, to which he had been introduced by Charles W. Morris in the United States, Carnap describes semiotics as the general theory of signs, which is divided into three parts based on the three factors involved in language. These factors are (1) the expression, (2) the designatum, and (3) the speaker. The part of semiotics that deals with all three of these factors is called pragmatics. The second part of semiotics, called semantics, abstracts from the speaker, and contains a theory of the meaning of expressions, which leads to the construction of a dictionary for translating the object language into the metalanguage. Finally the third part of semiotics is called syntax, which abstracts from both the speaker and the designata of the signs, in order to consider only the expressions. Carnap further distinguishes between descriptive semantics and syntactics on the one hand, and pure semantics and syntactics on the other. The former are included in pragmatics because they are empirical, while the latter are not because they are analytic. In pure semantics and syntactics the philosopher lays down definitions for certain concepts in the form of rules, and he studies the analytic consequences of these definitions. Nearly all of Carnap's work is in pure semantics and pure syntactics, and the terms "semantics" and "syntactics" mean pure semantics and pure syntactics in his texts, unless otherwise noted; Carnap's interest is typically more in constructional systems than in empirical linguistics.

A semantical system presupposes a syntactical system. A syntactical system or calculus, denoted K, consists of rules that define syntactical concepts, such as "sentence in K" and "provable in K". The smallest units of syntax in the system is called a "sign". Signs are combined into "expressions" according to the formation rules for the calculus. The most important type of expression is the "sentence". Sentences are derivable from other sentences, i.e. are "proved", in accordance with the transformation rules of the calculus. Transformation rules are also called the system's "logic", and for purposes of illustration Carnap typically utilizes Russell's first-order predicate calculus. All the rules of the syntactical system are analytical rules, and are expressed in a metalanguage; the defined language system is the object language.

Carnap defines a semantical system as a system of rules formulated in a metalanguage and referring to an object language, which rules determine a truth condition for every sentence of the language, i.e. a sufficient and necessary condition for each sentence's truth. The semantical system supplies an interpretation of the sentences of the syntactical system or calculus, because to understand a sentence is the same as to know under what conditions it would be true. It may be noted that truth conditions are not truth values. The semantical rules do not determine whether or not a sentence is true; the truth value of the sentence must be determined empirically. The truth condition need not be satisfied for the semantical rule to state it. As a set of definitions, a semantical system denoted S must set forth certain things. It must define:

(1) the classifications of the signs in *S*,

(2) the classifications of the expressions in *S*, such as "term in *S*" and "sentence in *S*",

(3) the meaning of "designation in *S*", and

(4) the meaning of "true in *S*".

These definitions may be enumerations or they may be recursive definitions. The meanings of expressions that are smaller than sentences are given by statements of designation, such as the rule for designation for predicates which may include " '*H*' denotes the property human". The meanings of sentences are given by statements of truth conditions called Tarski sentences, such as " 'The moon is round', if and only if the moon is round." The sentence in double quotes is in the metalanguage consisting of English, and the symbol or clause in the single quotes is in the object language. The truth condition statement could also be " 'The moon is round' is true, if and only if the moon is round", since to assert that a sentence is true with the predicate "is true" is to assert the sentence. These statements in the metalanguage are called "radical" concepts for the semantical system.

In the *Introduction to Semantics Carnap* describes L-semantics, which consists of L-concepts. In L-semantics an L-term applies whenever the term "true" applies on the basis of merely logical reasons in contrast to factual reasons. This truth is called L-truth or logical truth. These L-concepts are the same as those occurring in syntax, and Carnap states that logic is part of semantics even though it may also be dealt with in syntax. Corresponding to the L-concepts in semantics, there are identical

C-concepts in syntax. The relation between syntax and semantics is such that the sentences of a calculus denoted *K* are interpreted by the truth conditions stated in the analytic semantical rules of the semantical system, denoted *S*, provided that *S* contains all the sentences of *K*. However, not all possible interpretations of the calculus *K* are true interpretations. A semantical system *S* is a true interpretation of *K*, if the C-concepts of *K* are in agreement with the corresponding radical concepts in *S*. Furthermore not all true interpretations of the calculus *K* are L-true. The semantical system *S* is called an L-true interpretation for the calculus *K*, if the C-concepts in *K* are in agreement with the L-concepts in *S*.

Later in his *Meaning and Necessity* (1947) Carnap develops a definition of L-truth in terms of his concept of state description. A state description in a semantical system denoted *S*, is a class of sentences in *S* which contains for every atomic sentence either the sentence or its negation but not both. Such a sentence is called a state description, because it gives a complete description of a possible state of the universe of individuals with respect to all the properties and relations expressed by the predicates of the system. It thus represents one of Leibniz's possible worlds or Wittgenstein's possible states of affairs. To say that a sentence holds in a state description means that it would be true if the state description were true, i.e. if all the atomic sentences belong to it were true. Thus the L-concepts are precisely those that are true in all state descriptions, because they are true in all possible worlds, even though there is only one factually true state description.

Carnap further elaborates on L-truth in his "Meaning Postulates" (1952) reprinted in the appendix of the 1956 edition of *Meaning and Necessity*. His theory of L-truth and state descriptions initially applied to cases where the logically true statement is true only by virtue of the meanings of the logical terms in the statements, as in "Every *x* is either *P* or not *P*". But there are cases such as "If *x* is a bachelor, then *x* is not married", which are true by virtue of the meanings of the descriptive terms. Meaning postulates are object-langauge sentences introduced into a semantical system, that define the relations among descriptive terms in the sentence in addition to the meanings assigned by rules of designation expressed in the metalanguage. These meaning postulates are not said to be factually true by virtue of empirical investigation, but are true by a decision of the architect of the semantical system, who uses them as semantical rules. Carnap then introduces a modification of his concept of state description to include another kind of statement, that is the conjunction of all meaning postulates in the semantical system. Then he says that a sentence in a given semantical system is L-true, if it is L-implied by this conjunction of meaning postulates. This expanded notion of L-truth with meaning postulates is Carnap's explication of analyticity, by which is meant statements whose truth is known by virtue of the logical form and sometimes by the descriptive terms in the statement. Later he refers to this expanded idea of L-truth as A-truth.

Using his concept of state description Carnap defines the concept of ranges: the range of a sentence is the class of all state descriptions in which a sentence holds. Rules of ranges in turn determine the range of any sentence in the semantical system *S*. These rules are semantical rules that determine for every sentence in *S*, whether or not the sentence holds in a given state description. By determining the ranges, these rules together with the rules of designation for the component predicates and individual variables give an interpretation for all the sentences in *S*. This amounts to saying that to know the meaning of a sentence is to know in which of the possible cases it would be true. Carnap thus describes a semantical system in terms of four types of semantical rules: (1) rules of formation for sentences, (2) rules of designation for descriptive constants, (3) rules of truth, (4) rules of ranges.

Semantical Systems: Ontological vs Linguistic Issues

Meaning and Necessity has a more specific purpose than the earlier *Introduction to Semantics*. The former is the development of a new method of semantical analysis, which Carnap calls the method of extensions and intensions, and which is based on the customary concepts of class and property respectively. Carnap maintains that these concepts of extension and intension should be substituted for the idea of naming of an abstract entity. In his autobiography he notes that some philosophers [who happen to include Quine] reject this way of speaking as the "hypostatization of entities". In their view it is either meaningless or at least in need of proof, to say that such entities as classes and properties actually exist. But Carnap argues that such terms have long been used in the language of empirical science and mathematics, and that therefore very strong reasons must be offered, if such terms as "class" and "property" are to be condemned as incompatible with empiricism or as unscientific. He says furthermore that to label the use of such terms as "Platonistic" or as "Platonistic realism", as is done by these philosophers, is misleading, because these labels neglect the fundamental distinction between, say, physical laws containing real number variables, and ontological theses affirming or denying the reality of universals. Carnap dislikes the term "ontology", and he maintains that the issue between nominalists and realists regarding universals is a pseudo problem, which is devoid of cognitive content.

Carnap's thesis is that his method of extension and intension is a superior basis for semantical analysis than an alternative method based on the naming relation, because the latter leads to contradictions, when the names are interchanged with one another in true sentences. He thus refers to the "antinomy of the name relation", which is due to the fact that a predicate viewed as a name is ambiguous, since it can refer either to a class or to a property. Some systems avoid this ambiguity by rejecting properties, and Carnap rejects this loss. Others avoid the antinomy by having different names for properties and their corresponding classes, thus resulting in a higher degree of duplication of expressions. In Carnap's method of extension and intension the expressions for properties and for their corresponding classes have the same intension and extension. Thus both classes and properties are admitted without the inelegant duplication and without the antinomy; only one predicate is needed to speak about both a certain property and about its corresponding class.

The antinomy can be avoided by Carnap's method by prescribing for expressions with the same extension a principle of interchangeability, which is distinctive of extensional contexts. This prescription is achieved by means of the L-equivalence relation, such that extensions are defined in terms of intensions. The extension of a given intension is defined as the one L-determinate extension, that is equivalent to the given intension. Extensions are thus reduced to intensions. The result is what Carnap calls a "neutral metalanguage". While the metalanguage for an object language based on the name relation will contain such terms as "the class human" and "the property human", the neutral metalanguage for an object language based on the method of extension and intension contains only the neutral expression "human".

In "Meaning and Synonymy in Natural Language" (1955), also reprinted in the appendix to the 1956 edition of *Meaning and Necessity*, Carnap describes how his method of extension and intension is applicable in pragmatics as well as in pure semantics. The purpose of this paper is to give a behavioristic operational procedure for determining intension in natural language. This procedure is problematic, because unlike the construction of an artificial language, in which extension can be defined on the basis of intensions, the empirical investigation of an unknown natural language by the field linguist must begin with the identification of extensions. But the field linguist's determination of extensions is not problematic. On the basis of either spontaneous or elicited utterances of a native speaker of the unknown natural language, the field linguist can ascertain whether or not the native is willing to apply a given predicate to a thing. By such investigation the linguist determines firstly the extension of the predicate, the class of things to which the native is willing to apply the predicate, secondly the extension of the contradictory class of things to which the native will not apply the predicate, and thirdly the class of things for which the native will neither affirm nor deny the applicability of the predicate. The size of the third class indicates what Carnap calls the degree of extensional vagueness of the predicate. Carnap admits that this determination of extension involves uncertainty and possible error, either due to a failure to recognize an individual case or because of a failure to make the correct inductive inference to the intended thing. But he says that these hazards apply to all concepts in science, and they offer no reason to reject the concepts of the theory of extension.

Carnap's thesis is that the analysis of intension for natural language is a scientific procedure, which is methodologically just as sound as the field linguist's method of determining extension. And he notes his disagreement with Quine about this thesis. Carnap postulates the case in which two linguists agree on the extension of a native's use of a predicate, but not on the intension. Carnap maintains that in pragmatics the assignment of an intension is an empirical hypothesis, which like any other hypothesis can be tested by observation of linguistic behavior. In the empirical investigation of the native speaker's linguistic behavior, the linguist looks for what Carnap calls intensional vagueness. Extensional and intensional vagueness are related such that a decrease in one produces a decrease in the latter. This search is directed to finding out what variations of a given specimen are admitted within the range of the predicate, where "range" in the context of a discussion of natural languages means those possible kinds of objects for which the predicate holds. These are cases for which the native has never made a decision about the applicability of the predicate under investigation. The description of these possible cases is the intensional vagueness of the predicate. The linguist can therefore describe to the native speaker various imaginary cases, until he hits upon one that differentiates the otherwise co-extensive predicates. Carnap adds that rules of intension are necessary for the language of empirical science, because without them intensional vagueness would remain, and therefore prevent mutual understanding and communication. Carnap apparently believes that all vagueness can be removed from a predicate, when the predicate is taken from everyday discourse into scientific language. Carnap also elaborates his discussion to include intension for a robot. He maintains that from a logical point of view the pragmatical concept for a robot is the same as that for a human. If the internal structure of the robot is not known, however, the same empirical method that is used to determine intension for a human speaker can be used for a robot. In both cases the intension for a predicate for a speaker is the general condition that an object must satisfy for the speaker to apply the predicate to it. And if the intensional structure of the robot is known, the intension of a predicate can be known even more completely.

In his "Empiricism, Semantics and Ontology" (1950) also in *Meaning And Necessity* (1956) Carnap deals further with the problem of classes and properties, which some philosophers refer to as abstract "entities". Again he notes that in the language of physics it is hardly possible to avoid abstract entities, and that using a language referring to them does not imply embracing a Platonistic ontology. He views such language as perfectly compatible both with empiricism and with strictly scientific thinking. In this paper he explains further why this compatibility is possible. Firstly he notes that there are two kinds of questions concerning the existence or reality of entities. One kind is addressed by creating a system of new ways of speaking, which system is subject to new rules in the construction of a linguistic "framework", a whole semantical system, for the new entities in question. This first kind of question pertains to the existence of the entities referenced by the system as a whole, and Carnap calls these "external" questions. The other kind of question is appropriately called an "internal" question, since it pertains to the existence of a new kind of entity within the framework. Internal questions are resolved by scientific procedures either logical or empirical. The question of the reality of a kind of entity described by a theoretical term might serve as an example of an internal question. The problem of abstract entities, however, is an external question, and it is this latter type of question that concerns Carnap in this paper.

Carnap maintains that the introduction of a new language framework with its new linguistic forms does not imply any assertion of reality, but rather is merely a new way of speaking. Therefore, because the new language framework is not a new metaphysical doctrine, the acceptance of a linguistic framework containing terms referring to abstract entities does not amount to the acceptance of Platonism. Carnap then invokes his "principle of tolerance", which he had firstly expressed in his *Logical Syntax* many years earlier. The criteria he appeals to as a semanticist is not an ontological one, but rather is a pragmatical one. The relevant criterion is whether abstract linguistic forms of variables is expedient or fruitful for purposes for which the semantical analysis is made, such as clarification or for the construction of languages for communication, especially the language of science.

Semantical Systems: Physics and the Reduction of Theories

Even before Carnap had written his *Introduction to Semantics*, he had formulated his concept of science as a semantical system, and this concept did not change fundamentally for the duration of his contributing career. The early statements of this concept are set forth in his "Logical Foundations of the Unity of Science" and "Foundations of Logic and Mathematics" in *International Encyclopedia of Unified Science* (1938). In these works he asserts that philosophy of science is not the study of the activities of scientists, i.e. the pragmatics of science, but rather is the study of the results of the activity, namely the resulting linguistic expressions, which constitute semantical systems. More specifically the philosopher treats the language of science as an object language, and develops a metatheory about the semantics and syntax of this object language. The metatheory is expressed in a metalanguage.

A physical theory is an interpreted semantical system. Customarily a calculus is firstly constructed, and then semantical rules are laid down to give the calculus factual content. The resulting physical calculus will usually presuppose a logical mathematical calculus as its basis, to which there is added primitive signs and also the axioms which are the specific primitive sentences of the physical calculus in question. For example a calculus of mechanics of mass points can be constructed with the fundamental laws of mechanics taken as axioms. Semantical rules are laid down stating that the primitive signs designate the class of material particles, the three spatial coordinates of a particle x at time t, the mass of a particle *x,* and the class of forces acting on a particle *x* or on a space *s* at time *t.* Thus by semantical interpretation the theorems of the calculus of mechanics become physical laws, that constitute physical mechanics as a theory with factual content that can be tested by observations. Carnap views the customary division of physics into theoretical and experimental physics as corresponding to the distinction between calculus and interpreted system. The work in theoretical physics consists mainly in the essentially mathematical work of constructing calculi and carrying out deductions with the calculi. In experimental physics interpretations are made and theories are tested by experiments.

Carnap maintains that any physical theory and even the whole of physics can be presented in the form of an interpreted system consisting of a specific calculus, an axiom system, and a system of semantical rules for its interpretation. The axiom system is based on a logico-mathematical calculus with customary interpretation for the nondescriptive terms. The construction of a calculus supplemented by an interpretation is called a "formalization". Formalization has made it possible to forgo a so-called intuitive understanding of the theory. Carnap says that when abstract, nonintuitive formulas such as Maxwell's equations of electromagnetism were first proposed as new axioms, some physicists endeavored to make them intuitive by constructing a "model", which is an analogy to observable macroprocesses. But he maintains that the creation of a model has no more than aesthetic, didactic, or heuristic value, because it offers nothing to the application of the physical theory. With the advent of relativity theory and quantum theory this demand for intuitive understanding has waned. This issue leads to the topic of "abstract" terms and of logical reduction, one of the central questions in Carnap's treatment of physical theories as semantical systems.

A more adequate and mature treatment of physics as a semantical system, and especially of the problem of abstract or theoretical terms in the semantical system can be found in Carnap's "The Methodological Character of Theoretical Concepts" (1956) and in his *Philosophical Foundations of Physics: An Introduction to the Philosophy of Science* (1966). Firstly some preliminary comments about terms and laws: All the descriptive terms in the object languages used in science may be classified as either prescientific or scientific terms. The prescientific terms are those that occur in what Carnap calls the physicalist or thing-language. This language is not the same as the phenomenalist language advocated by Mach; Carnap had earlier in his career attempted to apply constructionalist procedures to the construction of a phenomenalist language in his *Logical Structure of the World* (1928), but later decided to accept a language in which the idea of a physical thing is not linguistically constructed out of elementary phenomena, because he came to believe that all science can be reduced to the thing-language. This thing-language refers to things and to the properties of things; in Russell's predicate calculus things and properties are symbolized as two distinct types of signs: instantiation signs and predicate signs. But the thing language is also expressible in a natural language such as English. The predicates or other descriptive signs referring to properties are of two types: observation terms and disposition terms. Observation terms are simply names for observable properties such as "hot" and "red". These words are called "observable thing-predicates". Disposition terms express the disposition of a thing to a certain behavior under certain conditions. They are called "disposition predicates" and are exemplified by such words as "elastic", "soluble", and "flexible". These terms are not observable thing-language properties, but by use of conditional reduction sentences they are reducible to observation predicates. Opposed to prescientific terms are scientific terms. Carnap classified all scientific terms as "theoretical terms" in a broad sense, even though physicists, as he notes, customarily refer to such terms as "length" and "temperature" as observation terms, because their measurement procedures are relatively simple. More abstract theoretical terms are exemplified by "electron" or "electrical field". A discussion of theoretical terms requires some further discussion of semantical rules in physical theory. There are two types of semantical rules: definitions and conditional reduction sentences. A reduction sentence for a descriptive sign is a conditional statement that gives for the sign the conditions for its application by reference to other signs. The reduction sentence does not give the complete meaning for the descriptive sign, but it gives part of its meaning. It is a "method of determination" enabling the user to apply the term in concrete cases. A definition is a special case of a reduction sentence, that gives all of the meaning of a descriptive term, because it is an equivalence or biconditional sentence. There is never more than one definition for a univocal term, but there may be many reduction sentences for a univocal term, each of which contributes to the term a part of its meaning. Both types of semantical rules introduce new terms into an object language. If one language is such that every descriptive term in it is expressible by reduction sentences in terms of another language, then the second language is called a "sufficient reduction basis" for the first language. For all scientific terms the scientist always knows at least one method of determination, and all such methods always either are reduction sentences or are introduced into an axiomatic system of physics by explicit definition in the axiomatic system.

Carnap states that he disagrees with the philosophy of the physicist Paul W. Bridgman, who stated in his *Logic of Modern Physics* (1927) that every quantitative concept must be defined uniquely by the procedures for measuring it. This principle is called "operationalism", and it implies that there are as many different concepts of temperature or length as there are different ways of measuring temperature or length. Carnap maintains that these different operational rules for measurement should not be considered as definitions giving the complete meaning of the quantitative concept. He prefers his idea of reduction sentences in which statements of operational procedures are semantical rules giving only part of the meaning of the theoretical term. In Carnap's philosophy what distinguishes theoretical terms from observation terms is precisely the fact that the meanings of theoretical terms are always partial and incomplete. This view distinguishes Carnap from Heisenberg and from other Positivists such as Nagel, who prefer equivocation to partial meanings. In Carnap's view statements of operational rules understood as reduction sentences together with all the postulates of theoretical physics function to give partial interpretations to quantitative concepts. These partial interpretations are never final, but rather are continually increased or "strengthened" by new laws and new operational or measurement rules that develop with the advance of physics. Such in brief is Carnap's taxonomy of terms.

Carnap classifies scientific laws as empirical laws and theoretical laws. This division does not correlate exactly to the division between observation terms and theoretical terms in the broader and less abstract sense of his meaning of "theoretical term". The distinction is based on how the laws are developed. Empirical laws are also called empirical generalizations, because they are developed by inductive generalization, which to Carnap means recognition of regularities by observation of repeating instances. The empirical laws contain observation predicates or magnitudes that are measured by relatively simple procedures that can be expressed in reduction sentences or definitions. Empirical laws therefore may contain theoretical terms, such as "temperature", "volume", and "pressure", as occur in Boyle's gas laws, as well as observation terms as may occur in such universal generalizations as "all ravens are black". The scientist makes direct observations or repeated measurements, finds certain regularities, and then expresses the regularities in an empirical law. Theoretical laws on the other hand cannot be made by inductive generalization, because they contain theoretical terms in the narrower or more abstract sense; these theoretical terms are too abstract for making laws by generalization. Examples of these terms are "electron", "atom", "molecule", and "electromagnetic field". These are the descriptive terms that the physicists also call theoretical and unobservable, and measurements associated with these theoretical terms cannot be acquired in simple or direct ways. The development of theoretical laws proceeds by the physicists' imaginative construction of theories in the object language of their science.

Having discussed Carnap's classification of the types of terms and of scientific laws, it is now possible to discuss the construction of physical theories. Logically there is firstly a calculus. Conceivably the calculus might be completely uninterpreted, but most often the calculus is supplied by what Carnap calls the logico-mathematical calculus with its semantical rules for its logical terms supplying the "customary" interpretations. In other words the physicist seldom develops his own logic or mathematics, although he may use a pre-existing mathematics that had never previously been used in physics, e.g. a non-Euclidian geometry. The physicist then postulates certain axioms, and the descriptive terms in the axiomatic system will either be primitive terms or will be completely defined by reference to primitive terms given in the axioms. In the axiom system the primitive terms may be classified either as elementary terms or as theoretical terms in the narrow or more abstract sense. Elementary terms are either observation terms or are simple magnitudes, which are theoretical terms in the less abstract sense. The elementary terms are given their semantical interpretation by semantical rules that either define them or give methods of determination by conditional reduction sentences.

The aim of the early Positivists was to make all the primitive terms elementary terms. In this way the semantics of the primitive terms would be given by semantical rules that would either designate them as observation predicates, or designate them by reference to experimental measurement procedures. And since none of the abstract theoretical terms are primitive in the axiomatic system, any such terms would have to be defined by reference to the primitive terms. This method would completely satisfy the early Positivist requirement that all the semantics in the physical theory be supplied by semantical rules that constitute an effective reduction of the theory to observations or to experimentally based measurements. This would surely insure that there would be no contamination of science by metaphysical "nonsense".

However, there is a problem with this approach, even though it would satisfy the requirements of the early Positivists. The theories actually constructed by physicists contain abstract theoretical terms that cannot be defined by reference to elementary descriptive terms which have semantical rules directly giving them their empirical meanings. As Carnap states, what physicists actually do is not to make all the primitive terms elementary terms, but rather to make the abstract theoretical terms primitive in the axiomatic system and to make the axioms of the systems very general theoretical laws. In this constructional procedure the semantical rules initially have no direct relation to the primitive theoretical terms. Carnap borrows Hempel's metaphorical language describing the axioms with their primitive terms as "floating in the air", meaning that the theoretical hypotheses are firstly developed by the imagination of the physicist, while the elementary terms occurring in the empirical laws are "anchored to the ground". There remains to connect the theoretical laws with the empirical laws. This is achieved by a kind of reduction sentence that relates the abstract theoretical terms in the theoretical laws with the elementary terms in the empirical laws. This reduction sentence is called the "correspondence rule", and it is a semantical rule that gives a partial and only a partial interpretation to the abstract theoretical terms; the axiomatic system is left open to make it possible to add new correspondence rules when theories are modified and as physics develops, until one day the theory is completely replaced in a scientific revolution by a newer one with different axioms. The new correspondence rules add more empirical meaning to the theoretical terms as theory is developed, and they also enable the physicist to derive empirical laws from the theoretical laws. The logical connection between the two types of laws enables the theoretical laws to explain known empirical laws, but Carnap maintains that the supreme value of a theory is its power to predict new empirical laws; explaining known laws is of minor importance in his view. He observes that every revolutionary theory has predicted new empirical laws that are confirmed by experiment.

But there still remains a problem for the Logical Positivist. In this more complicated relationship between theory and experiment, there is a question of how abstract theoretical terms can be distinguished from metaphysical "nonsense". Many philosophers of science, such as Popper, maintain that this is a pseudo problem that cannot be solved. But it was resolved to Carnap's satisfaction by the Ramsey sentence. The Cambridge logician and economist, Frank P. Ramsey, proposed that the combined system of theoretical postulates and correspondence rules constituting the theory be replaced by an equivalent sentence, which does not contain the theoretical terms; in the Ramsey sentence the theoretical terms are eliminated and are replaced by existentially quantified dummy variables. The Ramsey sentence has the same explanatory and predictive power as the original statement of the theory, but without the metaphysical questions that are occasioned by the original formulation with its theoretical terms. Carnap reports that Ramsey did not intend that physicists should abandon their use of theoretical terms; theory is a convenient "short hand" that is useful to the physicist.

Finally mention must be made of another application of the reductionist logic, the unity of science. Both Mach and Duhem expressed the belief that there is a basic unity to all science. In the Vienna Circle the principal advocate of using constructional methods for advancing the unity of science was Otto Neurath, a sociologist who was interested in the sociology of science as well as its linguistic analysis. In his autobiography Carnap stated that Neurath's interest in this effort was motivated by the belief that the division between natural sciences and sociocultural sciences, a division that is characteristic of the Romantic tradition, would be a serious obstacle to the extension of the empiricological method to the social sciences, and Neurath expressed a preference for the physicalist or thing language rather than the phenomenalist language, since the former is easier to apply in social sciences. His own views are given in his "Foundations of the Social Sciences" in the second volume of the *International Encyclopedia of Unified Science* (1944). But before Neurath had published his views, Carnap had published his "Logical Foundations of the Unity of Science" in the first volume of the *Encyclopedia* (1938), where he set forth the constructionalist procedures for the logical reduction of the descriptive vocabulary of the empirical sciences to the observational thing language. The use of the thing language presumes in Carnap's view a philosophical thesis called physicalism, the view that the whole of science can be reduced to the physical language, the language of physical things. Carnap says that the physiological and behavioristic approaches in psychology and social science are reducible to the observational thing language, but that the introspective method may not be reducible. Carnap's constructionalist program aims only to the logical reduction of descriptive terms in science to the observational thing language; this effort is not a reduction of the empirical laws of the sciences to one another. The reduction of laws occurs as a part of the development of the sciences themselves and is the task of the empirical scientist, not of the philosopher of science. The constructionalist procedures for the reduction of descriptive terms for the unity of science are the same as those that Carnap had developed for the reduction of theoretical terms.

Semantical Systems: Probability and Induction

In his article "Testability and Meaning" in *Philosophy of Science* (1936) Carnap abandoned the idea of verification, because he concluded that hypotheses about unobserved events in the physical world can never be completely verified by observational evidence, and he proposed instead the idea of confirmation. He became interested in the philosophy of probability in 1941, when he considered that the concept of logical probability might supply an exact quantitative explication of the concept of confirmation of a hypothesis with respect to a given body of evidence, such that it would become possible to speak of a degree of confirmation in a measurable sense. Up to that time there were fundamentally two kinds of concepts of probability, which were proposed by their advocates as alternatives. The earlier view is the frequency concept advanced by Richard von Mises and Hans Reichenbach. The other view is the logical concept advanced by John Maynard Keynes in 1921 and by Harold Jeffreys in 1939, and also considered by Ludwig Wittgenstein in his *Tractatus*, where he defined probability on the basis of the logical ranges of propositions. Wittgenstein's concept made a probability statement analytic instead of synthetic or factual as does the frequency concept. Carnap believed that the logical concept of probability is the basis for all inductive inference, and therefore he identifies the concept of logical probability with the concept of inductive probability.

In 1950 Carnap published *Logical Foundations of Probability*. This work on probability is not a development in the calculus of probability or in the techniques of statistical inference, but is a contribution to the interpretation of probability theory with the constructionalist approach; it is a further development of his metatheory of semantical systems. Here his distinction between object language and metalanguage serves as the basis for his relating the concepts of logical and statistical probability. Statements of statistical probability occur in an object language and are empirical statements about the world. Statements of logical probability occur in the metalanguage and are about the degree of confirmation of statements in the object language. Carnap also refers to the statements in the metalanguage for scientific theory as "metascientific" statements. However, for Carnap metascientific statements are not empirical, but rather are analytic or L-true; he does not recognize an empirical metascience. He accepts the frequency interpretation for the statistical probability asserted by statements in the object language; statistical probability therefore is the relative frequency of an occurrence of an event in the long run. Logical probability is the estimate of statistical probability, and it is the measure of the degree of confirmation. Symbolically he expresses this logical probability as:

$$c(h,e) = r,$$

which means that hypothesis h is confirmed by evidence e to the degree r. The variable r is the measure of the degree of confirmation, such that r can take values from 0.0 to 1.0; it is the estimate of the relative frequency and is expressed as:

$$r = m(e*h)/m(e),$$

where $m(e*h)$ is the number of observation sentences describing observed confirming instances e of hypothesis h, and $m(e)$ is the number of observation sentences e describing the total number of observed instances, both confirming and disconfirming. He calls m a measurement function.

In Carnap's view the logical foundation of probability is logic in the sense of L-truth, and he therefore draws upon his metatheory of semantical systems, in which his ideas of state description and range have a central role. A state description is a conjunction containing for every atomic sentence that can be formed in a language, either its affirmation or its negation but not both. Thus every sentence that is L-true, is true in all the state descriptions, and every sentence that is L-false or self contradictory, is false in every state description. F-true or factually true sentences, are true in only some state descriptions but are not true in others. When the idea of state description is related to the concept of logical probability, the L-true sentences are those that have a degree of confirmation on 1.0, the L-false sentences are those that have a degree of confirmation of 0.0, and the F-true sentences are those that have a degree of confirmation between 1.0 and 0.0. The closely related concept of the range of a statement is the class of all state descriptions in which an empirical statement is true, and it may also be defined as those state descriptions that L-imply the statement. Using the concept of range the equation $r = m(e*h)/m(e)$ may be said to be the partial inclusion of the range of *e* in *h* as measured by *r*. Thus the equation $c(h,e) = r$ is analogous to the statement that *e* L-implies *h* except that the range of *e* is not completely contained in *h*. Both types of statements are analytical or L-true statements in the metalanguage, because both are statements in logic, one in inductive logic and the other in deductive logic. In Carnap's view the logical foundations of probability is logic in the sense of L-truth.

In 1952 Carnap published *The Continuum of Inductive Methods,* which was to be the volume on the theory of induction that followed *Logical Foundations of Probability,* but he became dissatisfied with this treatment. For many years he continued to work on induction, and at the time of his death in 1970 he had completed "Inductive Logic and Rational Decisions" and "A Basic System of Inductive Logic, Part I" which were published in S*tudies in Inductive Logic and Probability,* Volume I (ed. Carnap and Jeffrey, 1971). Carnap did not complete Part II of "A Basic System", and it was edited for publication in 1980 by Jeffrey in *Studies in Inductive Logic and Probability*, Volume II. In "Inductive Logic and Rational Decisions" Carnap is concerned with Bayesian decision theory. In this context the term "probability" does not mean relative frequency, but rather means degree of belief. He distinguishes the psychological concept of actual degree of belief from the logical concept of rational degree of belief. The former is empirical and descriptive, while the latter is normative for rational decision making. Carnap considers the former to be subjective, since it differs from one individual person to another, while the latter is objective. Carnap maintains that contrary to prevailing opinion relative frequency is not the only kind of objective probability. He also calls the former "actual credence" and the latter "rational credence". Rational credence is the link between descriptive theory and inductive logic, and like inductive logic it is formal, deductive and axiomatic. The concepts of inductive logic and of normative decision theory are similar but not identical. The latter are quasi psychological, while the former have nothing to do with observers and agents, even as these are generalized so that the decision theory is not subjective. Hence there are separate measure functions and confirmation functions for rational decision theory and for inductive logic. In his "A Basic System of Inductive Logic" Carnap develops a set-theoretic axiomatic system, which uses set connectives instead of sentence connectives, and which is equivalent to the customary axiom systems for conditional probability.

Semantical Systems: Information Theory

In 1953 Carnap and Yehousha Bar-Hillel, professor of logic and philosophy of science at the Hebrew University of Jerusalem, Israel, jointly published "Semantic Information" in the *British Journal for the Philosophy of Science.* A more elaborate statement of the theory may be found in chapters fifteen through seventeen of Bar-Hillel's *Language and Information* (1964). This semantical theory of information is based on Carnap's *Logical Foundations of Probability* and on Shannon's theory of communication. In the introductory chapter of his *Language and Information* Bar-Hillel states that Carnap's *Logical Syntax of Language* was the most influential book he had ever read in his life, and that he regards Carnap to be one of the greatest philosophers of all time. In 1951 Bar-Hillel received a research associateship in the Research Laboratory of Electronics at the Massachusetts Institute of Technology. At the time he took occasion to visit Carnap at the Princeton Institute for Advanced Study. In his "Introduction" to *Studies in Inductive Logic and Probability*, Volume I, Carnap states that during this time he told Bar-Hillel about his ideas on a semantical concept of content measure or amount of information based on the logical concept of probability, in contrast to Shannon's statistical concept of the amount of information. Carnap notes that frequently there is a confusion between these two concepts, and that while both the logical and statistical concepts are objective concepts of probability, only the second is related to the physical concept of entropy. He also reports that he and Bar-Hillel had some discussions with John von Neumann, who asserted that the basic concepts of quantum theory are subjective and that this holds especially for entropy, since this concept is based on probability and amount of information. Carnap states that he and Bar-Hillel tried in vain to convince von Neumann of the existence of the differences in each of these two pairs of concepts: objective and subjective, logical and physical. As a result of the discussions at Princeton between Carnap and Bar-Hillel, they undertook the joint paper on semantical information. Bar-Hillel reports that most of the paper was dictated by Carnap, and it was originally published as a *Technical Report* of the MIT Research Laboratory in 1952.

In the opening statements of "Semantic Information" the authors observe that the measures of information developed by Claude Shannon have nothing to do with what the semantics of the symbols, but only with the frequency of their occurrence in a transmission. This deliberate restriction of the scope of mathematical communication theory was of great heuristic value and enabled this theory to achieve important results in a short time. But it often turned out that impatient scientists in various fields applied the terminology and the theorems of the theory to fields in which the term "information" was used presystematically in a semantic sense. The clarification of the semantic sense of information is very important, therefore, and in this paper Carnap and Bar-Hillel set out to exhibit a semantical theory of information that cannot be done with the concepts of information and amount of information used by Shannon's theory. Notably Carnap and Bar-Hillel's equation for the amount of information has a mathematical form that is very similar to that of Shannon's equation, even though the interpretations of the two similar equations are not the same. Therefore a brief summary of Shannon's theory of information is in order at this point before further discussion of Carnap and Bar-Hillel's theory.

Claude E. Shannon published his "Mathematical Theory of Communication" in the *Bell System Technical Journal* (July and October, 1948). The papers are reprinted together with an introduction to the subject in *The Mathematical Theory of Communication* (Shannon and Weaver, 1964). Shannon states that his purpose is to address what he calls the fundamental problem of communication, namely, that of reproducing at one point either exactly or approximately a message selected at another point. He states that the semantical aspects of communication are irrelevant to this engineering problem; the relevant aspect is that actual message is the one selected from a set of possible messages in a system that is designed to operate for all possible selections. If the number of messages in the set of all possible messages is finite, then this number or any monotonic function of this number can be regarded as a measure of the information produced, when one message is chosen from the set and with all choices being equally likely. Shannon uses a logarithmic measure with the base of the log serving as the unit of measure. His paper considers the capacity of the channel through which the message is transmitted, but the discussion is focused on the properties of the source. Of particular interest is a discrete source, which generates the message symbol by symbol, and chooses successive symbols according to probabilities. The generation of the message is therefore a stochastic process, but even if the originator of the message is not behaving as a stochastic process, the recipient must treat the transmitted signals in such a fashion. A discrete Markov process can be used to simulate this effect, and it can approximate an English-language message. The approximation to English language is more successful, if the units of the transmission are words instead of letters of the alphabet. During the years immediately following the publication of Shannon's theory linguists attempted to create constructional grammars using Markov processes. These grammars are known as finite state Markov process grammars. However, after Noam Chomsky published his *Syntactical Structures* in 1956, linguists were persuaded that natural language grammars are not finite state grammars but are potentially infinitestate grammars.

In the general case of the Markov process there exists a finite number of possible states of the system together with a set of transition probabilities, such that for any one state there is an associated probability for every successive state to which a transition may be made. To make a Markov process into an information source, it is necessary only to assume that a symbol is produced in the transition from one state to another. There is a special case called an ergodic process in which every sequence produced by the process has the same statistical properties. Shannon proposes a quantity that will measure how much information is produced by an information source that operates as a Markov process: given n events with each having probability $p(i)$, then the quantity of information H is:

$$H = \sum_{i=1}^{n} p(i) \log p(i).$$

In their "Semantic Information" Carnap and Bar-Hillel introduce the concepts of information content of a statement and of content element. The content of a statement is what is also meant by the Scholastic adage, *omnis determinatio est negatio*; it is the class of those possible states of the universe which are excluded by the statement. Expressed in terms of state descriptions, the content of a statement is the class of all state descriptions excluded by the statement. The concept of state description had been defined previously by Carnap as a conjunction containing as components for every atomic statement in a language either the statement or its negation but not both, and no other statements. The content element is the opposite in the sense that it is a disjunction instead of a conjunction. The truth condition for the content element is therefore much less than that for the state description; in the state description all the constituent atomic statements must be true for the conjunction to be true, while for the content element only one of the constituent elements must be true for the conjunction to be true. Therefore the content elements are the weakest possible factual statements that can be made in the object language. The only factual statement that is L-implied by a content element is the content element itself. The authors then propose an explicatum for the ordinary concept of the "information conveyed by the statement *i*" taken in its semantical sense: the content of a statement *i,* denoted *cont(i)*, is the class of all content elements that are L-implied by the statement *i.*

The concept of the measure of information content of a statement is related to Carnap's concept of measure over the range of a statement. Carnap's measure functions are meant to explicate the presystematic concept of logical or inductive probability. For every measure function a corresponding function can be defined in some way, that will measure the content of any given statement, such that the greater the logical probability of a statement, the smaller its content measure. Let *m(i)* be the logical probability of the statement *i.* Then the quantity *1-m(i)* is the measure of the content of *i*, which may be called the "content measure of *i*", denoted *cont(i).* Thus:

$$cont(i) = 1 - m(i).$$

However, this measure does not have additivity properties, because *cont* is not additive under inductive independence. The *cont* value of a conjunction is smaller than the *cont* value of its components, when the two statements conjoined are not content exclusive. Thus insisting on additivity on condition of inductive independence, the authors propose another set of measures for the amount of information, which they call "information measures" for the idea of the amount of information in the statement *i*, denoted *inf(i)*, and which they define as:

$$inf(i) = log \ \{1/[1\text{-}cont(i)]\}$$

which by substitution transforms into:

$$inf(i) = - \ log \ m(i).$$

This is analogous to the amount of information in Shannon's mathematical theory of communication but with inductive probability instead of statistical probability. They make their use of the logical concept of probability explicit when they express it as:

$$inf(h/e) = - \ log \ c(h,e),$$

where $c(h,e)$ is defined as the degree of confirmation and $inf(h/e)$ means the amount of information in hypothesis h given evidence e. Bar-Hillel says that *cont* may be regarded as a measure of the "substantial" aspect of a piece of information, while *inf* may be regarded as a measure of its "surprise" value or in less psychological terms of its "objective unexpectedness". Bar-Hillel believed that their theory of semantic information might be fruitfully applied in various fields. However, neither Carnap nor Bar-Hillel followed up with any investigations of the applicability of their semantical concept of information to scientific research, and Bar-Hillel's interests have since turned to the analysis of natural language.

Shreider's Semantic Theory of Information

Carnap's semantic theory of information may be compared to a more recent semantic information theory proposed by the Russian information scientist, Yu A. Shreider (also rendered from the Russian as Ju A. Srejder). In his "Basic Trends in the Field of Semantics" in *Statistical Methods in Linguistics* (1971) Shreider distinguishes three classifications or trends in works on semantics, and he relates his views to Carnap's in this context. The three classifications are: ontological semantics, logical semantics, and linguistic semantics. He says that all three of these try to solve the same problem: they try to ascertain what meaning is and how it can be described. Firstly ontological semantics, which is unrelated to Carnap's or Shreider's work, is the study of the various philosophical aspects of the relation between sign and signified. He says that it inquires into the very nature of existence, into the degrees of reality possessed by signified objects, classes and situations, and that it is closely related to the logic and methodology of science and to the theoretical foundations of library classification. The second classification is logical semantics, which studies formal sign systems as opposed to natural languages. This is the trend in which he locates Carnap, as well as Quine, Tarski, and Bar-Hillel. The semantical systems considered in logical semantics are basic to the metatheory of concrete sciences. The meaning postulates determine the class of permissible models for a given system of formal relations. A formal theory fixes a class of syntactical relations, whence there arises a fixed system of semantic relations between a text or a description of a possible world. The third classification is linguistic semantics, which seeks to elucidate the inherent organization in a natural language, to formulate the inherent regularities in texts and to construct a system of basic semantic relations. The examination of properties of extralinguistic reality, which determines permissible semantic relations and the ways of combining them, is carried considerably farther in linguistic semantics than in logical semantics, where the question is touched upon only in the selection of meaning postulates. However, linguistic semantics is still rather vague and inexact, being an auxiliary investigation in linguistics used only as necessity dictates. Shreider locates his work midway between logical and linguistic semantics, because it involves the examination of natural language texts with logical calculi.

Shreider's theory is a theory of communication which explains phenomena not explained by Shannon's statistical theory. Bibliographies in Shreider's English-language articles contain references to Carnap's and Bar-Hillel's 1953 paper, and Shreider explicitly advocates Carnap's explication of intensional synonymy in terms of L-equivalence, but Shreider's theory is more accurately described as a development of Shannon's theory, even though Shreider's theory is not statistical. English language works by Shreider include "On the Semantic Characteristics of Information" in *Information Storage and Retrieval* (1965), which is also reprinted in *Introduction to Information Science* (ed. Tefko Saracevic, 1970), and "Semantic Aspects of Information Theory" in *On Theoretical Problems On Information* (Moscow, 1969). Furthermore comments on Shreider and other contributors to Russian information science or "informatics" can be found in "Some Soviet Concepts of Information for Information Science" in the *American Society for Information Science Journal* (1975) by Nicholas J.

Belkin.

Like many information scientists who take up semantical considerations, Shreider notes that there are many situations involving information, in which one may wish to consider the content of the message signals instead of the statistical frequency of signal transmission considered by Shannon's theory. But Shreider furthermore maintains that a semantical concept of information implies an alternative theory of communication in contrast to Shannon's "classical" theory. Shannon's concept pertains only to the potential ability of the receiver to determine from a given message text a quantity of information; it does not account for the information that the receiver can effectively derive from the message, that is, the receiver's ability to "understand" the message. In Shreider's theory the knowledge had by the receiver prior to receiving the message is considered in order to determine the amount of information effectively communicated.

More specifically, in Shannon's probability-theoretic approach, before even considering the information contained in a message about some event, it is necessary to consider the *a priori* probability of the event. Furthermore according to Shannon's first theorem, in the optimum method of coding a statement containing more information requires more binary symbols or bits. In Shreider's view, however, a theory of information should be able to account for cases that do not conform to this theorem. For example a statement describing a newly discovered chemical element, which could be coded in a small number of binary symbols, and for which it would be meaningless to speak of an *a priori* probability, contains much information. On the other hand a statement describing the measurements of the physicochemical properties of some substance may be considerably less informative, while it may need a much more extensive description for its coding. The newly discovered element will change our knowledge about the world much more than measurement of known substances. A theory of information that can take into account the receiver's ability to "understand" a message, must include a description of the receiver's background knowledge. For this purpose Shreider's theory includes a thesaurus, by which is meant a unilingual dictionary showing the semantic connections among its constituent words. Let **T** denote such a thesaurus to represent a guide in which there is recorded our knowledge about the real world. The thesaurus **T** can be in any one of various states, and it can change or be transformed from one state to another. Thus let **M** represent a received message, which can transform the thesaurus **T.** Then the concept of amount of information, denoted **L(T,M)**, may be defined as the degree of change in the thesaurus **T** under the action of a given statement **M.** And for each admissible text **M** expressed in a certain code or language, there corresponds a certain transformation operator **q**, which acts on thesaurus **T.** The salient point is that the amount of information contained in the statement **M** relative to the thesaurus **T** is characterized by the degree of change in the thesaurus under the action of the communicated statement. And the understanding of the communicated statement depends on the state of the receiver's thesaurus. Accordingly the thesaurus **T** can understand some statements and not others. There are some statements that cannot be accepted by a given thesaurus at all, and the information for such a thesaurus is zero, which is to say **L(T,M)**=0, because the thesaurus **T** is not transformed at all. Such is the case of a student or a layman who does not have the background to understand a communicated message may be redundant to what the receiver already knows, such that there is no information communicated. Again **L(T,M)**=0, but in this case the thesaurus **T** has been transformed into its initial state. The interesting situation is that in which the receiver's thesaurus is sufficiently developed, that he understands the message but still finds his thesaurus transformed into a new and different state as a result of receipt of the new information. If the rules of construction of the transformation operator **q** is viewed as external to the thesaurus **T**, then the quantity **L(T,M)** depends on these rules. Since the transformation operator **q** may also therefore be revised, this implies that a preliminary increase of the knowledge stored in the thesaurus **T** not only can decrease the quantity of information **L(T,M)**, but can also increase it. Thus someone who has learned a branch of a science will derive more information from a special text in the branch than he would before he had learned it. This peculiar property of the semantic theory of information basically distinguishes it from the Shannon's classical theory, in which the increase in apriori information always decreases the amount of information from a message statement **M.** In the classical theory there is no question of a receiver's degree of "understanding" of a statement; it is always assumed that he is always "tuned". But in the semantic theory the essential role is played by the very possibility of correct "tuning" of the receiver.

In his 1975 article Belkin reports that Shreider has since further developed his theory of information to include the idea of meta-information. Meta-information is information about the mode of coding information, i.e. the knowledge about the relation between information and the text in which it is coded. In this sense of meta-information the receiver's thesaurus must contain meta-information in order to understand the information in the received message text, because it enables the receiver to analyze the organization of the semantic information, such as that which reports scientific research findings. Shreider maintains that informatics, which is the Russian equivalent to information science, is concerned not with information as such, but rather with meta-information, and specifically with information as to how scientific information is distributed and organized. Therefore, with his concept of meta-information Shreider has reportedly modified his original theory of communication by analyzing the thesaurus **T** into two components, such that **T=(Tm,To)**. The first component **Tm** consists of the set of rules needed for extracting elementary messages from the text **M**, while the second component **To** consists of the factual information that relates those elementary messages systematically and enables the elements to be integrated in **T.** The relationship between **Tm** and **To** is such that a decrease in the redundancy of coding of **To** requires an increase of the meta-information in **Tm** for the decoding of the coding system used for **To**. Hence the idea of meta-information may be a means of realizing some limiting efficiency laws for information by analyzing the dependency relation between information and the amount of meta-information necessary to comprehend that information.

It would appear that if the coding system is taken as a language, then Shreider's concept of meta-information might include to the idea of meta-language as used by Carnap and other analytical philosophers, or it might be incorporated into the metalanguage. Then the elements **Tm** and **To** are distinguished as metalanguage and object language respectively, although the philosophers have had little interest in examining the inverse dependency between them.

The Philosophy of Science

Carnap makes no explicit statement of the aim of science. A discussion of the aim of science is a discussion in the pragmatics of science, that is, what the scientist does as a user of scientific language when he does research. But Carnap identifies the pragmatics of language with the empirical investigation of historically given natural languages. He admits that a study of pragmatics may be useful, as for example to find out the effective form for a language system to be used in a branch of empirical science, as he says in his "Meaning and Synonymy in Natural Languages" (1955) in his *Meaning and Necessity* . In his "On Some Concepts of Pragmatics" (1955) in *Meaning and Necessity* he affirms the need for a pragmatics using theoretical constructs instead of just an empirical, behavioristic investigation. He proposes constructing pragmatical systems using the concept of assertion, in which he includes the idea of purpose or intention as a theoretical construct. However, Carnap did not develop either a behavioristic or a theoretical pragmatics for the language of science. Instead he always constructs his own languages usually using Russell's symbolic logic, and then uses these artificial languages to address the philosophical problems of interest to the Positivist program for philosophy, namely, the reduction of theoretical terms to demonstrate their meaningfulness and the reduction of the vocabulary of science to advance its unification. He gives no systematic account of the aim of science, that the scientist might use for his own constructional work in empirical science.

However, Carnap had elaborate and explicit views on the nature and function of scientific explanation. Explanation always involves laws, and laws are classified as either empirical laws or theoretical laws. Empirical laws explain facts, which are statements that describe individual instances. The explanation has the logical structure of a deduction. The premises of the deduction consist of at least one law and of facts that describe individual instances in the same terms as those occurring in the antecedent sentences of the conditional law. The conclusion is also a factual sentence that describes the individual instances in the same terms as those in the consequent sentence of the conditional law. In this manner empirical laws explain observed instances described by factual statements. Theoretical laws are related to empirical laws in a way that is analogous to the way that empirical laws are related to facts. The theoretical law is more general. It helps to explain deductively empirical laws that are already known and to permit the derivation of new empirical laws, just as the empirical laws help to explain facts that have been observed and to predict new facts. Furthermore the theoretical law puts several empirical laws into an orderly pattern, just as the empirical generalization puts many facts into an orderly pattern. The supreme value of theory is its power to predict new empirical laws; explaining known laws is of minor value. Every revolutionary theory in the history of science has predicted new empirical laws, that are confirmed by empirical tests.

Unlike Duhem, Carnap does not stratify the semantics of physics. To say that theoretical laws explain empirical laws is not for Carnap to say as Duhem did, that the theory is an axiomatic system with conclusions that are statements which parallel the empirical laws, and that have their own semantics that in turn refers to the empirical laws. In Carnap's view the theoretical terms receive all their semantics from the observation terms by means of reduction sentences known as correspondence rules. When Carnap says that theoretical laws explain empirical laws, he means that a deductive relationship is established between the axioms of the theory and the empirical laws, and that the relationship is mediated by the correspondence rules; the postulated axioms, which are the theoretical laws, together with the correspondence rules enable the physicist to explain empirical laws by logical deduction. In Carnap's philosophy the numerical approximation that Duhem saw existing between the solution sets for the equation deduced from the axioms on the one hand and the solution sets for the equation the empirical laws on the other hand, has no semantical implications and is not problematic. The post-Positivist philosophers agree with Duhem, and maintain that while the numerical difference between theoretical and empirical laws are experimentally indistinguishable because of measurement error, nonetheless the solution sets from the two types of laws are logically distinguishable, such that it is incorrect to say that experimental laws are logically derived from theoretical postulates.

Carnap's philosophy of scientific criticism is his thesis of confirmation. Both theoretical and empirical laws may be more or less confirmed, but empirical laws are confirmed directly by observation or measurement, while theoretical laws are confirmed indirectly through the confirmation of the empirical laws derived from them. Both empirical and theoretical laws may be classified as either universal or statistical laws. Most of Carnap's discussions of this distinction is in the context of empirical laws. All empirical laws are statements expressing observed regularities as precisely as possible. If a certain regularity is observed at all times and all places, then the regularity is expressed in the form of a universal law. But if the law asserts that an event occurs in only a certain percentage of cases, and if the percentage is specified or if in some other way a quantitative statement is made about the relation of one event to another, then the statement is called a statistical law. Both types of laws occur in the object language of science, and both are empirical statements. Statements about either universal and statistical laws occur in the meta-language, that refers to the object language of science, and for either types the statements in the metalanguage may refer to the degree of confirmation of the laws. Statements of the degree of confirmation are statements of logical probability associated with both universal and statistical laws. Logical probability is an estimate of the long-term relative frequency stated by the statistical laws, and takes values between zero and one inclusively. The statements associating the degree of confirmation to a statement in the object language, are statements in the metalanguage. The metalanguage is a language of the philosopher of science, and philosophy is not in Carnap's view an empirical or factual science. Philosophy of science is the logic of science, and the statements in the metalanguage are L-true or analytic. Logical probability is the logical relation similar to logical implication. By a logical analysis of a stated hypothesis *h* and the stated evidence *e*, one may conclude that *h* is not deductively implied but is partially implied by *e* to the degree *r*. For any pair of sentences *e* and *h* inductive logic assigns a number giving the logical probability of *h* with respect to *e*. In this way Carnap views the metalanguage to consist of analytic statements as opposed to the synthetic statements in the object language consisting of laws of nature.

Carnap's philosophy of scientific discovery gives different accounts for the discovery of empirical laws and the discovery of theoretical laws. His philosophy of discovery of empirical laws is inductivist; induction is the measurement of the degree of regularity in observed instances known either passively by casual observation or actively by experimentation. His philosophy of discovery of theoretical laws recognizes the role of the creative imagination. He gives consideration to the use of computers. He expresses doubts that rules can be established to enable a scientist to survey millions of sentences giving various observational reports, and then by a mechanized procedure applying these rules to generate a general theory consisting of a system of theoretical laws that would explain the observed phenomena. This is because theories deal with unobservables and use a conceptual framework that goes far beyond the framework used for the description of observations. Creative ingenuity is needed to create theories. Therefore Carnap concludes that there cannot be an inductive machine, a computer system into which the scientist can input all the relevant observation sentences, and then get an output consisting of a system of laws that explain the observed phenomena. He only believed that there can be an inductive machine that given observation *e* and hypothesis *h* will mechanically determine the logical probability or degree of confirmation of *h* on the basis of *e*. The post-Positivist philosophers of science rejected the Positivist's strong distinction between theory and observation. Like Einstein and Heisenberg, they maintained that theory determines what is observed. Therefore no theory-independent framework for observation exists.

Hempel's Critique of Analyticity

Carl G. Hempel was one of Carnap's more sympathetic colleagues and had been Carnap's assistant just after immigrating to the U.S. from Nazi Germany. But after reading Quine's "Two Dogma's of Empiricism" (1952) in which Quine criticized Carnap's concept of analyticity, Hempel gave serious reconsideration to Carnap's analyticity thesis. Hempel does not reject Carnap's concept of L-truth; his disagreement is only with the concept of A-truth, the truth that Carnap calls meaning postulates, which are known to be true by virtue of the meaning relations among the descriptive terms in the sentence.

Hempel's critique of A-truth is set forth in "Implications of Carnap's Work for the Philosophy of Science" in Schilpp's *The Philosophy of Rudolf Carnap* (1963) and relevant comments are to be found in his earlier work, "Theoretician's Dilemma" in *Minnesota Studies* (1958). Firstly Hempel considers problems of empirical significance presented by analyticity. After contrasting Carnap's concept of reduction sentences with the idea of definition, taking note that the reduction sentence offers convenient schema for a partial operational meaning, Hempel states that contrary to Carnap the reduction type of sentence does not eliminate all dependency on general empirical laws in these sentences. He says that P.W. Bridgman had advocated operational definitions one definition for every method of measurement because defining any measurement concept by more than one method of measure incurs the risk of an invalid empirical generalization, even if the different methods yield the same measurement value. The reduction type of sentence eliminates this risk, because in it only one generalization is used. However, Hempel says that an inductive risk is still incurred even for reduction sentences, since even if only one operational criterion is used, as in the reduction sentence, any application of a term requires a generalization. Therefore reduction sentences "fuse" together two functions of language, which had traditionally been thought to be totally different. These are firstly the specification of meanings and secondly the description of contingent fact. He maintains that the fruitful introduction of new concepts in science is always intimately bound up with the establishment of new laws.

Hempel then generalizes on his thesis that reduction sentences have the two functions of meaning specification and empirical law, to produce his own general conception of a semantical or "interpretative" system. Firstly he distinguishes an observational and a theoretical vocabulary. Then he states that a theory *T* characterized by a set of postulates with primitive theoretical terms constituting the theoretical vocabulary, is made an interpreted system by the set of sentences *J* satisfying three conditions: (1) *J* is logically compatible with *T*; (2) *J* contains no extralogical (descriptive) terms that are not an element of the observational or theoretical vocabulary; (3) *J* contains elements of the observational and theoretical vocabulary in an essential way, i.e. in a manner that does not make *J* logically equivalent to some set of sentences in which neither the observational or the theoretical terms occur. Interpretative systems so conceived share the same two characteristics that distinguish reduction sentences from definitions. Firstly they give only partial definitions of the theoretical terms they specify, and secondly they are not purely stipulative in character, but imply certain statements containing only observational terms. However unlike Carnap's concept of a semantical system with reduction sentences, Hempel's general concept of an interpretative system does not provide an interpretation, complete or incomplete, for each theoretical term individually in the whole system. Therefore in the interpretative system *J* the theoretical terms are not dispensable, and Hempel argues that in his definition of an interpretative system, the distinction between the theory and its interpretative sentences is arbitrary, because these two types of sentences have the same status and function. It is only in conjunction with the interpretative sentences that the theory can imply observational sentences, and the interpretative sentences no less than the theory may be theoretical laws. Furthermore, when discrepancies between predictions and experimental data call for modification of the predictive apparatus, suitable adjustments may be made not just by changing the theory but alternatively by changing the interpretative sentences. Therefore interpretative sentences must have the same status as the sentences constituting the theory, thus making it difficult to identify either theory or interpretative sentences as analytic. Following a similar line of argument Hempel rejects Carnap's proposal of introducing predicates by means of meaning postulates, which purport to separate the meaning specification function from the empirically descriptive function. Hempel questions the rationale for separating these two functions. He asks what distinctive status is conferred on a meaning postulate, since any statement once accepted in empirical science may conceivably be abandoned for the sake of resolving a conflict between theory and the stated body of available evidence. He says that apart from logical and mathematical truths, there can be no scientific statements that satisfy conditions for analytic meaning postulates.

In addition to discussing problems for empirical significance of analytical sentences, Hempel also discusses problems of empirical testing. Hempel references Carnap's *Logical Syntax of Language*, where Carnap references Poincare and Duhem, saying that no statement accepted in empirical science is taken to be immune from criticism and revision. Carnap furthermore stated that a statement in a scientific theory cannot be tested in isolation, but must be tested with other accepted statements, such that it is the entire theoretical system that is tested. And this is what Quine also maintains in his "Two Dogmas", which Hempel references in this context. Hempel relates that on Carnap's view of a semantical system, in which theoretical terms are viewed as being introduced by reduction sentences based on an observation vocabulary, it is possible to speak of individual sentences containing theoretical terms as being confirmable by reference to observation sentences. But Hempel notes that in his general concept of an interpreted theory, this idea has no useful counterpart, because one would have to say that the experimental import of a sentence relative to an interpreted theory is expressed by the class of nonanalytic observation sentences implied by the sentences and the theory. His view renders the notion of testability and experiential significance relative to a given theory, assigning all sentences of the theory the same experiential import represented by the class of all observation sentences implied by the theory. This is because testability and empirical significance are attributable not to scientific statements in isolation, but only to interpreted theoretical systems.

Hempel concludes that these considerations make it doubtful that the basic tenants of Positivism and empiricism can be formulated in a clear and precise way. The circumstance that empirical significance and testability requirements are applicable to entire theoretical systems, make these requirements extremely weak. For the Positivist that weakness permits the disturbing possibility of adding to contemporary physical theory an axiomatized metaphysics of Being and Essence, that would be an empirically significant system. One alternative is to exclude theoretical terms altogether, but Hempel invokes the criterion of simplicity. He concludes that the problem of giving a precise explication of this aspect of scientific theories presents a new and challenging task for the philosophy of science.

Carnap's Reply to Hempel

Carnap replies to Hempel's attach on the analytic-synthetic distinction both in the Schilpp volume containing Hempel's critique and in the concluding two chapters of his *Philosophical Foundations of Physics* (1963). He maintains that the analytic-synthetic distinction is of supreme importance for philosophy of science, and that the theory of relativity could not have been developed had Einstein not recognized the sharp dividing line between pure mathematics with its many types of logically consistent geometries, and physics in which only experiment and observation can determine which of these mathematical geometries can be applied most usefully to the physical world.

Firstly however Carnap takes up the identification of the analytic-synthetic distinction in natural language. He notes that natural language is sufficiently imprecise, that not everyone understands every word in the same way, such that some sentences may be ambiguous as to whether they are analytic or factual. The division depends on what characteristics described by the predicate terms, are taken to be essentially or definitively related to one another. For example is a redheaded woodpecker defined by its red colored head plumage? If not, than a green headed bird may be classified as a redheaded woodpecker, if it has other characteristics deemed definitive of the species. Carnap maintains that while certain statements may be ambiguous because of the vagueness of the predicates, the analytic-synthetic distinction as such is not therefore problematic for the same reason.

Carnap next turns to the analytic-synthetic distinction in an artificial observation language. In this case the distinction is determined by laying down precise rules, which are the meaning postulates or A-postulates. These rules determine what characteristics described by predicate constants are essential to their subjects. To the extent that these rules are vague, there will be sentences that are vague with respect to the analytic-synthetic status.

But Carnap says that in such cases the distinction between analytic and synthetic is not as such vague.

Then he turns to the determination of the distinction in an artificial theoretical language, where the fact that theoretical terms cannot be given complete interpretations causes special difficulties. He takes as an example a vapor track in a Wilson cloud chamber, which can be observed and can be explained in terms of an electron passing through the chamber. Such observations provide only a partial and indirect empirical interpretation of the entity referenced by the theoretical term "electron" to which the observed track is linked by correspondence rules. The problem is to find a way to distinguish in the linguistic network of correspondence postulates and theoretical postulates, those sentences that are analytic and those that are synthetic. It is easy to identify the L-true sentences, because descriptive terms are not involved in determining L-truth. But A-truth, the truth of analytic sentences, is problematic in this case. To recognize analytic statements in a theoretical language, it is necessary to have A-postulates that satisfy the meaning relations holding among the theoretical terms. But the theoretical postulates alone cannot serve as A-postulates, since without the correspondence rules the theoretical postulates have no interpretation at all. Yet the theoretical postulates together with the correspondence postulates cannot be analytic, because then the theory would have no empirical content.

Carnap notes Hempel's proposal that there is a double role for the theoretical and correspondence postulates, that defies the analytic-synthetic distinction, such that these postulates both stipulate meaning and also make empirical assertions. But Carnap proposes another way that preserves the empirical content of scientific theories while admitting the analytic-synthetic distinction. His proposal utilizes the Ramsey sentence, but without Ramsey's final step of eliminating the theoretical terms from the semantical system, since he believes that eliminating theoretical terms is too inconvenient for the scientists, who find that theoretical terms simplify their work enormously. Instead of splitting an interpreted theory into theoretical postulates and correspondence rules, Carnap proposes splitting it into analytic and factual sentences with the factual part consisting of a Ramsey sentence equivalent to the empirical content of the interpreted theory. The Ramsey sentence therefore implies the whole interpreted theory, and this implication is itself analytic; it is the analytic part of the theory. Carnap maintains that this analytic implication provides a way to distinguish between analytic and synthetic statements in the theoretical language, because the analytic implication is that if there exist entities that are referenced by the existential quantifiers of the Ramsey sentence, that are of a kind bound together by all the relations expressed in the theoretical postulates of the theory, and that are related to observed entities by all the relations specified by the correspondence postulates of the theory, then the theory itself is true.

In his "Theoretician's Dilemma" Hempel had criticized the Ramsey sentence as avoiding reference to theoretical entities only in Greek variables rather than in spirit. The Ramsey sentence still asserts the existence of certain entities of the kind postulated by a physical theory without guaranteeing any more than does the physical theory, that those entities are observable or at least are fully characterizable in terms of observables. Therefore, the Ramsey sentence provides no satisfactory way of avoiding theoretical concepts.

In his replies to Hempel in Schilpp's book Carnap says that he agrees with Hempel that the Ramsey sentence does refer to theoretical entities by the use of abstract variables. But he argues that these entities are not unobservable physical objects like atoms or electrons, but rather are purely logico-mathematical entities such as natural numbers, classes of such numbers, or classes of classes. The Ramsey sentence for a physical theory is a factual statement that says that the observable events in the world are such that there are natural numbers, classes of such numbers, or classes of classes, that are correlated with the events in a prescribed way, and which have among themselves certain relations.

Quine's Pragmatist Critiques

Willard Van Orman Quine was born in Akron, Ohio, in 1908. In 1930 he graduated *summa cum laude* in mathematics from Oberlin College, and then entered Harvard University's graduate school of philosophy. He wrote his doctoral dissertation under the direction of Alfred North Whitehead, the co-author with Bertrand Russell of the *Principia Mathematica*, and he published it as *A System of Logistic* in 1934. Quine became a faculty member of Harvard's department of philosophy in 1936, where he remained for the duration of his long career. He enjoyed traveling, and wrote an autobiographical travelogue as *The Time of My Life* in 1985. Quine described his long acquaintanceship with Carnap in

"Homage to Rudolf Carnap" (1970), a memorial article published in the year of Carnap's death, and reprinted later in Quine's *Ways of Paradox* (1976). Quine met Carnap during his European travels in the 1930's, and their dialogues continued after Carnap relocated to the United States in 1935. While Quine might be regarded as Carnap's principal protagonist, their philosophies are much more similar than different. In the memorial article Quine refers to Carnap as a towering figure, who dominated philosophy in the 1930's as Russell had in previous decades, and he also refers to Carnap as his greatest teacher. Their private correspondence has been published under the title *Dear Carnap, Dear Van* (ed. Creath, 1990), but the correspondence reveals nothing about their philosophical views that is not already known from their published works.

Quine's best known criticism of Carnap's philosophy is his rejection of the analytic type of statement. This criticism together with several others have their basis in Quine's Pragmatist view of empiricism. Quine published a brief statement of his own doctrine of empiricism as "The Pragmatist's Place in Empiricism" (1975), later appearing in his *Theories and Things* (1981) as "Five Milestones of Empiricism". This paper is ostensibly a history of empiricism in terms of five historical turning points, but miraculously the five historical milestones also happen to be the central theses of Quine's own Pragmatist philosophy. He summarizes these five turning points as follows:

1. The shift from ideas to words.
2. The shift of semantic focus from terms to sentences.
3. The shift of semantic focus from sentences to systems of sentences.
4. The abandonment of the analytic-synthetic distinction.
5. The abandonment of any first philosophy prior to natural science.

Quine's several criticism's of Carnap's Positivist version of empiricism may be viewed as having a basis in these five distinctive aspects of his Pragmatist version of empiricism. The first two of the five points are the basis for Quine's criticism of Carnap's doctrine of intensions, as well as a critique of the idea of propositions. The third point, sometimes known as the Duhem-Quine thesis, is the basis for Quine's critique of logical reductionism and for his wholistic thesis of semantical indeterminacy and his thesis of ontological relativity. The fourth is his rejection of analyticity, which follows from the third point. And the fifth and final point is Quine's critique of Carnap's doctrine of "frameworks" and of the distinction between "internal" and "external" questions. Each of these criticisms are considered in greater detail below.

Quine's Critique of Intensions and Propositions

At the close of his "Foreword" to Quine's *A System of Logistic* Whitehead commented that logic prescribes the "shapes" of metaphysical thought. The logic under consideration of course was that in Whitehead and Russell's *Principia Mathematica*, and there was probably no expositor of this logic who both illustrated and advocated Whitehead's comment more consistently than Quine; the comment both affirms the system-determined nature of semantics and suggests Quine's doctrine of ontological relativity. For more than a decade after *System of Logistic* Quine published a number of articles which describe how the Russellian symbolic logic and specifically how its theory of quantification enables the user of the logic to exhibit explicitly his ontological commitments, the shape of his metaphysics. The user's ontological commitment to the kinds of things he believes exists, is exhibited by the variable, the symbol that is bound by either the existential or the universal quantifier. The term "variable" in this context has a distinctive meaning that it does not have in mathematics. In his "A Logistical Approach to the Ontological Problem" (1939) reprinted in *Ways of Paradox* (1966) Quine expresses the role of logical quantifiers with the memorable refrain: To be is to be the value of a variable. This means that what entities there are from the viewpoint of a given discourse in the logic depends on what symbols are accessible to binding by quantifiers to become variables in the symbolic logic, and a shift from one discourse to another may involve a shift of ontology.

In 1947 Quine published "On Universals" in *Journal of Symbolic Logic* and "Logic and the Reification of Universals" in *From A Logical Point of View* (1953). In these papers he describes how the nominalist and realist views toward the historic problem of universals are expressed in the Russellian notation. The nominalist view is that only individuals exist, and it is expressed in the Russellian notation by limiting the quantifiers to ranging only over symbols referencing individual entities. On the other hand the universalist view affirms that attributes or properties exist. In the Russellian notation the existence of attributes is expressed by placing predicates within the range of quantifiers. For this reason Quine calls the universalist view the "Platonist" view, and he calls the attributes "abstract entities". Or when the abstract entities are said to exist in the human mind as meanings or concepts, Quine calls them "mental entities". The notational role of the quantifier is referential, such that whatever type of symbol may assume the role of a variable bound by a quantifier, thereby assumes the role of referencing an entity. Ostensibly Quine's purpose is not to advocate one or the other ontological thesis, but to advocate the role of the quantifiers as making a philosopher's ontological commitment explicit.

Quine has his own view on the issue of universals, and it might best be described as near-nominalist. In 1947 he co-authored with Nelson Goodman "Steps Toward A Constructive Nominalism" in *The Journal of Symbolic Logic*. Unlike most papers appearing in academic journals, this article was not so much an analytical paper, as it was a kind of nominalist manifesto advocating a programme for applying the symbolic logic. Quine has denied that he is a nominalist, because he accepts the existence of classes, which he views as a kind of abstract entity. And he accepts the existence of classes, because he could not eliminate them from the logistic reductionist programme. Except for his acceptance of classes, however, Quine is a nominalist, and he explicitly rejects the existence of such abstract entities or mental entities as properties or attributes, and intensions as propounded by Carnap. One reason that he rejects these latter types of abstract entities, is that they can be eliminated from the logistic reductionist programme, as he construes it. The second reason is that he denies that Carnap's intensions can be treated extensionally, as Carnap attempts to treat them by relating them to classes by analytical statements, a type of statement that Quine rejects.

In "Five Milestones" Quine notes that the first of the five turning points in the history of empiricism, the shift from ideas to words, had its origin with the medieval nominalists. Quine argues against the reification of universals, and he says that affirming the existence of abstract or mental entities is due to a common confusion, in which descriptive predicates are given a referential function that is properly had by bound variables. In "Ontological Relativity" he describes this error as a case of the copy theory of knowledge, which he says is an uncritical semantics. He ridicules this error as the "myth of the museum" and the "fantasy of the gallery of ideas", by which he means that words are mistakenly understood to be labels for ideas or meanings, as though they were exhibits. He views the confusion between names and descriptions to be a particularly pernicious philosophical error, and he maintains that Russell's theory of descriptions offers the way to avoid it. This is the technique used by Russell in his "On Denoting" in Mind (1908). In his "On What There Is" (1948) reprinted in *Logical Point of View* Quine says that Russell's theory of descriptions enables the philosopher to transform names into predicates, such that names should not be taken as an ontological criterion for deciding what is real. The correct criterion for determining the ontology of a language is the use of the quantified symbol or variable, so that predicates are not confused with names, and no claims are made to the effect that predicates name entities, unless the predicates are explicitly quantified.

Closely related to the first milestone, the second is the shift of semantic focus from terms to sentences. In "Five Milestones" Quine explains that the meanings of words are abstractions from the truth conditions of the sentences that contain them, and that it was the recognition of this semantic primacy of sentences that gave us contextual definition. Quine traces the development of contextual definition, which he calls a revolution in semantics, to Jeremy Bentham's technique of "paraphrasis", which is a kind of paraphrasing or circumlocution. If Bentham found some terms convenient but ontologically embarrassing, contextual definition enabled him in some cases to enjoy the services of the term, while disclaiming its denoting. In "Russell's Ontological Development" (1966) reprinted in *Theories and Things* (1981) Quine joins Ramsey's characterization of Russell's theory of descriptions as a paradigm of philosophical analysis, and he says that our reward for the paraphrasis technique is the recognition that the unit of communication is the sentence and not the word.

In his *Meaning and Necessity* Carnap explicitly affirms that intensions are not names either of concepts or of abstract entities. He maintains that like physical properties intensions may be said to be objective without invoking any hypostatization, and that they are indifferent to either concrete or abstract objects. Carnap's intensions are reminiscent of the Scholastic logicians' distinction between *suppositio* and *significatio* for terms, although Carnap never makes this comparison. According to the theory of *suppositio* a univocal term's *significatio* or meaning is the same whether the term occurs either as a subject or as a predicate in an affirmative categorical proposition. But its *suppositio* or supposition as a subject is called "personal", because it references the individual members of the class according to its associated quantifier, while its supposition as a predicate is called "simple", because no reference is made to the members of the class it signifies, and it's meaning is used indifferently with respect to instantiation. When Quine attacks Carnap's admission of attributes and intensions, as he does in "On the Individuation of Attributes" (1975) in *Theories and Things*, he attacks Carnap's use of analytic statements and does not claim that Carnap has confused names and predicates. But even apart from the issue of analyticity, Carnap's theory of intensions is inconsistent. In the section of *Meaning and Necessity* in which he discusses variables, Carnap explicitly agrees with Quine's view that the ontology to which one's use of language commits oneself comprises simply of the objects that one treats as falling within the range of values of one's variables, and he explicitly accepts Quine's refrain that to be is to be the value of a variable. Quine and Whitehead recognized, as Carnap had not, that one's logic shapes one's metaphysics, and Quine's papers on theory of reference had as their basis the thesis that the Russellian logic expresses existence exclusively by means of the instantiating quantifiers.

The Russellian manner of expressing ontological committment has its peculiar and controversial aspects, that are best seen by way of contrast with the earlier Aristotelian logic. In the Aristotelian logic existence is not affirmed by the quantifier, but rather by the copula term "is", as in "Every raven is black". The noteworthy difference is that in the Russellian notational conventions the only existence that can be affirmed is that of entities, namely those entities referred to by the quantified variable, such that any attempt to affirm the reality of attributes or properties must describe them as entities referred to by a quantified predicate. In the Aristotelian logic, however, the reality of what may be called an attribute signified by the predicate, need not be hypostatized as some kind of Platonic entity. Quine is therefore consistent in his use of the Russellian logic, when he describes the reality status of red, the property, as an abstract entity, and when he describes the reality status of red, the meaning, as a mental entity. According to the syntactical categories admitted by the Russellian logic all philosophers are either nominalists or Platonists, since they must either deny attributes as real by not quantifying the predicate, or they must affirm them as Platonic entities by quantifying over the predicate. In the Russellian logic attributes, properties, aspects, and accidents have no reality status apart from that of subsisting entities. Carnap's attempt to admit intensions or meanings and properties that are not hypostatized, is inconsistent with his use of the Russellian logic and with his agreement with

Quine that ontology is described by means of bound variables. And his complaint about erroneously labeling philosophers "Platonists" is similarly inconsistent. Other and more consistent philosophers have recognized the Russellian logistic to be an Orwellian-like "newspeak" for advocating a nominalist agenda hidden in its notational conventions, which the pontificating Quine would enforce as a "canonical notation".

In his *Medieval Logic and Metaphysics* (1972) the University of Manchester British philosopher, David P. Henry, asks how modern logic, caught as it is in the "entanglement" of the expression of existence in the quantifiers, can recapture the untrammeled approach to existence enjoyed by its medieval predecessors. He proposes reconsideration of modern formal logic of the Polish logician S. Lesniewski (1886-1939), which is unfamiliar to most modern logicians. In his autobiography Quine recounts his arguing with Lesniewski about "abstract entities" (Quine's characterization) while visiting Warsaw in the 1930's. Henry notes that Liesniewski's logic employs an interpretation of the quantifiers, which allows their dissociation from its currently conventional entanglement with the notion of existence, and he gives examples of how Lesniewski's interpreted system with its ontology may be used in the analysis of medieval themes including *suppositio* with an artificial language designed by Henry. In the present context the significance of Henry's work is that it shows how Quine's ontological agenda does not imply a simplistic dichotomy between modern mathematically expressed logic and antiquated colloquially expressed Aristotelian logic, but rather depends on very specific notational conventions distinctive of the Russellian logistic, to which there can and do exist alternatives. Quine's world view through the lenses of Russellian logistic with its ontological agenda reducing attributes either to "abstract entities" or to unreality, invites comparison with the obliviously contemplative noblemen of the airborne floating island of Laputa in Swift's satirical *Gulliver's Travels*. The Laputians viewed the world through the lenses of Cartesian geometry with Descartes' ontology of primary and secondary qualities. In Descartes' philosophy only geometrical or "primary" qualities have objective reality, while all others are "secondary", i.e. subjective and unreal. The Laputian noblemen were so obliviously faithful to their distorted Cartesian view of the real world, that they saw all reality as geometrical figures including even their wives, who were not similarly faithful to the Cartesian ontology and therefore felt so neglected that they were inclined to be unfaithful to their husbands. Comparison of Quineian and Cartesian ontologies with the aid of Gulliver's travelogue is not rhetorical; Quine's rejection of properties, attributes and qualities makes impossible any qualitative differentiation based on reality, and renders his real world as starkly extensional as Descartes'. This extensionalism built into the Russellian notational conventions is an ontological commitment that is as inconsistent with Quine's ontological relativity as his behaviorism.

Quine's Critique of Reductionism

The third milestone in "Five Milestones" is the semantical shift from sentences to whole systems of sentences. This shift to a wholistic (or holistic) view of the semantics of language is a central characteristic of Quine's philosophy, and it went through some evolution. He came to think that his earlier and more radical Pragmatism implies an unwanted cultural relativistic view of truth, and consequently in the 1970's he attempted to restrict the extent of his semantical wholism, so that the semantics of theory is not viewed as contributing to the semantics of observation language.

His first statement of his wholistic thesis is a highly metaphorical statement given in "Two Dogmas of Empiricism" (1951), one of his best known papers, reprinted in his *Logical Point of View* as well as in several anthologies. The two dogmas he criticizes in this paper are the Logical Positivist theses of analyticity and reductionism. He defines the reductionist thesis as the belief that each meaningful sentence is equivalent to some logical construct based on terms referring to immediate experience, and he notes that Carnap was the first empiricist who was not content with merely asserting the reducibility of science to terms of immediate experience, but who actually took steps toward carrying out the reduction in the *Aufbau*. Then Quine says that while Carnap later abandoned this radical reductionist effort, the dogma of reductionism continues in the idea that to each synthetic (i.e. empirical or nonanalytic) statement there is associated a unique range of possible sensory events, such that the occurrence of any of them would add to the likelihood of truth of the statement. Similarly for each synthetic statement there is associated another unique range of possible sensory events whose occurrence would detract from that likelihood. This dogma is implicit in the verificationist theory of meaning, and it survives in the thesis that each statement taken in isolation can admit of either confirmation or "infirmation", which is to say, either verification or falsification.

The view of empiricism that Quine advocates as his alternative to reductionism is the thesis that statements about the external world face the tribunal of sense experience not individually, but only as a corporate body. Quine references Duhem in this context, and his alternative view of empiricism has since come to be known as the "Duhem-Quine Thesis". However, Quine's wholistic view is more radical than that of Duhem. Quine purges Duhem's philosophy of physical theory of its Positivism by ignoring Duhem's two-tier semantics, that led to Duhem's distinction between "practical facts" and "theoretical facts". Quine's treatment of the difference between observation and theory is not a Positivist semantical metatheory. Furthermore, Quine's radical wholism does not admit a distinctive semantical status even for pure mathematics and formal logic. Speaking metaphorically Quine says that the totality of our beliefs including mathematics and logic is a man-made fabric which impinges on experience only along the edges. Then mixing metaphors he describes total science as a field of force whose boundary conditions are experience in which the laws of logic and mathematics are simply statements in the field that are more remote from experience. Any conflict with experience at the periphery occasions adjustments in the interior of the field, such that truth values must be redistributed over some statements, and a re-evaluation of some statements entails reevaluation of others because of the logical connections among them.

The enabling feature of Quine's wholistic doctrine of empiricism is his thesis that the total field is so empirically "underdetermined" by its boundary conditions, which are experience, that there is much latitude for choice as to what statements to re-evaluate in the light of any single contrary experience. And the criterion governing the choice of beliefs in the underdetermined system is entirely Pragmatic, where the objective is a relatively simple conceptual scheme for predicting future experience in the light of past experience. The thesis of the empirical underdetermination of language can be traced to Duhem's view of scientific theory. Duhem said that there could be many theories, all equally empirically adequate, that explain the same phenomenon. But Quine furthermore extends Duhem's thesis to include not just theory but all of language including observation language. He maintains that no statement is immune from revision, and he notes that revision even of the law of the excluded middle has been proposed as a means of simplifying quantum physics. Quine notes that there is a natural tendency when making revisions to disturb one's existing system of beliefs as little as possible, with the result that those statements that we are least likely to revise are those that have sharp empirical reference, while those that we are most likely to revise are those more theoretical statements that are relatively centrally located within the total network or web of beliefs. Later in his *Philosophy of Logic* (1970) this natural tendency becomes the "maxim of minimum mutilation", an idea similar to James' thesis of "minimum disturbance" in his *Pragmatism* (1907, Pp. 5960).

Quine's most elaborate statement of his wholistic thesis is set forth in his first full-length book, *Word and Object* (1960). Instead of the metaphorical statement of his view in "Two Dogmas" a decade earlier, here he expresses his thesis in the literal vocabulary of behavioristic psychology. Much of the book is an exposition of his thesis of semantic indeterminacy as it is manifested in translation between languages, and thus appears as his indeterminacy of translation thesis. In the translation situation he portrays the field linguist in the same situation as Carnap considers in "Meaning and Synonymy in Natural Language", where Carnap attempted to describe how the field linguist can ascertain a term's intension by identifying its extension from the observed behavior of native speakers of an unknown language. Carnap admitted that this determination of extension involves uncertainty and possible error due to vagueness, but he excused this uncertainty and risk of error because it occurs even in the concepts used in empirical science. While this admission of extensional vagueness in science made the fact unproblematic for Carnap, it had just the opposite significance for Quine. For Quine extensional vagueness is an inherent characteristic of language that he calls "referential inscrutability", and which he later calls "ontological relativity". And what Carnap called the intensional vagueness, Quine prefers to consider as a semantical indeterminacy in stimulus-meaning but without admitting intensions.

Quine rejects Carnap's thesis of intensions, explicates his own theory of meaning in terms of behavioristic psychology, and proposes his doctrine of "stimulus-meaning". Stimulus meaning is a disposition by the native speaker of a language to assent or dissent from a sentence in response to present stimuli, where the stimulus is not just a singular event but rather a "universal", a repeatable event form. Stimulus meaning is the semantics of those sentences that Quine had earlier described metaphorically as positioned at the edge of the system of beliefs viewed as a force field, as opposed to the more theoretical sentences that are in the interior of the field.

In Quine's philosophy the idea of stimulus meaning is not a special semantics, but rather is an attempt to isolate the net empirical content of each of various single observation sentences without regard to the theory that contains them yet without loss of what the sentence owes to that containing theory. This attempt to isolate the semantics of observation language is a move away from his earlier critique of reductionism, where reductionism is understood as statements having a unique range of possible sensory events, such that the statements can be criticized in isolation. But at this stage Quine still retains his original thesis of empirical underdetermination, in which empirical underdetermination is integral to his wholistic thesis of semantical indeterminacy or vagueness.

The underdetermination thesis admitting multiple and alternative observation sentences for the same stimulus situation presents a question: how can the same stimuli yield alternative stimulus meanings? One of Quine's answers is that the alternative theories or belief systems in which the stimulus situation is understood supply different significant approximations. But there still remains the question of how stimulus meanings are to be construed as approximations. In answer, Quine has a theory of vagueness which he sets forth in the third and fourth chapters of *Word and Object*, which resembles the latter Wittgenstein's thesis of paradigms, except that Quine explicitly invokes the behavioristic stimulus-response analysis of learning. On this analysis Quine rejects the view that stimulations eliciting a verbal response "red" are a well-defined or neatly bounded class. He maintains that the stimulations are distributed about a central norm, which when a language is initially being learned, may be a very wide distribution. The penumbral objects of a vague term are the objects whose similarity to ones for which verbal response has been socially rewarded in the learning process, is relatively slight. The learning process is an implicit induction on the part of the subject regarding society's usage, and the penumbral cases are those words for which that induction is most inconclusive for want of evidence, because the evidence is not there to be gathered. And society's members have had to accept similarly fuzzy edges when they were learning. There is an inevitability of vagueness on the part of terms learned by ostension, and it carries over to other terms defined by context on the basis of these ostensively learned terms.

Since the writing of N.R. Hanson in 1958 the participation of theoretical concepts in the semantics of observation language is often expressed by saying that observation is "theory-laden". And this semantical participation of theory in observation has made problematic the objectivity of observation, and therefore the decidibility of scientific criticism. In 1968 in "Epistemology Naturalized" in *Ontological Relativity* Quine states that Kuhn and Hanson among others have tended to belittle the role of evidence in science and to accentuate cultural relativism, and that such philosophers represent a wave of epistemological nihilism. He reports that Hanson maintains that observations vary from observer to observer according to the amount of knowledge that the observers bring with them, such that one man's observation is another man's closed book or flight of fancy, with the result that observation as the impartial and objective source of evidence for science is bankrupt. At this stage of Quine's thinking the semantical contribution of theory to observation is still problematic for Quine, but he continued to characterize observation language in terms of behavioristic theory of learning. In the chapter titled "Observation" in his *The Web of Belief* (1970) Quine says that an observation sentence is a sentence that can be learned ostensively by the association of heard words with things simultaneously observed, an association which is conditioned and reinforced by social approval or successful communication, and which becomes habitual. And because of the social character of its learning, the observation sentence must be understandable by all competent speakers of the language who might be asked to assent to it. Thus according to Quine the sentence "That is a condenser" is not an observation sentence, even if experts agree to it. Quine maintains contrary to the Positivists, that what qualifies a sentence as observational is not the lack of theoretical terms that may occur in theory formulations, but just that the sentence taken as an individual whole commands assent consistently or dissent consistently when the same global sensory stimulation is repeated. This behavioristic characterization initially enabled Quine to evade reference to semantics in his identification of observation language, and thereby to separate his view from that of the Positivists, who defined observation language in semantical terms. But in attempting to avoid a cultural relativist view of truth he thought he found in the likes of Hanson, Quine found himself getting back into the semantics of observation with the very Positivist objective of keeping the semantics of observation uncontaminated by that of theory.

After *Word and Object* and *Web of Belief* Quine further developed the Duhem-Quine thesis in his "On Empirically Equivalent Systems of the World" in *Erkenntnis* (1975), which as it happens had in 1930 been made the official journal of the Vienna Circle. This development of the Duhem-

Quine thesis represents a further restriction on Quine's earlier version on his wholistic semantical thesis of observation. Previously he had viewed empirical underdetermination as integral to semantical indeterminacy or vagueness in his semantical wholism. But in this paper he revises the concept of empirical underdetermination of language, and separates it from the wholistic view of the Duhem-Quine thesis. The scientific hypotheses that purport to describe things beyond the reach of observation are related to observation sentences by a kind of one-way implication, such that many alternative hypotheses may imply the same set of observation sentences, but not vice versa; observation sentences do not uniquely imply just one theory purporting to explain the observable events. It now is in this sense that natural science is "empirically underdetermined" by all possible events. Quine says that underdetermination lurks where there are two irreconcilable theory formulations each of which implies exactly the desired set of observation conditionals plus extraneous theoretical matter, and where no formulation affords a tighter fit. In Quine's vocabulary the phrase "observation conditional" is an empirical generalization expressed in conditional form and implying an observation sentence describing an individual event, and the phrase "theory formulation" is a conjunction of the axioms of a deductive theory, which implies observation conditionals. This is a different sense of "empirical underdetermination" than what Quine meant in "Two Dogmas", because it resurrects the idea of a semantically neutral observation language, which philosophers such as Kuhn and Feyerabend reject. These philosophers find a phrase such as "same observation sentences" when speaking of sentences implied by alternative theories to be very problematic; they deny that different theories can have the same set of observations because of the contribution of the semantics of theory to the semantics of observation language. Having revised "empirical underdetermination", Quine then distinguishes his revised concept from the wholistic doctrine of the Duhem-Quine thesis. He reiterates that the wholistic doctrine says that scientific statements are not separately vulnerable to adverse observations, since it is only jointly as a theory that they imply their observable consequences, with the result that any one of the statements can be adhered to in the face of adverse observations by revising others. Then he states that wholism lends credence to the underdetermination thesis, because in the face of adverse observations we are free always to choose among various adequate modifications of our theory, and all possible observations are insufficient to determine theory uniquely.

Also in this work Quine considers several criticisms or "reservations" about the wholism of the Duhem-Quine thesis, and in his defenses he will pick and choose between underdetermination (revised) and wholism (unrevised). The first criticism is that some statements closely linked to observation are separately susceptible to tests of observation, while at the same time these statements do not stand free of theory because they share much of the vocabulary of the more remote theoretical statements. Quine answers that the Duhem thesis does not imply equal status for all statements. He says that the Duhem thesis applies even for observation statements, since scientists do occasionally revoke observation statements when these statements conflict with a well attested body of theory, and when the experiment yielding the observation cannot be replicated. This is such a weak concession to semantical wholism and the indeterminacy of observation, that it effectively limits wholistic theory participation in the semantics of observation language to the status of errors of observation.

A second reservation pertains to the breadth of the theory: If it is only jointly as a theory that scientific statements imply their observable consequences, then how inclusive does that theory have to be? Does the wholistic scope have to include the whole of science taken as a comprehensive theory of the whole world? Quine sees science as an integrated system of the world as science exists at any point in its historical development, but he does not view it as integrated by reductionism into a single unified science. He says that Duhem wholism admits that science is neither discontinuous nor monolithic, but as variously joined and loose in its joints in varying degrees. Later in "Five Milestones" Quine elaborates on this idea by saying that all sciences interlock to some extent not only because of a common logic and mathematics, but also because small "chunks" may be ascribed their independent empirical meaning nearly enough, since some vagueness in meaning must be allowed for. This defense based on vagueness calls upon the semantical indeterminacy that enables wholism.

A third reservation is that the semantical and ontological wholism may imply a cultural relativistic view of truth. Quine denies that his wholism implies a cultural relativistic view of truth. His first argument is external to the wholistic thesis. He finds a paradox in the thesis of cultural relativism: if truth were culture bound, then the advocate of cultural relativism ought to see his own culture-bound truth as absolute. The cultural relativist can not proclaim cultural relativism without rising above it, and he cannot rise above it without giving it up. Quine then turns to the issue of irrationality of theory choice, the argument for cultural relativism that is internal to wholism. He argues that the choice between empirically equivalent alternative systems need not be irrational; he says he will settle for a "frank dualism". He says that oscillation between rival theories is standard scientific procedure, because it is thus that one explores and assesses alternative hypotheses. In this defense Quine switches between underdetermination and wholism. Rationality of theory choice is based on comparability of theories permitted by a neutral observation language. This neutrality is admitted by Quine's revised underdetermination thesis, since it is only theories and not observations that are incompatible. The dualism is therefore merely one due to empirical equivalence. The idea of empirical underdetermination as newly revised in this article is not the context in which the issue of irrationality of theory choice emerges. It emerges in the context of wholism where theory participates in the semantics of observation language. Quine switches to the wholistic context, when he says that whatever we affirm, we affirm as a statement within our aggregate theory of nature as we now see it, and that there is no extratheoretic truth. Quine's frank dualism has not been very frank in this defense. Quine's revised concept of empirical underdetermination is not consistent with his semantical wholism. The revised concept of underdetermination permits a neutral observation language, while the Duhem-Quine wholism continues to permit theory to resolve the vagueness in the semantics of observation language.

Quine eventually recognized this inconsistency. Just as he imposed logical one-way restrictions for his revised concept of empirical underdetermination, he found that he must impose semantical one-way restrictions in the semantical wholism of the Duhem-Quine thesis. In his "Empirical Content" (1981) in *Theories and Things*, which he notes contains "echoes" from "Empirically Equivalent Systems of the World", Quine explicitly uses Hanson's terminology saying that observation sentences are "theory-laden", but he reconstrues the intended meaning of Hanson's phrase to mean that the terms embedded in observation sentences may recur in theory formulations. Thus while Quine here says that observation sentences are theory-laden, he denies to the semantics of theory any participating role in the semantics of observation. In fact in Quine's construing of "theory-laden" it is not observation language that is theory-laden, but rather theory that is observation-laden. Still later in "Truth" in his *Quiddities* (1988) he is explicitly reconciled about refusing to admit theory any resolving function in the semantics of observation. There he says that we work out the neatest world system, and we tighten the squeeze by multiplying the observations. Tightening the squeeze in observation sentences is the progressive reduction of vagueness but only by the addition of information in additional observation sentences. Quine's limitation on which contexts may resolve vagueness and which ones may not, is arbitrary and *ad hoc*. His wish to make observation sentences semantically uncontaminated by theory is a Positivist atavism, even though his motivation is not characteristically Positivist. His point of departure was not a preconceived semantics for observation; he attempted a behavioral (behavioristic) characterization of observation language instead. Still, he believed that an unrestricted wholistic, theory-dependent, context-determined semantics encompassing both theory and observation language implies a relativistic and subjectivist philosophy of truth. And this fear of a relativistic view of truth led him to revise his original version of his Duhem-Quine thesis.

Quine the logician always saw theory language as an axiomatic system with observation language serving as its derived theorems. For Quine, Isaac Newton's mechanics is still "theory" today. On a thoroughly Pragmatic concept of scientific theory, however, theory language is identified not by contrast to an observation semantics or by semantics at all, but by reference to its function or pragmatics in science: it is that discourse that is proposed for testing in contrast to that which is presumed for testing. Thus, observation language need not be exclusively identified as either theory or nontheory language (unless the Pragmatist simply chooses to define "observation" correlatively to his functional definition of "theory"), and all contexts consisting of explicitly or implicitly universally quantified sentences believed to be true operate to resolve the vagueness in the meanings of their common univocal terms. Quine's view is not a Pragmatist view of theory based on the function of theory in empirical basic science, but is better characterized as an archival concept of theory, or what N.R. Hanson called an "almanac" view. Correspondingly his concept of observation language is an archival concept of observation language. Quine believed that this archival view would enable him to make observation language a repository of permanent truth. And his motive is his wish to evade the relativistic view of truth, that he believed is implied by the unrestricted context determination of semantics.

As it happens, Quine is mistaken in his belief that a relativistic, context-determined semantics implies a relativistic view of truth. Regardless of how culture-bound and context-determined may be the semantics of a language, it is not possible either to affirm or to deny truthfully just anything expressed by sentences made with those concepts. The empirical underdetermination of language implies that many alternative sentences can be said which are consistent with the same observations. Still, the empirical constraint imposed exogenously on sentences by the real world forbids any arbitrary distribution of truth values over a set of semantically interpreted grammatical sentences. When any subset of these sentences is given definitional force to specify its semantics, then only some of the remainder sentences containing the same descriptive terms can also be true. Truth is always relative to what is said, but the real world in which all language users live forbids ingenuously asserting just anything in the semantically interpreted language. Therefore, semantical relativity does not imply relativism of truth, but just the opposite: with an adequate metatheory of semantical description semantical relativity explains the partial equivocation that makes it impossible for the same sentences occurring in two different belief systems, to be true in one belief system and false in another, because it explains how the same sentence is not simply and completely the same statement in each system. Thus it is not necessary for the semantics of observation language to be quarantined from the semantics of theory, in order to assert the objectivity of truth.

Each person acquires the semantics of what Quine calls observation sentences from his own personal experiences, and he acquires it publicly and ostensively in the circumstances of his learning situation. There is a wide variation among people between what is learned ostensively and contextually, but even for language learned ostensively by most people, intersubjectivity is increased with successive approximation, as the web of belief grows and imposes increasingly more shared truth conditions on the ostensively acquired semantics. The entire web of beliefs may be viewed on analogy with an underdetermined system of conditional equations, in which the addition of a new equation further restricts the range of numeric values that the set of variables may accept as solution sets. A noteworthy difference in this mathematical analogy is that the mathematical variables can take only one numeric value at a time without becoming ambiguous, while each of the descriptive terms, including those used as mathematical variables in applied mathematics in empirical science, take on the semantic values distinguishable in the explicitly related universal statements in the system of beliefs simultaneously, subject only to the preservation of univocity. Thus all the terms explicitly related by the sentences in the web of beliefs may participate in one another's univocal semantics, and thereby resolve one another's vagueness in relation to each other. And as implicit statements are made explicit by deduction, the vagueness in the meanings of the terms of the system is even further resolved.

But Quine viewed meanings as abstract or mental "entities", and then developed his behavioristic theory of stimulus meanings, which he called "behavioral dispositions" to evade the representative function of language. He could not be expected to have developed a metatheory of semantical description enabling him to describe how meanings participate in one another. The closest Quine came to the idea of semantical participation was the idea of the resolution of vagueness. His rejection of the dichotomous analytic-synthetic distinction is a worthy start toward such a metatheory, but his rejection of the distinction was actually a rejection of analyticity as such, except in the cases that he called "analytical hypotheses" used for translations. As it happens, rejection of the analytic-synthetic dichotomy does not imply the rejection of analyticity as such. Universally quantified statements believed to be true for empirical reasons may also be used analytically to exhibit the complexity in the meanings of their constituent terms by displaying their component semantic values that constitute the discriminating capability in the descriptive function of the language. In other words all universal empirical statements in the web of beliefs are analytical hypotheses.

Quine's Critique of Analyticity

The fourth of the five milestones that Quine finds in the history of empiricism is the abandonment of analyticity in the traditional analytic-synthetic dichotomy. He calls his exclusive acceptance of synthetic statements "methodological monism". The rejection of analyticity is one of the earliest theses that distinguish Quine's philosophy of language. His first published statement of the rejection is his "Truth by Convention" (1936) originally in *Philosophical Essays for A.N. Whitehead*, later reprinted in his *Ways of Paradox*. Analytic statements are those that are true by linguistic convention, and they include the propositions of logic and mathematics.

Essentially his argument in this paper is based on the rejection of an infinite regress; he argues that some logic is needed and is presupposed to develop logic. Thus he asks whether or not it makes any sense to say that the truths of logic and mathematics are destined to be maintained independently of our observation of the world, so that truth by convention may apply. Fifteen years later his critique of analyticity took a different tack in "Two Dogmas", where he drew upon the Duhem-Quine thesis of semantical wholism, in order to attack linguistic synonymy upon which analyticity is based. The statement "No bachelor is married" is made analytic by substitution of synonyms "bachelor" and "unmarried man" in the statement "No unmarried man is married", because the latter statement is true in all interpretations of its nonlogical or descriptive terms. Quine notes that Carnap explained analyticity by appeal to state descriptions; a statement is analytic if it is true in all state descriptions. Quine says that appeal to state descriptions works only if the atomic statements of the language are mutually independent, i.e. if the language has no extralogical synonym pairs such as "bachelor" and "unmarried man". Thus on Quine's thesis, Carnap's criterion for analyticity in terms of state descriptions is a reconstruction at best of logical truth, not of analyticity. Quine argues that all instances of synonymy except those occurring in purely stipulative definitions introducing notational abbreviations, are based on observed synonymy occurring in natural language. These include synonymies occurring in reduction sentences, analytic sentences and Carnap's semantical rules; and they all depend on the thesis contrary to Duhem's thesis, that it is possible to determine the truth or falsehood of sentences in isolation from one another. Invoking Duhem's thesis Quine rejects the distinction between a factual component and a linguistic component in the truth of any individual statement, which is the basis for the analytic-synthetic distinction.

Shortly after writing "Two Dogmas" Quine wrote "Carnap and Logical Truth" (1954) in *Philosophy of Rudolf Carnap* (1963). This critical essay's most distinctive characteristic relative to Quine's prior essays is its treatment of the effects of linguistic and scientific change on analyticity and logical truth. Carnap's interest in philosophy was originally inspired by Einstein's use of non-Euclidian geometry and by Hilbert's formalistic approach to mathematics. Quine says that the initial tendency to treat geometries as true by convention together with the tendency toward formalization were extended to mathematical systems generally. But Quine maintains that formalist mathematics has been "corrupted" by supposing that postulates are true by convention, and he rejects the idea of semantically uninterpreted postulates. Quine treats the subject of postulates in a manner similar to his earlier treatment of definitions in "Two Dogmas". He distinguishes two types of postulates: "legislative" and "discursive". The former type is a stipulative definition that merely introduces previously unused notation, and it initiates truth by convention. Discursive postulation on the other hand is a selection from a preexisting body of truths, of certain ones for use as a basis from which to derive others initially either known or unknown. Most notably what discursive postulation fixes is not truth, but only some particular ordering of the truth. All postulation may be said to be conventional, but only legislative postulation admits to truth by convention. The importance of the distinction, however, is that it refers to an act and not to any enduring consequences. The conventionality in postulation is a passing trait, which is significant at the moving frontier of science, but which is useless in classifying the sentences behind the lines. Conventionality is a trait of events and not of sentences. And if legislative postulates are subsequently singled out in some later exposition, they have the status of discursive postulates in the subsequent exposition. The artificiality of legislative truth does not linger as a localized quality, but suffuses with the corpus and becomes integral with it. Quine does not explicitly reference Duhem in this context, but Duhem's wholism is clearly operative. Quine says that legislative postulation occurs continually in the theoretical hypotheses of natural science. The justification of any theoretical hypothesis can at the time of hypothesizing consist in no more than the elegance or convenience which the hypothesis brings to the containing body of laws and data. There is indirect but eventual confrontation with empirical data, but this can be remote. Furthermore, some such remote confirmation with experience may be claimed even for pure mathematics and logic. A self-contained theory that can be checked with experience includes not only its various theoretical hypotheses of so-called natural science, but also such portions of logic and mathematics as it makes use of. There is no line to be drawn between hypotheses which confer truth by convention and hypotheses that do not; logic and mathematics are not qualitatively different from the rest of science.

Quine elaborates by illustration: Suppose a scientist introduces a new term for a certain substance or force by an act of legislative definition or postulation. Progressing, he then evolves hypotheses regarding further traits of the named substance or force. And then further progressing he identifies this substance or force with one named by a complex term built up of other portions of his scientific vocabulary. This new identity will figure in the ensuing developments quite on a par with the identity which first came by the act of legislative definition, or on a par with the law which first came by the act of legislative postulation. And revision in the course of further progress can touch any of these affirmations equally. Quine says that scientists proceeding in this way are not slurring over any meaningful distinction. Legislative acts occur routinely. Carnap's dichotomy between analytic and synthetic, between truth by meaning postulate and truth by force of nature, has no clear meaning, even as a methodological ideal. The fabric of our sentences, our web of beliefs as Quine calls them later, develops and changes through more or less arbitrary and deliberate revisions and additions of our own, more or less directly occasioned by the continuing stimulation of our sense organs. Carnap replies at the end of the volume in which Quine's critique was published. He emphasizes that his explication of "analytic" has always been for a formalized language, one for which explicit semantical rules are specified and that lead to the concept of truth. He rejects Quine's demand that semantical concepts such as analyticity and synonymy must also be explicated pragmatically by an empirical criterion in behavioristic terms applicable to natural language. He therefore maintains that Quine's objections are not directed against his semantical *explicata*, and that A-truth is not objectionable.

Carnap then turns to Quine's critique of analyticity in situations where there is a change in artificial language, from *L(n)* to *L(n+1)*. Firstly Carnap agrees with much of what Quine says in "Two Dogmas", where Quine sets forth his neo-Duhemist wholistic thesis. Carnap agrees that a scientist who discovers a conflict between his observations and his theory and who must therefore make a readjustment somewhere in the total system of science, has much latitude with respect to the places where a change is to be made. Remarkably Carnap also agrees that in this procedure of readjustment, no statement is immune to revision, not even statements of logic or mathematics. But Carnap rejects Quine's characterization of an analytic statement as one held true come what may. And Carnap furthermore denies that a change in language invalidates the analytic-synthetic distinction. In defense of analyticity Carnap distinguishes two types of linguistic change. The first type is a change of language from *L(n)* to *L(n+1)*. He says that this type constitutes a radical alteration and perhaps a revolution. It occurs only at certain historically decisive points in the development of science. The second type is a mere change in or an addition of a truth value ascribed to an indeterminate statement. An indeterminate statement is one having a truth value that is not fixed by the rules of the language, i.e. by postulation of logic, mathematics, or perhaps physics. This second type of change occurs "every minute" according to Carnap. He says that his concept of analyticity has nothing to do with the first type of transition; his concept of analyticity refers only to some given language, *L(n)*. The truth of a sentence, *S*, in *L(n)* is based on meanings in *L(n)* of the terms occurring in *S*. In *L(n)* analytic sentences cannot change their truth value, and furthermore neither can the synthetic postulates of physics and their logical consequences.

Quine's critique of analyticity is principally directed against what Carnap called A-truth, which is truth based on the semantics of the descriptive vocabulary in the sentence, a lexical basis. As a symbolic logician Quine continues to rely on logical truth, on the kind of sentence that Carnap calls L-truth, but his reasons are different than Carnap's. In "The Ground Of Logical Truth", the eighth chapter in his *Philosophy of Logic,* Quine admits to an acceptable sense of logical truth, the truth that is evident because of the grammatical structure of the logically true sentence. But Quine rejects Carnap's doctrine of linguistic truth, the thesis that language alone can make logical truth independently of the nature of the world. In view of Carnap's defense of analyticity, it is doubtful that Carnap continued to maintain such a view. In any event, Quine maintains that the validity of logical truth depends on the relation of grammatical structure to the structure of the real world. He argues that the distinction between the lexical and the grammatical is variable not only among different languages, but also within the same language.

Quine's Rejection of First Philosophy

The last of the five milestones in Quine's history of empiricism is what he calls the abandonment of the goal of a first philosophy. By first philosophy he means any philosophy that is prior to natural science. Traditionally metaphysics and epistemology are considered to be first philosophy. Quine calls his position "naturalism". The term "naturalism" has meant many different things in the history of philosophy. A term that Quine does not use is "scientism". In "Five Milestones" Quine defines his naturalism as the view that natural science is an inquiry into reality, a fallible and corrigible inquiry, but not answerable to any suprascientific tribunal, and not in need of any justification beyond observation and the hypothetico-deductive method. This statement by Quine is not merely an affirmation of the autonomy of empirical science from metaphysics, as may be found in Duhem's philosophy of science. Quine rejects the view that there is any philosophical tribunal for science, by which he means any knowledge separate from empirical "common sense", that he views to be continuous with science in his wholistic philosophy of language. Furthermore, Quine maintains that epistemology is an empirical discipline that he assimilates into empirical psychology, which for him is behavioristic psychology. He describes the scientific epistemologist as asking how animals, presumably human, can have managed to have arrived at science from the limited information from surface stimulations, and as pursuing this inquiry to yield an account that pertains to the learning of language and the neurology of perception.

Quine gives two reasons for his naturalism by which he rejects all first philosophy. One reason is what he calls an "unregenerate" realism, the robust state of mind of the natural scientist who has never felt any qualms beyond the negotiable uncertainties internal to his science. He expresses his realism even more emphatically in his "Scope and Language of Science" (1954) reprinted in *Ways of Paradox,* where he states that we cannot significantly question the reality of the external world or deny that there is evidence of external objects in the testimony of our senses; for to do so is to dissociate the terms "reality" and "evidence" from the very application which originally did most to invest these terms with whatever intelligibility they may have for us. He maintains that the notion of reality independent of language is derived from our earliest impressions, and then carried over into science as a matter of course. The second reason for Quine's realism is what he calls the despair of being able to define theoretical terms generally in terms of phenomena even by contextual definitions. This is a rejection of the Logical Positivist problem for which reductionism of theoretical terms was thought to provide an answer. On the Positivist philosophy there is no justification for affirming the reality of theoretical entities, unless these terms are firstly established as semantically meaningful. The purported solution is the reduction of theories to observation sentences, which are the source for the semantics and ontology of theories. Quine rejects the Positivists' problem, because it involves a prior ontology or first philosophy consisting in the Positivists' observation language. In Quine's view

Positivism is a kind of metaphysics, Positivists' antimetaphysical rhetoric not withstanding.

Fundamental to Quine's second reason for rejecting first philosophy is his thesis of ontological relativity. This thesis can be found in Quine's literary corpus even before he came to call it "ontological relativity" in the mid-1960's. In "Two Dogmas" after rejecting the dogma of reductionism, he says that physical objects are conceptually imported into the linguistic system as convenient intermediaries, as irreducible posits comparable epistemologically to the gods of Homer. What he calls the "myth" of physical objects is epistemologically superior to others including the gods of Homer, in that it has proved to be more efficacious than other myths as a device for working a manageable structure into the flux of experience. Microphysical entities are posited to make the laws of macroscopic objects and ultimately to make the laws of experience more manageable. Science is a continuation of common sense, and it continues the commonsense expedient of swelling ontology to simplify theory. Shortly later in "Posits and Reality" (1955) Quine says that if we have evidence for the existence of bodies of common sense, we have it only in the way in which we may be said to have evidence for the existence of molecules. All science is empirically underdetermined, and the only difference between positing microphysical and macrophysical entities is that the theories describing the former are more underdetermined. In this context Quine is using the term "underdetermined" in same sense as he used it in "Two Dogmas" to express his neo-Duhemist wholistic view of language.

The thesis of ontological relativity is also prefigured in *Word and Object*. Just as Carnap recognized extensional vagueness, Quine recognized referential indeterminacy which he calls referential "inscrutability". Inscrutability of reference is due to the semantic indeterminacy of direct ostension. This indeterminacy occurs in translation of an unknown language by the field linguist, but it occurs in all language, and is not distinctive of the translation situation. It permits reference and ontology to be completely system determined in linguistic context. In chapter six of *Word and Object* Quine says that everything to which we concede existence is a posit from the standpoint of the theory-building process, and is simultaneously real from the standpoint of the theory that is built. The explicit statement of ontological relativity itself is in "Ontological Relativity" (1968) in *Ontological Relativity*. Quine uses the phrase "ontological relativity" explicitly on analogy with Einstein's relativity theory in physics. He maintains that reference is nonsense except in relation to a coordinate system, where the coordinate system is some background language. Asking for ontological reference in any more absolute way than by reference to a background language is like asking for absolute position or absolute velocity, rather than for position or velocity relative to a frame of reference. The ultimate background language to which we take recourse in practice is our mother tongue, in which we take words at face value with their primitively adopted and ultimately inscrutable ontology. Any subordinate theory must be interpreted by reference to this home language. This thesis would seem to be contrary to Heisenberg's doctrine of closed-off theories, since on Quine's view no theory can be semantically closed off from the rest of language. Quine opposes his thesis of ontological relativity to Carnap's thesis of the distinction between external ontological questions and internal factual questions set forth in "Empiricism, Semantics and Ontology". In Quine's view there can be nothing like Carnapian external questions which are external to the home language. In "Carnap's Views on Ontology" (1951) reprinted in *Ways of Paradox* Quine maintains that ontological questions are on a par with questions in natural science. Within science there is a continuum of gradations from the statements that report observations to those that reflect basic features of quantum theory and relativity theory. Similarly statements of ontology and even of mathematics and logic form a continuation of this continuum, though these are more remote from observations than the central principles of quantum theory or relativity theory. Quine says that the differences along this continuum are only differences of degree and not differences in kind.

While the semantical wholism of the Duhem-Quine thesis has received much attention, it is seldom realized that Quine's rejection of all first philosophy is one of its most consequential implications for philosophy of science. When the Duhem thesis of physical theory is extended to the whole of language, not only is all semantics made context-determined, but also all ontologies described by the semantics are made vulnerable to empirical criticism; there are no longer any protected ontologies. Quine's thesis of ontological relativity has the historic and revolutionary effect of excluding all ontological considerations from the criteria for scientific criticism. In his philosophy it is empirical adequacy that decides ontological questions, rather than prior ontological commitments that decide the acceptability of scientific theories. Quine subordinates all questions of ontology to the empirical adequacy of the theory affirming the ontological claims in question. He maintains that the human knower can never do better than to occupy the standpoint of one or another theory, whether the theory purports the existence of either macrophysical or microphysical entities. All entities are "posits" affirmed by one or another theory, and all are worthy of our patronage just to the extent that the theory positing them is empirically adequate. However detailed may be the relevant observation language, empirical underdetermination (in Quine's earlier sense) and its consequent semantical indeterminacy always admit alternative choices of theory. And the consequent referential inscrutability may admit to as many correspondingly alternative choices of entities. Quine's rejection of ontological criteria in scientific criticism is called "scientific realism". It is not only characteristic of the contemporary Pragmatist philosophy, but was also the practice of Galileo, Einstein and Heisenberg. In developing his theory of relativity Einstein posited relativistic time as real instead of Newton's absolute time, and he rejected Lorentz's relegation of relativistic time to the status of apparent time and Lorentz's retention of Newton's absolute time as real. A central thesis of the Copenhagen interpretation, or at least Heisenberg's noninstrumentalist version, is its realistic claims about the wave-particle dualism and the indeterminacy principle. Heisenberg referenced Einstein's realism in relativity theory as a precedent, when he posited the reality of the quantum theory's wave-particle duality and indeterminacy claims. The practice of letting the empirical adequacy of a theory operate as the criterion for the acceptability of its ontology did not begin with Einstein or Heisenberg. A historic and well known example is Galileo's realistic interpretation of the Copernican theory, which placed him in conflict with the Aristotelian ontology enforced by the Roman Catholic Papacy.

This is a distinctively and thoroughly Pragmatist view, that separates Quine from his Positivist and Romanticist predecessors. Ironically it also separates him from certain other aspects of his own philosophy, notably its behaviorism and its near-nominalism. The Romanticists insist upon and the Positivists insist against the introduction of "mentalism" in explanations in the social and behavioral sciences. But on a thoroughly Pragmatist philosophy of science, this ontological issue is decided by the empirical adequacy of the behavioral and social science theories. Different theories in different sciences at different times or even at the same time will admit different ontologies, some behavioristic and others nonbehavioristic. It may seem odd that Quine's behavioristic "naturalized" epistemology is actually an exception to his thesis of ontological relativity. But Quine is neither the first nor the last philosopher-king to exercise a sovereign's right of eminent domain in his own philosophy, and exempt himself from his own laws.

Comment and Conclusion

Mach and Duhem were not only Positivist philosophers of science, they were also practicing research physicists, who furthermore wrote histories of physics. Carnap on the other hand was neither a practicing research physicist nor a historian of physics. His philosophical work had a remoteness from the physicists' research practices, because the Vienna Circle had an epistemological (i.e. metaphysical) agenda for scientific criticism, that did not actually operate in research physics. Carnap aimed to construct a metalogic for science, but he did not apply his constructionalist techniques to the language used by scientists; instead he used the symbolic logic of Russell and Whitehead to substitute for the object language that he claimed he was investigating. But the symbolic logic is not useful to the physicist. The development of the Ramsey sentence, for example, was hailed by Carnap and others such as Russell and Braithwaite as a great philosophical achievement; but it would be a rare physicist who would consider the Ramsey sentence at all consequential to either the practice or the history of physics. The situation is aptly stated by Radnitzky in the "Epilogue" in the first volume of his *Contemporary Schools of Metascience* (1968), where he says that the logical empiricists had not produced any metascience at all, because they did not study the producers of scientific knowledge or the production or even the results. The post-Positivist philosophers rejected Positivism because they correctly recognized its irrelevance to research science and its inadequacy as a philosophy of science.

When the post-Positivist philosophers rejected Positivism, many of them also rejected its constructionalism. Many Pragmatists in particular found their wholistic concept of the semantics of language to be incompatible with the mechanistic and procedural character of logical constructionalism; in the wholistic view the semantics of science makes the development of science a nonlogical process. But they rejected too much, because the Logical Positivists' analytical approach is more valuable than the Russellian symbolic logic and the Positivist philosophy of science which used the logic. In this age of the computerized discovery system Carnap's constructionalism and his metatheory of semantical systems may with certain noteworthy modifications be carried forward into contemporary and future methodology of science.

Some such modifications are as follows: A first important modification is that the object language that is constructed by a discovery system is not the Russellian symbolic logic; it is the mathematical equations or other technical language actually used in the relevant science. Scientists never use the Russellian symbolic logic for the expression of their theories. Use of the symbolic logic to express empirical science was never more than a caricature. A second modification of Carnap's work is the use of a computer language for the metalanguage. The computer language gives the metalanguage a disciplined and procedural character that the colloquial metalanguage does not have. The computer language in which the discovery system is written operates as a metalanguage in which the formation rules of the object language are expressed in computer instructions; the discovery system in other words is a metalanguage expressing a mechanized generative grammar.

A third modification pertains to Carnap's concept of semantical rules that interpret a semantical system. The semantical rules for interpreting a mechanically generated semantical system might be viewed as analogous to Carnap's meaning postulates, in that all of them are stated in the object language instead of the metalanguage, and are not like Carnap's rules of designation, which occur in the metalanguage. Two relevant types of semantical rules may be distinguished. One type consists of those semantical rules that are the mechanically generated statements and equations. These consist only of the statements constituting a mechanically generated and empirically acceptable theory, the theory statements which are believed to be true. But not all the semantical rules occurring in the object language are mechanically generated. A second type consists of test design statements, which are accepted independently of any statements of theory generated by the system, so that the generated theory is not tautological and can be tested independently. But the semantical rules for mechanically generated semantical systems are unlike Carnap's meaning postulates, because they are not just analytical sentences. With Quine's rejection of any distinctively analytic truth it is possible to view sentences as both analytic and synthetic, and the semantical rules that describe the semantical interpretation of the object-language statements must be viewed as both analytic and synthetic sentences. They are more like Quine's analytical hypotheses or discursive postulates. These semantical rules might also be viewed as similar to Carnap's reduction sentences, which he says determine only "part" of the meaning of theoretical terms. But Carnap has never explained how it is possible for the meanings of terms to have parts. Viewing the sentences as both analytic and synthetic enables the empirical statements constituting the generated theory to exhibit the parts of the meanings of their constituent terms, just as analytic statements always have. Both test design statements and generated theory statements, both of which are believed to be true for empirical reasons and not because of the meanings of their constituent terms, are object-language statements functioning as semantical rules each of which contribute part of the meaning to each of their common descriptive terms.

A fourth modification pertains to Carnap's idea of a state description. The Carnapian state description is not a useful concept for describing the semantical systems generated by mechanized discovery systems. It consists of singular "atomic" statements expressed in Russellian logic, and was conceived with the intent of explicating precisely the idea of L-truth and A-truth. The semantical systems generated by the discovery systems contain only universal statements constituting the theories generated by with the formation rules in the computerized generative grammar. In contrast to the semantical systems in Carnap's philosophy, which were devised for static analyses, the semantical systems in metascience are intended to describe the semantical changes occurring in the development of new theories, which is a dynamic procedure. Accordingly the Carnapian idea of a state description must be revised for describing the computer system input and output object language, in order to reveal the semantical changes produced by the discovery system. The inputted information for the discovery system is drawn from the current state descriptions consisting of the several theories that have been advanced by the particular scientific profession. These theories supply the vocabulary inputted to the computerized discovery system. This vocabulary has its semantics specified by semantical rules consisting of test design statements, which are common to both input and output state descriptions. These test design statements are not changed by the discovery system, and they supply semantical continuity for identifying the subject of the theories independently of the theories. The computerized discovery system generates a set of outputted state descriptions consisting of alternative empirically adequate theories, which are semantical rules describing the semantics of the new theories.

A fifth modification consists of replacing Carnap's theory of information with Shreider's semantical metatheory, if the concept of state description as revised in the manner described above is identified with Shreider's concept of thesaurus. But unlike Shreider's theory there are actually two types of transformations involved. Firstly there is the mechanized syntactical transformation, the generation of new theories which are the output messages. And secondly there is also the semantical transformation on the part of the system users who communicate with the computer, when they attempt to interpret its output. The computer system is a transmitter and information source, that generates message texts consisting of new theories, and a user having a thesaurus consisting of one of the input semantical systems, i.e. an old theory, must transform his thesaurus to conform to one of the output semantical systems, a new theory. Thus the amount of information transmitted to a user depends on the degree of transformation between his initial thesaurus and the outputted theory that must transform his thesaurus for him to understand the new theory.

It might be added that there is in all cases a psychological resistance to the extent that the amount of information communicated is large. And there may also a philosophical resistance depending on the using-scientist's philosophy of science. If the scientist is a Romantic, he will be philosophically ill disposed to accept the newly generated theories containing large amounts of information, because he will find they are not "intuitively plausible" and do not "make sense". Romanticism retards the development of science, because it forbids the unfamiliar. The Positivist will be less affected by such philosophical cognition constraints. Positivists believe in the special importance of the familiar, which they call the "observable". But some, like Carnap, opposed "models" in which technologically less accessible microphysical processes are explained on analogy with more familiar macrophysical processes. The philosophy of science that offers the least impediment to the reception of new information is Pragmatism, according to which no ontology may serve as a criterion for scientific criticism.

WERNER HEISENBERG AND THE SEMANTICS OF QUANTUM MECHANICS

Werner Heisenberg (1901-1976) was born in Wurzburg, Germany, and studied physics at the University of Munich, where he wrote his doctoral dissertation under Arnold Sommerfeld in 1923 on a topic in hydrodynamics. He became interested in Niels Bohr's atomic theory and went to the University of Gottingen to study under Max Born. In 1924 he went to Bohr's Institute for Theoretical Physics in Copenhagen. He developed the quantum matrix mechanics in 1925, and then developed the uncertainty principle in 1927. From 1927 to 1941 he was a professor of physics at the University of Leipzig. In 1932 he was awarded the Nobel Memorial Prize for Physics. In the Second World War, he was the director of the Kaiser Wilhelm Institute for Physics in Berlin. After the war he established and became director of the Max Planck Institute of Physics initially at Gottingen, and then after 1958 at Munich. His principal publications in which he set forth his philosophy of physics consist of the "Chicago Lectures of 1930" published as *The Physical Principles of the Quantum Theory* (1950, [1930]), *Philosophical Problems of Nuclear Science* (1952) currently published under the title of *Philosophical Problems of Quantum Theory* (1971), *The Physicist's Conception of Nature* (1955), an interpretative history of physics, *Physics and Philosophy: The Revolution in Modern Science* (1958), his intellectual autobiography published as *Physics and Beyond* (1971), and *Across the Frontiers* (1974).

Heisenberg's mature philosophy of science was not significantly influenced by the doctrines of professional philosophers, although he was a Positivist early in his career and later rendered Bohr's view of observation in neo-Kantian terms, even though he was not a metaphysical idealist. The formative intellectual influences on his mature philosophical thinking were two of the great physicists of the twentieth century, Einstein and Bohr. As it happens, these two philosophical influences were contrary to each other, and each pulled Heisenberg's thinking in opposite directions. Therefore, some preliminary remarks about the philosophical views of both Einstein and Bohr are in order for understanding Heisenberg's own philosophy of science.

Heisenberg's Discovery and Einstein's Semantical Views

Reference was made above in the discussion of the philosophy of Mach, about the influence of Einstein's admonition on Heisenberg's development of the indeterminacy relations. This episode in the history of science, which Heisenberg relates in "Quantum Mechanics and a Talk with Einstein (1925-1926)" in *Physics and Beyond,* is a watershed event for the contemporary Pragmatist philosophy of science, and Heisenberg's description of his personal experience and thought processes deserve close examination. Firstly he discusses why he had been led to believe that he could develop a quantum theory exclusively on the basis of observed magnitudes. He writes that in the summer of 1924 he had attempted to guess the formula that might successfully describe the line intensities of the hydrogen spectrum using methods involving the idea of electron orbits which he thought would be successful because of earlier work by Kramers in Copenhagen. When use of these methods hit a dead end, he became convinced that he should ignore the idea of electron orbits. He decided instead that he should treat the frequencies and amplitudes associated with the spectral line intensities as substitutes, because the line intensities are observable directly, while the electron orbits are not. He was led to this approach because he recalled a conversation years earlier in which a friend told him that Einstein had emphasized the importance of observability in relativity theory. In May of 1925 Heisenberg suffered a severe hay fever attack and had to absent himself from his academic duties. While recuperating on the island of Heligoland he continued to work on the problem by considering nothing but observable magnitudes, and during this period of isolation he developed his matrix mechanics.

About a year later he was invited to give a lecture at the University of Berlin physics colloquium to present his matrix mechanics. Einstein was in the assembly, and after the lecture he asked Heisenberg to discuss his views with him in his home that evening. In that discussion Einstein argued that it is in principle impossible to base any theory on observable magnitudes alone, because in fact the very opposite occurs: it is the theory that decides what we can observe. He argued that when the physicist claims to have observed something new, he is actually saying that while he is about to formulate a new theory that does not agree with the old one, he nevertheless must assume that the new theory covers the path from the phenomenon to his consciousness and functions in a sufficiently adequate way that he can rely on it and can speak of observations. The claim to have introduced nothing but observable magnitudes, is actually to have made an assumption about a property of the theory that the physicist is trying to formulate. Heisenberg was using his idea of observation as if the old descriptive language could be left as it is. Heisenberg replied that Einstein was using language a little too strictly, and that until there is a link between the mathematical quantum theory and the traditional language, physicists must speak of the path of an electron by asserting a contradiction. Bohr's complementarity is such a contradiction.

Heisenberg also replied by referencing Mach's view that a good theory is no more than a condensation of observations in accordance with the principle of thought economy. Einstein explained that Mach thought that a theory combines complex sense impressions just as the word "ball" does for a child. But he also stated that the combination is not merely a psychological simplification but is also an assertion that the ball really exists, because it makes assertions about possible sense impressions that may occur in the future. Einstein thus affirmed a realistic philosophy, and criticized Mach for neglecting the fact that the real world exists, that our sense impressions are based on something objective, and that observation cannot be just a subjective experience. Heisenberg accepted Einstein's realism on these grounds, and admitted that theory reveals genuine features of nature and not just our of knowledge.

In the preface to *Physics and Beyond* Heisenberg notes that conversations cannot be reconstructed literally after several decades, and that the book is not intended as a collection of memoirs; but that careful attention has been paid to the precise "atmosphere" in which the conversations took place, because in that conversational atmosphere the creative process of science is made manifest. His book contributes to explaining how the cooperation of different people may culminate in scientific results of the utmost importance. Heisenberg stated that his purpose is to convey even to those remote from atomic physics some idea of the mental processes that have gone into the genesis and development of science. From these statements a reader of his book may reasonably conclude that whatever inaccuracies may have occurred in Heisenberg's reconstruction of his conversation with Einstein in 1926, or however he may have misunderstood Einstein's views, the importance of the recollections is not diminished thereby, since it had led him to his development of the uncertainty principle a year later.

In a chapter titled "Fresh Fields (1926-1927)" Heisenberg offers a description of his own mental processes in his development of the uncertainty relations. To the contemporary reader this description has value apart from his philosophy; just as Newton attempted to philosophize about his work with his denial that he created hypotheses, so too did Heisenberg attempt to philosophize about his work in his own systematic and explicit philosophy of language, his doctrine of closed-off theories and of perception. But these recollections of his cognitive experiences in "Fresh Fields" are not an attempt at a systematic philosophy; they are more simply his recollection of his own cognitive experiences as a central participant, and is valuable as a historical document. As it happens, in the contemporary philosophical perspective these recollections are more valuable than his explicit attempt to philosophize on the nature of language and perception. Heisenberg's writings reveal that his development of the uncertainty relation was occasioned by several historical circumstances. One of these that he discusses in "Fresh Fields" was the development of the wave mechanics by Schroedinger and its disturbing effects on the thinking of the physicists at Copenhagen. The wave equation did not contain Planck's constant as did Heisenberg's matrix mechanics, while Planck's constant was thought by Bohr and the Copenhagen physicists to be central and necessary for any modern microphysical theory. Then Max Born, formerly a teacher of Heisenberg, proposed a probability interpretation of the wave equation, such that each point in space and instant in time the equation gives the probability of finding an electron at a given point and instant. The upshot was that while neither the matrix mechanics nor the wave mechanics could be rejected for empirical reasons, they nevertheless seemed to be logically incompatible. In addressing this problem Bohr and Heisenberg took different approaches. Bohr attempted to admit simultaneously to the validity of both theories by maintaining that both the classical wave and particle concepts used to describe the experimental observations, are necessary for characterizing atomic processes, even though in the language of ordinary discourse and of classical physics, these two concepts are mutually exclusive. This semantic inconsistency became Bohr's complementarity principle. But Heisenberg relates that he did not like this approach, and that he wanted a unique, that is, a consistent and unequivocal, physical interpretation of the magnitudes in the mathematical formalism, one that is derivable from the matrix mechanics by strict logic. Heisenberg reports that this objective was one of the reasons that led him to derive the uncertainty relation.

A second reason leading him to the uncertainty principle was the fact that neither the wave mechanics nor the matrix mechanics seemed capable of explaining the observed phenomenon of the trajectory of the electron in the Wilson cloud chamber. Such ideas as trajectories and orbits do not figure in the mathematical formulations of the matrix mechanics, and the wave mechanics could only be reconciled with the existence of a densely packed beam of matter, if the beam spread over volumes that are much larger than the dimensions of an electron. This problem of the observed phenomenon in the cloud chamber led Heisenberg to reformulate the questions he was asking himself in his statement of the problem; he attempted to relate the observed path of the electron in the cloud chamber to the mathematics of the matrix mechanics. In February and March, 1927, while Bohr was vacationing in Norway and while Heisenberg was again alone with his thoughts, as he had been when he first developed the matrix mechanics, his attempts to relate the cloud chamber observations to the matrix mechanics bought to mind his discussion with Einstein the prior year in Berlin, and specifically it recalled to mind Einstein's statement that it is the theory that decides what can be observed. In "Fresh Fields" he describes his thinking processes as he attempted to employ Einstein's advise: Firstly he reconsidered the idea that what is observed in the cloud chamber is a trajectory. The idea of a trajectory is a concept in Newtonian physics. Therefore, when he thought that he was observing the trajectory of an election in the cloud chamber, the theory that was deciding what was being observed was the Newtonian theory, not his quantum theory. Then secondly after reconsidering the Newtonian observations and recognizing that it is not necessary to think in Newtonian terms, he viewed the phenomenon as merely a series of ill defined and discrete spots through which the electron had passed like the water droplets that are much larger than the dimensions of the electron. Then thirdly he reformulated the question, and asked how quantum theory instead of Newtonian theory can represent the fact that an electron finds itself approximately in a given place and that it moves approximately with a given velocity. Using Einstein's thesis that the theory decides what can be observed, Heisenberg concluded that the processes involved in any experiment or observation in microphysics must satisfy the laws of quantum theory. The magnitude of the water droplets suggested room for approximation for the minute electron, and Heisenberg asked whether it is possible to make these approximations so close that they do not cause experimental difficulties. He then derived the mathematics of the uncertainty principle in which the approximations are governed by a limit that is a function of Plank's constant.

Heisenberg's use of Einstein's thesis may be contrasted with Duhem's Positivist view. Unlike earlier Positivists, Duhem admitted that theory has a valid place in science, but he did not view observation as dependent upon theory, as did Einstein. Instead, Duhem maintained that while theory provides an interpretation for observation, observation without theory is not only possible but furthermore offers a certitude that theory cannot offer; theoretical interpretation is something added to fundamental and uninterpreted observation. Thus, the characterization of the vapor trail in the cloud chamber as a series of water droplets would on Duhem's thesis be an example of an atheoretical and uninterpreted observation. On Einstein's thesis, however, there is no observation without theory, and the characterization of the cloud chamber vapor trail as a series of water droplets is no less interpreted than the characterization of the phenomenon by means of Newtonian or quantum theoretical concepts. Even though the interpretation in terms of water droplets does not employ a mathematically expressed theory, the concepts of water droplet, of the cloud chamber, and even of the reflected light needed to view the water droplets in the cloud chamber, have built into them many hypotheses or as Einstein says, "expectations", that are tacitly assumed in order to make the observations.

Heisenberg had formulated his uncertainty principle by the time Bohr had returned to Copenhagen from his vacation in Norway. Initially Bohr objected to the idea, while at the same time Heisenberg disliked the complementarity idea that Bohr had developed while away. After several weeks of argument they finally agreed that the two approaches are related; the uncertainty principle reconciles at the microphysical level and in the mathematical formalism of quantum mechanics, what cannot be avoided and yet what cannot be stated consistently in the language supplied by classical physics and ordinary language, which is suitable only to describe phenomena at the macrophysical level. What is expressed consistently with the mathematical formalism of the uncertainty principle is the impossibility of measuring simultaneously both the position and the impulse of the electron with a degree of accuracy greater than the limit imposed by Planck's constant, a limit that is imposed by virtue of the nature of the microphysical phenomenon itself and not merely by the measurement technique. What is described inconsistently at the macrophysical level and in the language of classical physics by means of complementarity, is the wave and particle manifestations of the unitary phenomenon.

Heisenberg's description based on his own experience of the interpretative character of all perception and observation and of the role of scientific theory in determining the interpretation, articulates one of the most characteristic feature of the contemporary Pragmatist philosophy of science. It is more valuable than Duhem's exemplification of the theoretical interpretation of the laboratory apparatus in the opening passages of the chapter titled "Experiment in Physics" in *Aim and Structure of Physical Theory*, not only because Duhem's explanation is Positivist, but also because Heisenberg's description of his experiences is given in the context of his development of the uncertainty principle, one of the most noteworthy achievements of twentieth century theoretical physics. As it happens, Heisenberg did not like Pragmatism, or at least the Pragmatism he encountered during his visit to the United States and described in "Atomic Physics and Pragmatism (1929)" in *Physics and Beyond*. Even though his description of the interpretative character of perception and observation actually contributed to the contemporary Pragmatism, he was influenced by Bohr in ways that impeded his developing a philosophy of science that is consistent with Einstein's thesis that theory determines what is observed, and this influence places Heisenberg's explicit philosophy of science closer to the Positivist philosophy than either Einstein's or the Pragmatists' views. This influence of Bohr consisted of a naturalistic philosophy of the semantics of language, and the result is Heisenberg's neo-Kantian philosophy of perception and his doctrine of closed-off theories.

Heisenberg's Discovery and Einstein's Ontological Criteria

An ontology may be defined as those aspects of the real world that are described by the semantics of a discourse, such as a scientific theory, that is believed to be true. Philosophers and scientists have had different ontological commitments, because they hold different criteria for determining which among alternative descriptive discourses is true. It may be said that Einstein had two different ontological criteria for physics, one explicitly set forth by him, and another that he tacitly used and which therefore may be called his implicit criterion. In Newtonian physics and in relativity theory these two different criteria are not easily distinguished, because in each case they yield similar ontologies, but in quantum theory they yield fundamentally different ontologies.

Einstein's explicit ontological criterion for deciding what is physically real is set forth in his "Can Quantum Mechanical Description of Physical Reality be Considered Complete?" in *Physical Review* (1935), in his "Reply to Criticisms" in *Albert Einstein* (ed. Schilpp, 1949), and in his "Physics and Reality" in *The Journal of the Franklin Institute* (1936). There are several statements. One that he sets forth as his "programmatic aim of all physics" is his criterion of logical simplicity, which he sets forth as the aim of science. The aim of science in Einstein's view is a comprehension as complete as possible of the connections among sense impressions in their totality, and the accomplishment of this aim by the use of a minimum of primary concepts and relations. He goes on to say that the essential thing about the aim of science is to represent the multitude of concepts and theorems close to experience as theorems logically deduced and belonging to a basis, as narrow as possible, of axioms and fundamental concepts, which themselves can be chosen freely. The aim of science is the logical unity of the world picture. Einstein interprets the history of physics as an evolution under the direction of this aim of science. This criterion requires that microphysical and macrophysical theories affirm one single consistent ontology and use the same basic concepts of what is physically real. He also says that the conviction that field theory is unable to give a solution to the molecular structure of matter and to the quantum phenomenon, is a prejudice. He demands that the ontology of field theory supply the uniform fundamental ontology, and he uses this explicit ontological criterion to criticize the Copenhagen statistical interpretation of quantum physics. In a famous article titled "Can Quantum Mechanical Description of Physical Reality be Considered Complete?" in *Physical Review* co-authored with Podolsky and Rosen he describes the Copenhagen interpretation as "incomplete", by which he meant that further research is needed to make quantum theory consistent with the ontology of field physics, the ontology of deterministic causality and of the physical continuum in four dimensions. The argument in this paper, often called the "EPR argument" after the three co-authors, includes a thought experiment, which is based on explicit criteria for completeness and for physical reality. The completeness criterion says that a physical theory is complete only if every element of the physical reality has a counterpart in the physical theory. The criterion for physical reality is that if without in any way disturbing a system, one can predict with certainty the value of a physical quantity, then there exists an element of physical reality corresponding to this physical quantity. This criterion's reference to independence of any act of observation is repeated in a later statement of the programmatic aim of all physics in "Remarks" in *Albert Einstein.* The thought experiment in the EPR argument attempts to demonstrate that the statistical theory's satisfaction of the reality criterion does not result in its satisfaction of the completeness criterion.

The stated criteria for completeness and for physical reality are defined such that deterministic field theory satisfies both criteria while statistical quantum theory does not. The point of departure, the basic premises of the argument, are Einstein's ontological preferences. In an article with the same title appearing in the same book Bohr argued that the phrase "without in any way disturbing a system" in Einstein's criterion for physical reality is ambiguous, because its meaning in classical physics is not the same as that in quantum physics. Bohr maintained that in quantum measurements the object measured and the observing apparatus form a single indivisible system that defies any further analysis at the quantum level. A large literature developed around the technicalities of the physical thought experiment, but in practice the physicists chose their ontological premises according to their preferences about the ontological conclusions, depending on whether one agreed or disagreed about Einstein's view that quantum theory must have the same ontology as field physics. And for most of the next twenty years the preferred conclusion was the Copenhagen interpretation of the statistical quantum theory.

On the other hand Einstein's implicit ontological criterion was operative in his development of the special theory of relativity. This criterion (stated explicitly) is that the empirically adequate scientific theory must be interpreted realistically. Unlike his explicit criterion, which subordinates a scientific theory and its interpretation to a preconceived ontology, the implicit criterion subordinates ontological commitment to the outcome of empirical scientific criticism. Heisenberg used such an ontological criterion for developing the uncertainty principle, and he may have consciously been imitating Einstein's use of it in the latter's development of special relativity. Heisenberg unquestionably used this criterion to defend the Copenhagen interpretation of quantum theory against Einstein's criticism based on the latter's explicit ontological criterion for physical reality. In defense of the Copenhagen interpretation, Heisenberg referenced Einstein's realistic interpretation of the Lorentz transformation equation. In his discussions about Einstein's special theory of relativity in *Physics and Philosophy* and in *Across the Frontiers* Heisenberg describes as the "decisive" step in the development of special relativity, Einstein's rejection of Lorentz's distinction between "apparent time" and "actual time" in the interpretation of the Lorentz transformation equation, and Einstein's taking "apparent time" to be physically real time, while rejecting the Newtonian concept of absolute time as real time. In other words this decisive step consisted of taking the Lorentz transformation equation realistically, and of letting it define the ontology of the physically real because of its empirical adequacy. Nowhere does Heisenberg write that he was consciously imitating this discovery strategy at the time that he developed the uncertainty relation. But in his "History of Quantum Theory" in *Physics and Philosophy* he describes his use of the same strategy. In this description of his thought processes Heisenberg does not refer to his conversation with Einstein in Berlin in 1926. He states that his thinking in the discovery experience of the uncertainty principle consisted of turning around a question: Instead of asking himself how one can express a given experimental situation, the cloud chamber experiment, in the Newtonian mathematical scheme, he asked whether only such experimental situations can arise in nature as can be described in the formalism of the matrix mechanics. The new question is a question about what can arise or exist in reality. Years later in "The Uncertainty Principle and Foundations of Quantum Mechanics" (1976) he says explicitly that this meant that there was not a real (Newtonian) path of the electron in the cloud chamber. Heisenberg's strategic answer to the new question, the uncertainty relation, resulted from this realistic interpretation of the quantum theory. Similar remarks are to be found in "The Development of the Interpretation of the Quantum Theory" in Pauli's *Niels Bohr and the Development of Physics* (1955) where Heisenberg says that he inverted the question of how to pass from an experimentally given situation to its mathematical representation, by using the hypothesis that only those states which can be represented as vectors in Hilbert space can occur in nature and be realized experimentally. He immediately adds that this method of solution had its prototype in Einstein's special theory of relativity, when Einstein had removed the difficulties of electrodynamics by saying that the apparent time of the Lorentz transformation was the real time, that similarly it is now assumed in quantum mechanics that real states can always be represented as vectors in Hilbert space (or as mixtures of such vectors), and that the uncertainty principle was the simple expression for this assumption.

If at the time that he developed the uncertainty principle, Heisenberg was not consciously imitating the discovery strategy that Einstein used for development of special relativity, it is nevertheless not difficult to imagine how Heisenberg hit upon it independently. For the realist, it is a small step from Einstein's semantical thesis that it is the theory that decides what can be observed, to the ontological thesis that it is the theory that decides what is physically real, where the theory in question is the most empirically adequate one. This strategy in which the empirical adequacy of a scientific theory as revealed by scientific criticism decides the ontology to be accepted, is a reversal of the more traditional relation in which currently accepted ontological and metaphysical views are included among the criteria for scientific criticism. Heisenberg's approach is the contemporary Pragmatist thesis of scientific realism. Heisenberg explicitly compares his realistic interpretation of the statistical quantum theory to Einstein's realistic interpretation of the Lorentz transformation equation, when he defends the ontology of the Copenhagen interpretation against Einstein's explicit ontological criterion for physical reality. In "Criticism and Counter-proposals to the Copenhagen Interpretation of Quantum Theory" in *Physics and Philosophy* he characterizes the ontology advanced explicitly by Einstein as the ontology of "materialism", which he says rests upon the illusion that the kind of existence, the direct actuality of the world around us, can be extrapolated into the atomic range. In the closing paragraphs of this chapter of his book he states that all counterproposals offered in opposition to the Copenhagen interpretation must sacrifice what he calls the symmetry properties of the quantum theory, namely the wave-particle symmetry and the position-velocity symmetry. He explicitly states that like the Lorentz invariance in the theory of relativity, the Copenhagen interpretation cannot be avoided, if these symmetries are held to be genuine features of nature.

Contemporary philosophers and historians of science have learned to recognize the operation of the realistic interpretation of empirically successful theories in the history of science. As new and empirically superior theories are developed, their realistic interpretations produce new ontologies with new ideas and beliefs about what is real, including ideas of the nature of causality. And when the new theory with its new ontology is attacked by the establishment, the "authorities" of the particular scientific profession, it is invariably attacked with the ontological beliefs defined by a preceding and less empirically adequate theory. Einstein seems not to have been unaware of this historical of phenomenon. In his "Reply to Criticisms" he stated that the scientist cannot afford to carry his striving for epistemological systemic as far as will the philosopher, and that while the scientist gratefully accepts the epistemologist's analysis, nonetheless the facts of experience, by which he presumably means scientific evidence, do not let the scientist be too much restricted in the construction of his conceptual world by the adherence to an epistemological system. Einstein was faithful to this insight to the extent that he rejected the Positivist philosophy, but he did not follow through with it, when he functioned as his own epistemologist and attempted to impose the deterministic ontology of field theory upon quantum theory.

Bohr's Influence on Heisenberg and Issues with Einstein

Niels Bohr was one of the leading atomic physicists of the first half of the twentieth century. He had studied in England under J.J. Thompson and Lord Rutherford, and received the Nobel Memorial Prize for Physics in 1922 for his theory of the structure of the atom. He founded the Copenhagen Institute for Theoretical Physics in 1920, and as its director was actively recruiting talented staff members, when he accepted an invitation to deliver a series of lectures on atomic physics at the University of Gottingen in the summer of 1922. In "Quantum Theory and its Interpretation" in *Niels Bohr* (1963) Heisenberg reports that he first met Bohr at the Gottingen lectures, which he attended with his teacher, Arnold Sommerfeld. At the time Heisenberg was a twenty-two year old, fourth semester student at the University of Munich. Heisenberg came to Bohr's attention, because in the discussions following one of the lectures, he dissented from Bohr's optimistic assessment of a theory developed by Kramers at Copenhagen. Heisenberg relates that Bohr was sufficiently worried about the objection, that after the discussion he asked Heisenberg to take a walk with him for a conversation. During the walk Bohr talked about the fundamental physical and philosophical problems of modern atomic theory, and the encounter resulted in an invitation for Heisenberg to visit the Institute at Copenhagen for a few weeks, and later to hold a position. Heisenberg describes his impressions of Bohr as primarily a philosopher rather than a physicist, and he states that he found Bohr's philosophy to be fascinating, although he also states that he and Bohr had different views on the role of mathematics in physics.

Bohr's philosophy of atomic physics is set forth in his *Atomic Physics and the Description of Nature* (1934), "Discussions with Einstein" in *Albert Einstein* (ed. Schilpp, 1949), *Atomic Physics and Human Knowledge* (1958), and *Essays 1958/1962 on Atomic Physics and Human Knowledge* (1963). Bohr's philosophical views may have been influenced by some casual reading of the philosophical literature, but he never references any philosopher in his writings, and his views seem largely to be the product of his own reflections on his research in atomic physics and on the work of the staff at Copenhagen. In "Quantum Theory and Its Interpretation" Heisenberg states that Bohr had developed views on the semantics of language and scientific theory many years before he had met Bohr and before he developed his matrix mechanics. Bohr's mature philosophy of science included two theses: Firstly that the mathematical formalisms of microphysics cannot describe the microphysical domain that lies beyond ordinary experience. Secondly that the only language that is capable of a descriptive semantics is the language of ordinary discourse and its refinement in classical Newtonian physics. Heisenberg did not accept the first thesis, and had a different concept about the abstract nature of mathematics. But Bohr's second thesis had a lifelong influence on him, an influence that had a retarding effect on his development of his own philosophy.

Bohr gives various reasons why in his view the mathematical formalisms of microphysics have no descriptive semantics, and are only symbolic instruments for making calculations and predictions. One reason given in "Discussions with Einstein" is the occurrence of a complex number in the formalism. Apparently he believed that reality can be described only by equations that have only real numbers for values. Another reason given in "The Solvay Meetings and the Development of Quantum Theory" (1962) in his *Essays 1958/1962* is the interpretation of the statistical wave function in a configuration space of more than four dimensions; Bohr believed that real physical space has no more than four dimensions. But the basic reason why Bohr interpreted the mathematical formalism of quantum theory instrumentally is his belief that only the language of everyday discourse and its refinement in classical physics can have a descriptive semantics. He maintained that ordinary language and classical physics must be used to describe any experimental set up in physics, while at the same time he believed that classical physics is too limited to describe the microphysical domain beyond ordinary experience. It is limited not only because Newtonian physics is inadequate as a microphysical theory, but also because of the inherent nature of human cognitive perception. This is a philosophy of the semantics of language that is a variation on the naturalistic thesis. Because of Bohr's philosophy of perception, Einstein as well as many philosophers of science were led to conclude that Bohr's philosophy of science is Positivist.

If Bohr's philosophy of science is a Positivist philosophy, it is a peculiar one. His statements of his philosophy that are most often referenced in this connection by philosophers of science are those in *Atomic Physics and the Description of Nature.* In the opening "Introductory Survey (1929)" he states that both relativity theory and quantum theory are concerned with physical laws that lie beyond ordinary experience, and which therefore present difficulties to our "accustomed forms of perception". In quantum theory the limitations of these forms of perception are revealed by the need for the complimentary description, the inconsistent description of the phenomenon as both a wave and a particle. Both of these two forms based on classical physics are necessary for a complete description, even though they are inconsistent in classical physics. Yet these "customary" forms of perception cannot be dispensed with, since all human cognitive experience must be expressed in terms of them. The fundamental concepts of classical physics therefore will never become superfluous for the description of physical experience; they must be used to describe experiments and to relate the mathematical symbolisms to the data of experience.

In Einstein's attack on Bohr's philosophy of quantum theory the central issue is the ontology of the Copenhagen interpretation, which Einstein critiqued with his programmatic aim of all physics. The explicit criterion set forth in the programmatic aim of science is the "complete" description of any individual situation, as it supposedly exists irrespective of any act of observation or substantiation. Accordingly he characterized the Copenhagen interpretation as a version of Bishop Berkeley's idealist thesis "*esse est percipi*", a characterization that is not accurate, because Bohr did not maintain that the atomic phenomenon is produced by a cognitive process but rather by the physical processes of measurement in the experimental set up. In this matter Einstein seems to have confused an epistemological issue with a physical one. But Bohr is not blameless for the confusion. For example in "Introductory Survey (1929)" he opens with statements emphasizing the subjectivity of all experience and the difficulties in distinguishing between phenomena and their observation; and he concludes the chapter with the statement that "to be" and "to know" lose their unambiguous meanings. From an epistemological viewpoint some of Bohr's statements are ambiguous as to whether he is advancing a realist or an idealist philosophy. Some of Heisenberg's earlier statements are also suggestive of an idealist position, as for example when he says in the opening chapter of *The Physicist's Conception of Nature*, that since we can no longer speak of the behavior of the particle independently of the process of observation, the natural laws formulated in the quantum theory no longer deal with the elementary particles themselves, but only with our knowledge of them. But later Heisenberg is very clear about avoiding any metaphysical idealism; in "The Copenhagen Interpretation of Quantum Theory" in *Physics and Philosophy* he states explicitly that quantum theory does not contain genuinely subjective features, since it does not introduce the mind of the physicist as part of the atomic event, and that the transition from possible to actual in the act of observation is in the physical and not the psychical act of observation.

This metaphysical idealist/realist confusion notwithstanding, however, Einstein's central ontological thesis is that the statistical quantum theory is incomplete in the sense that further theoretical research is necessary, in order to develop a complete theory that would give the uncertainty relation a status in future physics, which he thought should be analogous to the status had by statistical mechanics. What is most noteworthy is that Einstein admits that the indeterminacy principle is not empirically incorrect, even as he rejects the Copenhagen ontology because it does not conform to his explicit ontological criterion. In the 1949 "Reply to Criticisms" Einstein conceded that his incompleteness thesis is the minority view among physicists; contemporary philosophers as well as physicists have accepted the indeterminacy thesis of the Copenhagen interpretation of the statistical quantum theory, and have rejected the deterministic ontology advocated by Einstein. When confronted with the dilemma of having to choose between an established ontological criterion and a new but empirically adequate quantum theory, both the contemporary physicists and the contemporary Pragmatist philosophers of science have opted for the latter, contrary to Einstein's arguments for the former.

In addition to the ontological issue between Bohr and Einstein about what is physically real, there is also a related epistemological issue about the relation between sense perception and intellectual concepts, which is also a semantical issue about the relation between observation language and theory language. Einstein had portrayed Bohr as a Positivist because of Bohr's views about perception and the semantics of language. This portrait is debatable, because Positivists do not usually speak of forms of perception, and particularly about the limitations of such forms of perception for physics. But in his 1934 book Bohr writes of the necessity of these forms of perception for science to reduce our "sense impressions" to order. Even though Einstein himself uses the phrase "sense impressions" in his statement of the aim of science in "Physics and Reality" in 1936, he seems to have taken Bohr's discussion referencing sense impressions to mean that these are no concepts or categories in perception. Einstein opposed this view, and states in 1949 in his "Reply to Criticisms" that thinking without positing categories and concepts is as impossible as breathing in a vacuum. He furthermore states that his philosophy differs from Kant's only by the fact that he does not view categories as unalterable and as conditioned by the understanding, but rather views them as "free conventions". The philosopher of science may ask whether Einstein's neo-Kantian views without Kant's idealism and *a priorism* is still recognizably Kantian. But the point to be emphasized is that Einstein's thesis that concepts are necessary for perception and that they are free conventions amounts to a restatement of what he told Heisenberg in 1926, when he said that it is the theory that decides what we can observe. In this earlier statement Einstein might consistently have told Heisenberg that observation without theory is as impossible as breathing in a vacuum. Perhaps it was in response to Einstein's criticisms in these matters that Bohr refrains in his later writings from using the phrase "sense impressions". Instead Bohr merely describes the concepts of classical physics as a refinement of the concepts of ordinary discourse, so that he is no longer mistakenly taken as saying that perception occurs without any concepts or forms.

Nonetheless there is still a fundamental difference between the semantical views of Bohr and Einstein. Einstein's thesis that concepts are free conventions is intended to mean that there are none of the inherent limitations in observation or in language, such as Bohr had maintained. In

Bohr's phrase "customary forms of perception", the term "customary" does not mean the same thing as the term "convention" in Einstein's phrase "free conventions". The limitation that Bohr said these customary forms of perception impose on descriptive language are not temporary limitations, that will be removed with the change in language customs resulting from the further development of theory. Rather these limitations are inherent in the nature of the human cognitive processes of perception and consequently in the semantics of descriptive language. They are therefore permanent. There is no such permanence according to Einstein's view; the free conventions of human thought, in the concepts and categories in language and scientific theory, are not only conventions that are free to change, but are destined to change with the advancement and further development of scientific theory. The difference between Bohr's and Einstein's semantical views is the difference between the naturalistic and the artifactual philosophies of the semantics of language.

A few more comments about the relation of the semantical issue to contemporary philosophy of science: Einstein's semantical views anticipated those of the contemporary Pragmatist philosophers of science in several respects, and his arguments against Positivism undoubtedly had an influence on them, even though he is seldom referenced in the philosophical literature. Einstein rejected the Positivist thesis that each individual concept in a theory requires specific justification of its meaningfulness, when the concept is indispensable for the theory, and when the theory in its entirety has been empirically validated. This is a rejection of the Logical Positivist problem of theoretical terms. He also rejected Bridgman's operationalist thesis, and its requirement that each of a theory's assertions must be independently interpreted and tested, because this procedure has never yet been accomplished for any scientific theory, and furthermore in Einstein's view, it cannot be accomplished. On Einstein's thesis a physical theory need only imply empirically testable assertions in general; there exists no logical path from the empirically given to the conceptual world. Both the individual concept and the individual assertion in a theory, confront the empirically given in connection with the entire system of assertions, because there is an element of arbitrary choice between the empirical and the conceptual world, that result in what Einstein calls an "embarrassment of riches" for the theorist. This element of the arbitrary in the relation between the empirical and the conceptual is the basis for the contemporary Pragmatist philosophers' thesis that the semantics of language is not predetermined by nature, as Bohr and the Positivists had maintained, but rather is a cultural artifact. Thus the meanings of individual terms and assertions are not determined by their relation to the empirical world individually, but by their relation to one another in the larger context of a discourse, such as a scientific theory. This much is at least consistent with Einstein's views. But the contemporary Pragmatists take a step that probably Einstein cannot be associated with; they equate the dependence of meanings upon context with a wholistic view of the semantics of language. But one cannot be certain about what Einstein might have said: while Einstein affirmed an artifactual theory of the semantics of language, he did not develop a theory of meaning description.

On the other hand Einstein took his views a step in another direction than the Pragmatists, when he advanced his explicit ontological criterion of logical simplicity for the whole of physics. This is a nonempirical criterion for scientific criticism, which Einstein used to argue that the concepts that are successful in field theory must also be used in quantum theory. It was this requirement that macrophysical and microphysical theories use the same ontological categories, that led Einstein to reject the ontology of the Copenhagen interpretation of the statistical quantum theory. Here the contemporary Pragmatist philosophers of science depart from Einstein's views; the element of arbitrariness that both they and Einstein admit in the relation between the empirical and the conceptual, leads the Pragmatists to admit a pluralism in empirical science, that both Einstein and most Positivist philosophers found to be scandalous. This pluralism is opposed to Einstein's explicit ontological criterion of simplicity for all of physics. The Pragmatists do not find the theorists' embarrassment of riches permitted by the artifactual character of the semantics of language to be as embarrassing as Einstein's nonempirical criterion for scientific criticism, which seeks to constrain the development of scientific theory by imposing a monolithic ontology. On the Pragmatist view pluralism is characteristic of the development of science, and some Pragmatist philosophers maintain that it is a condition for its advancement.

Semantical Revision and the Doctrine of Closed-off Theories

Heisenberg's response to the conflicting influences of Einstein and Bohr was his thesis of closed-off theories. In "Questions of Principle in Modern Physics" originally given as a lecture at the University of Vienna in

1935 and since published in his *Philosophical Problems of Quantum Physics* he sets forth the central questions that are addressed by his philosophy of physics. He firstly asks how it was possible for there to have occurred the "strange" revision of the fundamental concepts of physics during the preceding thirty years. Then secondly he asks what is the truth content of classical physics and of modern physics in view of this conceptual revision. He notes that these are also the questions that were posed and seriously discussed by Bohr, who approached them from the fundamental premises of quantum theory. It is noteworthy that Heisenberg's philosophy of science addresses questions formulated by Bohr. The formulation of the questions in terms of how a conceptual revision is possible suggests a naturalistic philosophy of the semantics of language as a point of departure, since on the artifactual thesis the possibility of a fundamental semantical revision is not problematic. When concepts and meanings are understood to be cultural artifacts, then semantical change may be expected as a matter of course.

As it happens, Heisenberg did not depart very far from the naturalistic thesis. He developed a theory of semantical revision, but it is also a theory of semantical permanence. His mature philosophy of science is not Positivist, but like the Positivists he maintains that classical physics is permanently valid, and that its concepts are necessary for experimentation in physics. He states that classical physics is based on a system of mathematically concise axioms, whose physical content is fixed by the choice of words used in them. These words determine unequivocally the application of the system of axioms to nature. Wherever concepts like mass, velocity and force can be applied, there Newton's law, $F=ma$, will be true; the validity of the claim of this law is comparable to Archimedes law of the simple lever, which today forms the theoretical basis for all load-raising machines, and which will be true for all time. Therefore in spite of the fact that there has been a revision of classical mechanics, the axiomatic system developed by Newton is still valid. The revision pertains to the limits encountered in the application of the axiomatized system of concepts of classical physics; it is not the validity but only the applicability of classical laws, that has come to be restricted by relativity theory and quantum theory. The experiences that provide the basis of relativity theory have demonstrated that the simple time concept of Newton's mechanics ceases to be of use, when dealing with bodies moving with a velocity approaching that of light. And in microphysics classical mechanics can predict the correct track of the electron in the Wilson cloud chamber, but if without observation of its track the electron is reflected at a diffraction grating, the basis for an unambiguous application of the space-velocity concept has disappeared, and classical laws cannot be applied to such a process.

Having described how the axiomatized mathematical system of classical physics is permanently valid, Heisenberg then describes how revision is possible. The revision of classical physics is possible because of a lack of precision in the concepts used in the system. While the quantities, the variables x, t, and m used in the Newtonian system are linked without ambiguity by the system of equations, which contain no degree of freedom apart from initial conditions, the words "space", "time", and "mass", which are attributed to the quantities are tainted with all the lack of precision that may be found in their everyday use. The validity of classical physics is limited by the lack of precision of the concepts contained in its axioms. As a result of this lack of precision science may be forced into a revision of its concepts as soon as it leaves the field of common experience; the concepts currently used may lose their value for the orderly presentation of new experience. But this revision cannot be known in advance. For example before the experiences of quantum theory the results of the Wilson cloud chamber experiments could unhesitatingly be expressed as "we see in the cloud chamber that the electron has described such and such a path", and this simple description could be accepted as an experimental fact. It was only later that physicists came to know from other experiments the problematic nature of the phrase "path of an electron". Scientific progress consist initially in the unhesitating use of existing terms for the description of experience, and then the revision of those terms as demanded by new experience. The lack of precision contained in the systems of concepts of classical physics is necessary, and therefore even the mathematically exact sections of physics represent only tentative efforts to find our way among a wealth of phenomena.

Classical concepts must be retained for experimentation in physics. So far as the concepts of space, velocity and mass can be applied unhesitatingly, as in everyday experiences, Newtonian principles still apply. The Newtonian laws represent an "idealization" achieved by taking into account only those parts of experience that can be ordered by the concepts of space, time and mass on the assumption of objective events in time and space, and therefore they always remain the basis for any exact and objective science. Since we demand of the results of science that they can be objectively demonstrated, we are forced to express these results in the language of classical physics. For example for an understanding of relativity theory, it is necessary to stress that the validity of Euclidian geometry is presupposed in the instruments that are used to show the deviation from Euclidian geometry, i.e. the measure of the deviation of sunlight [an apparent reference to Eddington's 1919 eclipse experiment to test relativity theory]. Furthermore the very methods used for the manufacture of these instruments enforce the validity of Euclid's geometry for these instruments within the range of their accuracy. Similarly we must be able to speak without hesitation of objective events in time and space in any discussion of experiments in atomic physics. Heisenberg concludes that while the laws of classical physics seen from the viewpoint of modern physics appear only as limiting cases of more general and abstract connection, the concepts associated with these laws remain an indispensable part of the language of science, without which it is not possible even to speak of scientific results. Therefore, while mathematically exact sections of physics are tentative, the classical concepts must nevertheless be used for the description of experiments.

Heisenberg further elaborates these views in several later articles and chapters in his books. In "The Notion of a 'Closed Theory' in Modern Science" in *Across the Frontiers* he discusses the criteria for scientific criticism and the evolution of the aim or goals of science. When Einstein developed his special theory of relativity, it was evident that Maxwell's theory of electromagnetic phenomena could not be traced back to mechanical processes that obey Newton's laws, and the inference seemed unavoidable that either Newtonian mechanics or Maxwell's theory must be false. Physicists concluded that Newton's theory is strictly speaking false. This mislead many scientists into unwittingly attempting to describe the phenomena of the world exclusively by means of the concepts of field theory. This represented a goal of science that was commonly accepted from Newton's theory, that science should proceed by means of a unitary conceptual scheme, except that now the concepts should be those of field theory instead of classical mechanics. But in both cases the concepts supplied an objective and causal description of the process involved, and were therefore thought to be universal. These common concepts were rejected by quantum theory for the description of the atom, although they must still be used to describe the results of an observation while standing in a complimentary relation to one another. Thus physicists no longer say that Newton mechanics is false and must be replaced by quantum mechanics, which is correct. Instead it is said that classical mechanics is a consistent self enclosed scientific theory, and that it is a strictly true and correct description of nature, whenever its concepts can be applied. Quantum theory has only restricted the applicability of Newtonian mechanics, and has made classical physics a "closed-off" theory. In contemporary physics there are four great disciplines that are closed-off theories. They are (1) Newtonian mechanics, (2) Maxwell's theory and special relativity, (3) the theory of heat and statistical mechanics, and (4) nonrelativistic quantum mechanics, atomic physics and chemistry. General relativity is not yet closed off.

Heisenberg then turns to a discussion of the properties of a closed-off theory and of its truth content. One property of a closed-off theory is its consistency as an axiomatized mathematical system. The most celebrated example is Newton's *Principia Mathematica.* Another property is that the concepts of the theory must be directly anchored in experience. Before the axiomatic system is developed, concepts describing everyday life remain firmly linked to the phenomena and change with them; they are compliant toward nature. But when they are axiomatized, they become rigid, and they "detach" themselves from experience. The system of concepts rendered precise by axioms is still very well adapted to a wide range of experiences, but axiomitization of concepts sets a decisive limit to their field of application. The discovery of these limits is part of the development of physics. Yet even when the boundaries of the closed theory have been encountered and overstepped, and new areas of experience are ordered by means of new concepts, the conceptual scheme of the closed theory still forms an indispensable part of the language in which the physicist speaks of nature. The closed theory is among the presuppositions of the wider inquiry; we can express the result of an experiment only in the concepts of earlier closed theories.

Heisenberg summarizes the properties of closed-off theories as follows: Firstly the closed-off theory holds true for all time. Whenever experience can be described by the concepts of the closed-off theory, even in the most distant future, the laws of this theory will always be correct. Secondly the closed-off theory contains no perfectly certain statements about the world of experiences; its successes is contingent. Thirdly even with the uncertainty of its contingency, the closed-off theory remains a part of scientific language, and therefore is an integrating constituent of our current understanding of the world. Heisenberg sees the evolution of modern science differently than Einstein's description in "Physics and Reality". The historical processes that have given rise to the whole of modern physics since the conclusion of the Middle Ages, is a developmental process consisting of a succession of intellectual constructs, which take shape as if from a "crystal nucleus", out of individual queries raised out of experience, and which eventually once the complete crystal has developed, again detach themselves from experience as purely intellectual structures that forever illuminate the world for us as closed-off theories. Thus the history of science is like the history of art, where the goal is to illuminate the world by means of intellectual constructs. In his "The End of Physics" in *Across the Frontiers* he adds that while physics consist of many closed-off systems, it is not possible to close off physics as a whole. Today it is necessary to seek out new and still more comprehensive closed-off theories, or "idealizations" as he also calls them, which will include both relativity theory and quantum theory as limiting cases.

Closely related to his thesis of closed-off theories is Heisenberg's theory of abstraction. In "Abstraction in Modern Science" in *Across the Frontiers* he defines abstraction as the consideration of an object or a group of objects under one viewpoint while disregarding all other properties of the object. All concept formation depends on abstraction, since it presupposes the ability to recognize similarities. Primitive mathematics developed from abstraction, e.g. the concept of the number three. Mathematics has formed new and more comprehensive concepts, and thereby ascended to ever higher levels of abstraction. The realm of numbers was extended to include the irrational and complex numbers. This view is quite different from Bohr's, who believed that the mathematical formalisms used in physics have no descriptive semantical value but are merely symbolic instruments for calculation and prediction, particularly if they contain complex numbers or represent more than four dimensions as in quantum theory. In Heisenberg's philosophy abstraction, the consideration of the real from a selective viewpoint, produces idealizations of reality which are axiomatic mathematical structures that become closed-off, as the historical development of science reveals the limitations of their applicability and occasions the creation of new theories.

In expounding his semantical doctrine of closed-off theories Heisenberg departed from Bohr. Comparison of their views reveals essential similarities, but it also reveals differences. Bohr's semantical views are stated in "Discussions with Einstein" where he says that Planck's discovery of the quantum of action makes classical physics an "idealization" that can be unambiguously applied only in the limit, where all actions involved are large in comparison with the quantum. A more elaborate statement is given in "The Solvay Meetings" in *Essays 1958/1962*. There he firstly says that unambiguous communication of physical evidence demands that the experimental arrangement and that the reading of observations be expressed in common language suitably refined by the vocabulary of classical physics. Then secondly he states that in all experimentation this demand is fulfilled by using as measuring instruments bodies like diaphragms, lenses, and photographic plates, which are so large and heavy that notwithstanding the decisive role of the quantum for stability and properties of such bodies, all quantum effects can be disregarded in the account of their position and motion. Finally and thirdly he says that in classical physics we are dealing with an idealization according to which all phenomena can be arbitrarily subdivided, and all interaction between measuring instruments and the object under investigation can be neglected or compensated for. Bohr seems to be using the term "idealization" as Heisenberg does, but he reserves it for the classical physics. He does not admit a separate set of distinctively quantum concepts because of his instrumentalist interpretation of the quantum theory formalism. In his view there are no quantum concepts defined by the equations of the quantum theory, but rather there are only classical concepts and the semantically uninterpreted mathematical formalism used for generating predictions expressed in classical terms.

Bohr's "Forms of Perception" and Neo-Kantianism

Having based his doctrine of closed-off theories on Bohr's philosophy of observation, Heisenberg attempts to relate Bohr's view to that of Kant. Heisenberg's statements are found in his "Recent Changes in the Foundations of Exact Science" (1934) in *Philosophical Problems of Quantum Physics*, in his "The Development of Philosophical Ideas Since Descartes in Comparison with the New Situation in Quantum Theory" in *Physics and Philosophy*, in his "Quantum Physics and Kantian Philosophy (1930-1932)" in *Physics and Beyond*, and in his "Planck's Discovery and the Philosophical Problems of Atomic Theory" in *Across the Frontiers*. Like Einstein, Heisenberg rejects the Positivist phenomenalism and advocates realism. In "Planck's Discovery" he states that quantum theory does not consider sense impressions to be the primary given, and that if anything is the primary given in quantum theory, it is the reality that can be described by the concepts of classical physics. And in "Development of Philosophical Ideas Since Descartes" he describes his revision of Kant's views with the phrase "practical realism", since in Heisenberg's view things rather than perceptions are the given for the human mind.

But while Heisenberg is opposed to Positivism as much as Einstein, his referencing the philosophy of Kant is not motivated by his anti-Positivism; Heisenberg is interested merely in relating Kantianism to his philosophy of observation. In "Changes in the Foundations of Exact Science" he says that in the field of the philosophy of perception, Kant's philosophy has been put into a new light as a result of the critique of absolute time and Euclidian space by relativity theory, and by the critique of the law of causality by quantum theory. In particular, the question of the priority of the forms of perception and of the categories of the understanding must be reconsidered. He states that there are two apparently contradictory propositions that must be reconciled: On the one hand relativity theory and quantum theory have shown that our space-time forms of perception and the category of causality are not independent of all experience in the sense that they must for all time remain essential constituents of every physical theory. On the other hand, as Bohr taught, the applicability of the classical (i.e. Kantian) forms of perception and the law of causality are the premises of every objective experience even for modern physics. The physicist can only communicate the course of an experiment and the result of a measurement by describing the necessary manual operations and instrument readings as objective events taking place in the space and time known to our intuition. And he could not infer the properties of the observed object from the result of measurement, unless the law of causality guaranteed an unambiguous connection between measurement and object. Heisenberg resolves the contradiction between the two statements as follows: Physical theories can have a structure differing from classical physics, only when their aims are no longer those of immediate sense perception; that is to say, only when they leave the field of common experience dominated by classical physics. In "Quantum Physics and Kantian Philosophy" Heisenberg views Kant's philosophy of perception as a closed-off theory, as he elsewhere describes closed-off theories in physics. He compares the validity of Kant's philosophy to the validity of

Archimedes' theory of the lever, and he states that Kant's theory is eternally true, just as Archimedes' theory is eternally true; Kant's analysis of perception represents true knowledge, that applies wherever thinking beings enter into the kind of contact with their environment called "experience". Relativity theory and quantum theory have defined the limits of the *a priori* in the exact sciences in ways that could not have been known to Kant. The *a priori* has not been eliminated from physics, and Kant's analysis of how we come by our experiences is essentially correct. But the *a priori* has become "relativised" in the sense that classical concepts are *a priori* conditions for relativity and quantum theory, since classical concepts are necessary for experiments. The progress of science in Heisenberg's view has changed the structure of human thought, and has taught us the meaning of "understanding".

In the closing paragraph of his "Quantum Physics and Kantian Philosophy" Heisenberg states that he has described the relationship between Kant's philosophy and modern physics from the perspective of Bohr's teachings. The importance of Heisenberg's discussion of Kant is that it treats the philosophy of perception. And the philosophy of perception in turn is important because it often serves as a philosophy of observation in philosophy of science and epistemology. In the context of the philosophy of modern physics the central problem is the semantics of modern physical theory. On the one hand where the semantics of the language of physics is said to be supplied by observation and perception, the usual assumption is that perception is a natural cognitive process that predetermines the semantical content of language in an invariant, objective, and atemporal manner. On the other hand the history of science is a history of change, of development of theory, that is not easily supposedly delivered to the human mind by perception. Traditional philosophy of perception suggests a naturalistic philosophy of the semantics of language, while history of science suggests an artifactual philosophy of semantics.

In anticipation of the discussions to follow about the views of other philosophers, it may be said that contemporary philosophers have addressed this semantical issue and specifically the topic of perception in a different manner than did Heisenberg or Bohr, or for that matter most earlier philosophers. Unquestionably there are limitations to what can be perceived; there are objects that are too small to be seen with the human eye, there are sounds that are pitched too high to be heard by the human ear, etc. Yet observation is permeated with learned ideas, permeated with interpretation, such that it has a variability, a subjectivity, and a historicity that are not invariably associated with the outcomes of the functioning of our natural faculties for perception. The Pragmatist philosopher of science, Norwood R. Hanson, addressed this difficult mixture of nature and culture in knowledge by reconsidering the concept of perception in observation. Hanson's answer is the same as Einstein's admonition to Heisenberg, that theory decides what the physicist can observe. But this view is not considered by Heisenberg in his discussion of his doctrine of closed-off theories or in his discussion of Kant.

On the other hand a theory of knowledge based on perception is sometimes called a "psychologistic" theory of knowledge, and some philosophers object to psychologism, even when the particular psychologistic position advocated does not assert that the laws of logic are "laws of thought" in the sense of psychological laws. The notable contemporary example is the philosopher of science, Karl Popper, who rejects psychologism, and makes a strong distinction between perception and observation with his distinction between "world two", the domain of subjective psychology including perception, and "world three', the domain of objective knowledge including observation ("world one" is the world of objective nature treated by natural sciences such as physics). In this way Popper separates the roles of nature and culture, so that following Einstein, who said that theory decides what we can observe, Popper says that there is no observation without theory. Whatever the strategies of the contemporary philosophers of science, they converge on the idea that the content of knowledge is not predetermined by nature in perception. In this respect they depart not only from Kant and the Rationalists and from Hume and the Positivists, they also depart from Bohr, Heisenberg and the other Copenhagen physicists.

On Scientific Revolutions

Heisenberg considers the development of modern quantum theory to be one of the two great scientific revolutions of twentieth century physics; the other in his view is relativity theory. The complete title of his 1958 book is *Physics and Philosophy: The Revolution in Modern Science*. But by the 1960's the term "revolution" as used in connection with the development of science had become what Heisenberg calls a "vogue word" because of some similarities between scientific revolutions and social revolutions.

Possibly the vogue status of the term was due to the popular monograph, *Structure of Scientific Revolutions*, written by Thomas Kuhn in the United States in 1962, but Heisenberg never references Kuhn. Heisenberg discusses his idea of revolution in science in a lecture delivered to the Association of German Scientists in Munich in 1969, which was published in English in 1974 as "Changes of Thought Pattern in the Progress of Science" in his *Across the Frontiers*. Heisenberg recognizes the operation of sociological forces in the scientific professions, but his views are different from those of Kuhn.

Heisenberg defines a "revolution" in science as a change in thought pattern, which is to say, a semantical change. He states that a change in thought pattern becomes apparent, when words acquire meanings that are different from those they had formerly, and when new questions are asked. He does not explicitly reference his semantical thesis of closed-off theories in this context. However, the episodes in the history of post-Newtonian physics that he cites as examples of scientific revolutions are the same as those that he also says resulted in new closed-off theories in the history of physics. And the semantical change that occurs in the transition to a new axiomatic theory and the closing off of the old one, is the change involved in the transition to a new thought pattern. The central question that Heisenberg brings to the phenomenon of revolution in science understood as a change in thought pattern, is how the revolution is able to come about. The occurrence of the revolution is problematic because of resistance to the change in thought pattern offered by the profession. Heisenberg also expresses the question in more sociological terms, when he asks how a small group of physicists are able to "constrain" other physicists to make the change in thought pattern in spite of the latter's resistance to do so. Firstly he discusses the reasons for resistance. Then he discusses various proposed explanations about how the resistance is overcome.

In his discussion of the reasons for resistance he states that there have always arisen strong resistances to every change in the pattern of thought. The progress of science proceeds as a rule without much resistance or dispute; the scientist has by training been put in readiness to fill his mind with new ideas. But the case is altered when new groups of phenomena compel changes in the pattern of thought. Here even the most eminent of physicists find immense difficulties, because a demand for change in thought pattern may create the perception that the ground is to be pulled from under one's feet. A researcher who has achieved great success in his science with a pattern of thinking he has accepted from his young days, cannot be ready to change this pattern simply on the basis of a few novel experiments. Heisenberg states that once one has experienced the desperation with which clever and conciliatory men of science react to the demand for a change in the pattern of thought, one can only be amazed that such revolutions in science have actually been possible at all. Undoubtedly the case in Heisenberg's experience is the desperation that he saw in Schroedinger's and especially Einstein's opposition to the new thought pattern represented by the Copenhagen interpretation of the quantum mechanics.

He then considers several possible answers to the question of how scientific revolutions can come about in spite of the resistances, of how the resistances are overcome. One answer that he rejects is that the revolution is due to a strong revolutionary personality. He maintains that no such strong personality could overcome the profession's resistance. Another answer that he rejects might be described as a variation on the conspiracy thesis, the view that a small group of physicists intended from the outset to overthrow the existing state of the science. He states that never in its history has there ever been a desire for any radical reconstruction of the edifice of physics; this is because at the onset of a revolution there is a very special, narrowly restricted problem, which can find no solution within the traditional framework. The revolution is brought about by researchers who are genuinely trying to resolve the special problem, but who otherwise wish to change as little as possible in the previously existing physics. It is precisely the wish to change things as little as possible, which demonstrates in Heisenberg's opinion, that the introduction of novelty is a matter of being compelled by the facts. The change of thought pattern is enforced by the phenomena. He concludes therefore that the way to make a scientific revolution is to try to change as little as possible: it is an error to demand the overthrow of everything existing because of the risk of attempting a change that nature makes impossible. Small changes on the other hand show what is compelled by the facts, and in the course of years or decades, they will enforce a change in thought pattern and shift the foundation of science. An example of such a small change is Planck's quantum of action, that years later resulted in the modern quantum theory.

Having rejected the view that scientific revolution occurs because of a conspiracy either with or without a strong revolutionary personality, Heisenberg then considers the answer that the resistances to revolution are overcome simply because there is a "right" and a "wrong" in physics, and the new theory is right while the old theory is wrong. It is noteworthy that Heisenberg does not reject the thesis that there is a right and a wrong in the sense of a correct and an incorrect, and in view of his thesis of closed-off theories, it would be remarkable if he did. Furthermore he had explicitly rejected historical relativism in his "Quantum Physics and Kantian Philosophy". Still he finds that there is a problem with this answer as an explanation for overcoming resistances, namely that the right theory has not always prevailed. He cites as an example the dominance of the geocentric theory of Ptolemy over the heliocentric theory of Aristarchus. Therefore, while there are absolute standards for criticism of scientific theories, there still remains the question of why some correct theories succeed in gaining acceptance over the strong forces of resistance, while others do not, even though the rejected theories may be correct. Heisenberg then proposes his own answer. Scientists perceive that with the new pattern of thought, they can achieve greater success in their science than with the old; the new system proves to be more fruitful. Heisenberg states that once anyone has decided to be a scientist, he wants above all to get ahead, to be on hand when the new roads open up; it does not satisfy him merely to repeat what is old and has often been said before. Consequently the scientist will be interested in the kind of problems where there is something to be done, where he has the prospect of successful work. That is how relativity theory and quantum theory came to prevail according to Heisenberg. He describes this as a "pragmatic criterion of value", and he states that while one cannot always be certain that the right theory will always prevail, nevertheless these are forces that are strong enough to overcome the resistances to a change in thought pattern.

Since Heisenberg was a principal participant in one of the great scientific revolutions in modern physics, his views based on his personal experience deserve singular consideration. He was undoubtedly impressed by the resistances offered to the Copenhagen interpretation by Schroedinger and especially by Einstein. While few contemporary philosophers of science accept his doctrine of closed-off theories with its Positivist view of observation, which he uses to interpret his experience of scientific revolution, they recognize the operation of sociological forces including the thrust of opportunistic careerism. And they also recognize that semantical change occurs in scientific revolutions, and that the adjustment it imposes on the affected profession operates as a cause of resistance within it, even though they do not accept Heisenberg's theory of semantical change and permanence. Unlike others such as Kuhn, Heisenberg does not identify the institutionalized criteria for scientific criticism with the existing thought pattern, and he does not maintain that the revolution is a change with no institutional framework controlling it. Heisenberg avoids the historical relativism found by many in Kuhn's thesis, and which is explicitly embraced by Feyerabend. And one would not expect the proponent of the doctrine of closed-off theories and the advocate of Bohr's theory of observation to find the process of scientific criticism to be very problematic; the scientist is simply compelled by the facts, and the facts are not a problematic matter. Failure of the correct theory to overcome the forces of resistance, and indeed the very existence of those resistances, is due to the professional failure of those who can not adjust to new thought patterns when the facts compel, and not to any inherently problematic character in the process of scientific criticism itself.

The contemporary Pragmatist philosopher of science can only wonder what Heisenberg might have said, were he to have followed through with Einstein's thesis that it is the theory that decides what the physicist can observe; how he would have addressed the consequent problem that the concepts used to describe the facts are supplied by the choice of thought patterns expressed in the theory. Facts are not unproblematic, as the Pragmatist N. R. Hanson attempted to demonstrate in his *Patterns of Discovery* (1958), where he considers at length the interpreted nature of the observations relevant in the choice between the geocentric and the heliocentric theories of the planetary motions. If what Heisenberg calls "the pragmatic criterion of value" determines the choice between the old and the new theories, and each theory determines the facts, then a problem arises as to how facts can have any independently compelling force. This is a problem that currently vexes the contemporary Pragmatist philosophers of science.

The Philosophy of Science

Heisenberg's philosophical writings can be related to the four basic questions addressed by contemporary professional philosophers of science. The question of the aim of science has a special importance in Heisenberg's philosophy, because it was explicitly developed to defend the Copenhagen interpretation of quantum theory against Einstein's explicitly formulated programmatic aim of all physics. Heisenberg's views are expressed in his "Notion of 'Closed Theory' in Modern Science" and in his "On the Unity of the Scientific Outlook on Nature" (1941) in *Philosophical Problems of Quantum Physics*. Einstein used his programmatic aim of physics to claim that the statistical quantum theory is "incomplete" in the sense that it does not represent an adequate explanation for the problem that it addresses, and that further research work is still needed. The reason that it is still incomplete, is that it is not consistent with the ontology of field physics, which describes physical reality as continuous in four dimensions and deterministic. Heisenberg denied Einstein's thesis that the microphysical theory must employ the same ontological concepts as those used in macrophysical field theory, and his doctrine of closed-off theories was motivated in part by the desire to show how multiple ontologies can co-exist in physics. This is Heisenberg's pluralism in science. The Copenhagen interpretation of quantum theory is complete in Heisenberg's view, because it is a closed-off theory, and like all closed-off theories it is not only a complete solution to the problem that it addresses, but it is also a permanently true solution. In Heisenberg's philosophy of science the aim of science is to progress through a sequence of closed-off theories, and it is not, as Einstein maintained, to progress toward a single and all-inclusive ontology. The result of physics pursuing its aim as Heisenberg views it, has been the architectonic scheme for physics, a scheme of closed-off theories which he delineates in his "Relation of Quantum Theory to Other Parts of Natural Science" in *Physics and Philosophy*.

In Heisenberg's treatment of the question of scientific discovery, two aspects may be distinguished: One may be called the syntactical or structural aspect, and the other may be called the semantical or the interpretative aspect, which also includes ontological considerations. The structural aspect pertains to the development of the new formal axiomatic system, the new mathematical theory. These new formal structures are the result of repeated failures in the conservative attempts by the researchers to extend a currently acceptable theory to explain phenomena in a new domain of experience, and eventually may result in a revolutionary development. Closely related to the first aspect is the second, the interpretative problem. When extension of Newtonian physics could not solve the problem of microphysics, and the matrix mechanics was eventually developed by Heisenberg, the interpretation of the new matrix mechanics still remained problematic. Using Einstein's thesis that the theory decides what the physicist observes, Heisenberg reinterpreted the relevant observations in the Wilson cloud chamber experiment, and developed the uncertainty relation and its nondeterministic ontology. The new interpretation was accomplished by taking the new quantum theory realistically, as a description of the ontology of the microphysical world. When Einstein attacked the statistical quantum theory, he attacked only the second aspect, the Copenhagen interpretation with its nondeterministic ontological claim; he rejected the indeterminacy claim as a true description of the real world.

The third question pertains to the nature of scientific explanation. Heisenberg's views on this question are implicit in his position against Einstein's objections to the Copenhagen interpretation. Einstein's objection to the Copenhagen interpretation was that it was incomplete as a scientific explanation. This objection is a very traditional type of objection, because historically the concept of scientific explanation has been defined in terms of one or another ontology, and Einstein demanded conformity to the ontology described by the concepts of field physics. Bohr placed himself and his Copenhagen colleagues at a disadvantage, when he employed the vocabulary of their critics by referring to the statistical quantum theory as "noncausal"; he accepted the definition of causality in terms of the ontology of classical physics and field theory. But Heisenberg maintained that the revolutionary developments in physics consisted in part of interpreting the new mathematical formalism realistically, such as accepting the field as a reality, accepting relativistic time as real time and abandoning the concept of absolute time, and most notably accepting the uncertainty relation as describing the real microphysical world as nondeterministic. This amounts to separating the concept of scientific explanation from any preconceived ontology, a view of scientific explanation that was quite radical in its time, even though it is now the common property of the contemporary philosophers of science.

In striking contrast to his radical concept of scientific explanation, Heisenberg's treatment of the fourth question, the question of scientific criticism, is very conservative: he believed that his doctrine of closed-off theories enables him to explain how scientific theories can be permanently true. His views of explanation and of criticism represent a very unusual combination of views; historically philosophers and scientists have maintained that scientific explanations are permanently true, because as explanations, they describe correctly the one and only actual ontology. Heisenberg's philosophy of scientific criticism includes a semantical thesis, which is a thesis of both semantical change and of semantical permanence. Whether or not this semantical thesis is a sustainable one is certainly questionable, particularly when it depends on such curious processes as the semantics of words becoming "detached" from the variables occurring in the closed-off axiomatic theories, when the theories encounter the limits of their applicability. A philosopher of science such as Popper would dismiss such a process as a "content-decreasing" stratagem: if when a theory is criticized by an experimental test, the words expressing the test outcome describe something contrary to what the theory had predicted, then the attempt to save its truth claim by equivocation, by the "detachment" of the meanings describing the experimental outcome from the terms in the theory, only makes the theory tautological. In other words Heisenberg's doctrine in effect says a theory is true where it is true, and that where it is not true, it is not falsified, because it becomes silent. But it must be remembered that the relevant situation in Heisenberg's view is that in which the theory has been applied to a new domain of experience, where it is used to address a new problem that it was not initially formulated to address. Therefore, when the meanings of the words expressing the apparently falsifying experimental outcome became "detached" from the variables in the theory, the reasonable question arises as to whether or not the equivocation does not actually produce two theories: one is the original theory that is adequate to the original problem it was initially formulated to solve; the other one is the extended version of the original that is used to address the new problem arising in the new domain of experience. Only the extended version is falsified by the new test; the original theory is still true. As it happens, Heisenberg was motivated in part to develop the thesis of permanent truth and the doctrine of closed-off theories, in order to be consistent with Bohr's belief that Newtonian physics must be permanently true, because it supplies physics with the language of observation. But even after discounting this motivation and the assumption about observation that gave rise to it, Heisenberg's thesis of permanent truth raises worthy questions about the criteria for individuating theories. And since the opposing view expressed by Popper would relegate the original theory to the status of a mere tautology resulting from the content-decreasing stratagem, instead of recognizing it as a scientific, empirical theory, Heisenberg's thesis of permanent truth and his doctrine of closed-off theories also raises worthy questions about the criteria for distinguishing theory from nontheory language in empirical science. It cannot be said that either Heisenberg or

Popper or any of the contemporary Pragmatists have formulated a metatheory of semantical description that adequately addresses these questions.

Comment and Conclusion

There exists an inconsistency between Heisenberg's systematic philosophy of science as set forth in his philosophical works and his practice of science as revealed in his autobiographical accounts. This inconsistency is attributable to the conflicting influences of Einstein and Bohr, and it has its basis in two fundamentally different philosophies of the semantics of the language of science, particularly where the language of science is used to describe observations. Einstein's philosophy of language is that the semantics of the language of scientific theory is an artifact, a "free convention", a cultural product instead of a natural product, such that concepts and categories used for observation in physics do not have their information content specifically determined by the natural processes of perception. Bohr's philosophy of language on the other hand is that the semantics of the language of scientific observation is not a cultural product, but rather is the natural product of perception. Concepts or what he calls the "forms of perception" have their information content determined by the natural processes of perception.

It was evident to Heisenberg as well as to nearly every other physicist at the time, that revolutionary revisions had been made in twentieth-century physics, and Heisenberg wanted to explain how such developments in the history of science could produce correspondingly revolutionary revisions in the semantics of the language of physical theory. Heisenberg's response was his doctrine of closed-off theories, and the philosophy of language that he used for his semantical theory was greatly influenced by Bohr. This doctrine restricts semantical revision to the description of phenomena that lay beyond ordinary perception and thereby retains semantical permanence for the description of phenomena which are accessible to everyday observation and which are described by the language and concepts of classical physics. From the point of view offered by the doctrine of closed-off theories Newtonian physics is permanently valid; it is the observation language for physics, and it is necessary for reporting and communicating experimental measurements and observations in microphysics. In Heisenberg's doctrine of closed-off theories, the theories of modern physics can exhibit semantical change while the language of scientific observation can exhibit semantical permanence. This is also characteristic of the Logical Positivist philosophy of science, which too assumes a naturalistic philosophy of the semantics of language. But in the Logical Positivist philosophy the relation between the language of theory and the language of observation is mediated by sentences variously called "reduction sentences" (Carnap), "correspondence rules" (Feigl and Nagel), or "bridge principles" (Hempel and Oppenheim), such that the sentences in the observation language are logically deduced from the sentences in the axiomatic theory. In Heisenberg's philosophy, however, the relation between observation and theory is not mediated by sentences and it is not deductive, although like Logical Positivism, it uses semantical equivocation to preserve semantical permanence for observation language while permitting meaning variance for theory language.

In Heisenberg's semantical theory all observation must be with concepts either of classical physics or of "everyday" language. These concepts are not exactly the same. The everyday concepts have a "lack of precision" or vagueness, while the concepts of classical physics have their content rigidly fixed by virtue of their occurrence in the laws and axiom system of Newtonian physics. The concepts of quantum physics also have their content fixed by virtue of their occurrence in the laws and axiom system of quantum physics. The quantum concepts are alternative concepts relative to classical concepts, because the laws of classical and quantum physics are mutually inconsistent. But the quantum concepts are not alternative resolutions of everyday concepts, because the quantum concepts cannot be used for observation. The fact that classical and quantum concepts occur in mutually inconsistent laws or axiom systems implies in Heisenberg's view that when they are associated with the same descriptive term or variable, they are alternative meanings making that term or variable equivocal. The equivocal relation between classical and quantum concepts is illustrated in the cases of the terms "position" and "momentum". These terms occur in both classical and quantum physics. On the one hand the meanings of these terms are used to describe observational measurements, and as such they must be classical concepts. This classical semantics includes the idea that in principle the measurements can be indefinitely accurate, even though in practice the degree of accuracy is limited, and the idea that nature is fundamentally continuous. On the other hand there are also meanings for these terms that are distinctive of quantum physics, and this semantics does not say that in principle measurements may be indefinitely accurate. Instead the quantum meanings include the idea of the indeterminacy relations, which say that the accuracy of their joint measurement is in principle limited to a function of the value of Planck's constant, and the idea that nature is fundamentally discontinuous. Therefore, in order for observation to be possible in quantum physics there must exist in quantum physical laws an equivocation such that for every quantum concept determined by the context of quantum laws, there must be a corresponding classical concept determined by the context of classical physics. The advocates of the Copenhagen interpretation of the modern quantum theory argue that in practice the concepts of classical physics actually do in fact operate in descriptions of the macrophysical experimental apparatus and measurement. Furthermore Bohr's semantically inconsistent complementarity thesis describes duality with these classical concepts, while for Heisenberg duality is expressed consistently with quantum concepts associated with the variables in the equations of the mathematical theory.

Yet Heisenberg's use of Einstein's admonition for describing the vapor trails in the Wilson cloud chamber, and his subsequent development of the indeterminacy relations, does not agree with his explicit doctrine of closed-off theories. Einstein's admonition consists of the semantical thesis that it is the theory that decides what the physicist can observe. According to Heisenberg's semantical doctrine of closed-off theories the quantum theory cannot decide what the physicist observes in the cloud chamber, because the macrophysical description of what is observed must contradict the quantum description. In spite of the suggestiveness of the discrete water droplets in the vapor trail, the classical concepts used for observation dictate that physical nature is fundamentally continuous.

Contrary to Heisenberg's semantical doctrine of closed-off theories, classical concepts are not necessary for observation, variables in the quantum laws are not equivocal, and all the concepts in the quantum theory are quantum concepts including the concepts used for observation. It is possible with a theory of semantical description to follow through with Einstein's admonition and to say that theory decides what the physicist can observe, because the concepts occurring in the quantum theory and that are used for observation are quantum concepts. Such a new semantical theory is needed, because Heisenberg had premised his doctrine of closed-off theories on the naturalistic philosophy of language. The twentieth century has been a century of historic scientific revolutions, and these dramatic revisions in physical theory have caused correspondingly profound revisions in the semantics of the language of theory. Attempts to preserve a permanent semantics for observation, while at the same time to explain the semantical revisions resulting from revolutionary developments in theory, have resulted in attributing equivocation to language that physicists none-theless are able in practice to use unambiguously. These historic scientific revolutions have motivated post-Positivist professional philosophers of science to reject the naturalistic philosophy of language, and to accept the artifactual philosophy of language.

It is necessary to consider further how to describe the semantics both of quantum theory and of experimental observation, in order to exhibit how concepts in this language are determined as human artifacts instead of predetermined as products of nature, and to explain why these concepts do not involve complete equivocation. The doctrine of closed-off theories contains certain basic assumptions that are in need of reconsideration. One basic assumption in the doctrine is the tacit assumption that all concepts are indivisible or simple wholes, that must be either completely different or completely the same. The doctrine maintains that any macrophysical apparatus must employ the classical concepts for describing the experimental apparatus and observed measurements, and it is assumed that those classical concepts are simply and wholly different from the quantum concepts that have their contents rigidly fixed by their mutual relations in the quantum microphysical theory. The other basic assumption is that observation language and theory language must be distinctively associated with macroscopic phenomena and microscopic phenomena respectively.

Both of these basic ideas contain errors. Firstly it is incorrect to assume that concepts in physics or anywhere else are simple wholes, that cannot be analyzed into component parts. And secondly it is necessary to reconsider the Copenhagen school's basis for dividing the relevant language into statements of experiment and statements of theory. Specifically rejection of the naturalistic philosophy of language implies rejecting two mental associations that occur in the doctrine of closed-off theories. One association is the classical-macroscopic-experimental-observation association. The other is the quantum-microscopic-theoretical association. Consider firstly an alternative to the wholistic view, and how it affects the thesis of equivocation. Reflection on the common occurrence of looking up a word in a unilingual dictionary or thesaurus reveals that the meanings of words are not simple wholes, but rather have component parts that are identified by the other words occurring in the dictionary definition or lexical entry. These definitions give semantical descriptions of the meanings they define, and in order to function in this way they always must have the force of universal affirmations accepted as true. Dictionary definitions are often viewed as describing the complete meaning of the term, but dictionary definitions are minimal statements. Usually an understanding of the meaning of a technical term requires a larger context consisting of a set of several statements containing the term. Since Quine rejected the analytic-synthetic distinction, all universal empirical or synthetic statements may be viewed as also analytic. Thus if one were to draw up a list of noncontradictory affirmative statements containing the term, with each statement having universal import and accepted as true, then each of the mutually consistent statements constituting the list would describe part of the meaning of the common subject term, and the entire list as well as each statement in it may be called a "semantical description" of the common subject term's univocal meaning. A term is equivocal if any of the universal affirmations in the semantical description are mutually inconsistent This equivocation is made explicit if the predicates of the inconsistent universal affirmations can be related to one another by universal negations accepted as true. This amounts to saying that none of the instances to which the common subject term refers are the same instances. The two meanings in the equivocation have separate semantical descriptions, when the original list is subdivided into subsets containing mutually consistent universal affirmations, such that each subset is a semantical description of the different meanings of the equivocal term instead of descriptions of different parts of the one meaning of a univocal term. The equivocations postulated by Heisenberg's doctrine of closed-off theories applied to microphysics is purported to be due to a logical inconsistency between the laws or axiomatic systems of classical and quantum physics.

In addition to the properties of equivocation and univocation there is another aspect of language called vagueness. Terms are univocal or equivocal relative to meaning, but meanings are clear or vague relative to one another. Meanings are clear in relation to one another, if they can be related to each other by universal affirmations or negations accepted as true, and they are vague in relation to each other if they cannot be so related by universal statements. Adding consistent, related, universal statements to a univocal term's semantical description list has the effect of resolving the vagueness in the concept associated with the term by adding or excluding meaning. All meanings are vague and admit to further resolution or clarification, because the semantical description can always be increased by additional universal statements, although responsible addition often requires additional scientific research. This residual vagueness has been referred to by Waismann as the "open texture" of concepts.

In the empirical sciences equations have the force of universal affirmations. Their universal logical quantification is changed to particular logical quantification, when the variables are given numerical values by measurement, or by calculation with the equation whose initial conditions have been satisfied by measurement. Each measurement performance is an experiment instance, the results of which are described by measurement values in the equation expressing particular logical quantification by the assignment of numerical values to the variables of the equation. Different measurement performances may result in the same or different measurement values. They will likely result in different measurement values, either because of measurement errors in the execution of the experimental design, or because of different initial conditions in the ranges of the variables. An equation is not given particular logical quantification merely by the range of possible values of its variables. Side conditions may severely restrict the range of applicable values, but there is always a universal claim made by the equation, because there exists an infinitely large number of possible measurement performances, each one of which could be described by a statement having particular logical quantification. Furthermore no mathematically expressed theory need be a more "general theory" relative to any other theory, in order to be a logically universal statement. In the case of quantum physics the microphysical quantum theory need not also be a macrophysical theory, in order to be a universal theory, just as it need not be a biological or a sociological theory, in order to be a universal theory.

Turn next to the relation between the language of experiment and the language of theory, the second basic assumption in the doctrine of closed-off theories. To begin with, consider the meaning of the word "theory". It is still conventional to refer to Newton's "theory" of gravitation, to Einstein's "theory" of relativity, and to the quantum "theory", even though the physics profession had decided decades ago either to accept or to reject them as physical explanations. In this conventional usage the term "theory" does not mean the same thing with respect to these explanations, as it did when they were firstly proposed for testing as explanations of problematic phenomena. When they were firstly proposed, these explanations represented language that had a much more hypothetical status in the judgment of the profession than they do today, and they were topics of controversy. There is, therefore, an ambiguity between "theory" understood as an accepted or rejected explanation on the one hand, and "theory" understood as a tentative proposal submitted for empirical testing on the other hand. Unlike the former archival understanding, which Russell Hanson calls "almanac" science, the latter understanding of "theory" relates to the function of theory language in the development of science, where the function under consideration is that of empirical testing. From this functional perspective theories are universal affirmations that are proposed for testing. And if a theory is not falsified by any test, it is accepted as having the status of an explanation, even though there may be other tested and nonfalsified theories addressing the same problem also having the status of explanations accepted by other scientists in the profession.

In an empirical test the semantics of the vocabulary in all the relevant discourse is controlled by a decision that is antecedent to the performance of the test. This is the decision as to what language is presumed true for testing and what language is proposed for testing. The former language is the statements of test design, and the latter is the theory. This decision is entirely pragmatic, since it is not based on the syntactical or the semantical characteristics of language, but rather is based on the use or function of the language, namely empirical testing. The test design statements are those that by prior decision and agreement among cognizant members of the profession, have the incorrigible force of definitions. These statements are presumed to be true regardless of the outcome of the test, and serve to identify the subject of investigation throughout the test. And the theory is the language that by prior decision and agreement has the status of an hypothesis. The hypothesis is believed to be true to the extent that it is considered worthy of testing, although the proponent may be quite convinced. If the test outcome is a falsification, then it is the theory and not the statements of test design, that is judged to have been falsified. However, a falsification may lead some interested scientists, such as the theory's proponent and its advocates, to reconsider the beliefs underlying the test design, even while admitting that the test was executed in accordance with its design. This reversal of test design and theory may result in productive research.

The decision distinguishing test design and theory language made prior to the experiment may but need not result in identifying mathematical equations as the statements of theory, and of identifying colloquial discourse as the statements of test design. The decision is not based on syntactical characteristics of the language or syntactical criteria, and the test design statements may include mathematically expressed statements. Even more relevantly the decision is not based on semantical criteria, as advocates of the naturalistic philosophy of the semantics of language purport. The decision is not based on any purportedly inherent distinction between observation and theory, whether or not, as in the case of quantum theory, the observation concepts are called "classical" or "macroscopic", and the theoretical concepts are called "quantum" or "microscopic". The distinction between statements of test design and statements of theory is neither syntactical nor semantical; it is distinctively pragmatic.

In order for the test design statements to function as independent evidence in relation to the theory proposed for testing, the test design statements and the theory statements must be logically independent. The one set of statements cannot be merely a transformation of the other. Nor can the test design statements imply the contradiction of the theory or any of its alternatives. The semantical consequence of this logical independence of a theory relative to the universal statements of test design, is that each of the terms common to both the test design statements and theory statements have their semantics defined by reference to the meanings of the other terms in the test design statements, but not by reference to the meanings of the other terms of the theory. The meanings of these common terms are therefore vague with respect to the meanings of the other terms of the theory, because the statements of the theory are too hypothetical to function as definitions, except perhaps for the proponent and other advocates of the theory, who believe in the theory as strongly as they believe in the truth of the test design statements. For this latter group the test design statements and the theory statements contribute parts to the complex meaning of the univocal terms common to both sets of statements. The most notable semantical consideration is that the two sets of statements are not mutually inconsistent while both are believed to be true, such that the terms common to both sets of statements are not equivocal.

Such is the semantical situation before the test is performed and before the profession knows of the test outcome. A test with a nonfalsifying outcome produces a semantical change. After the test the theory no longer has the hypothetical status that it formerly had merely as a proposal, but assumes a status that is neither more nor less contingent than other accepted universal statements including the test design statements. The semantical outcome is that both the test design statements and the theory statements (now elevated to the status of an explanation) are semantical rules exhibiting the composition of the meanings of the univocal terms common to both sets of statements. In Heisenberg's doctrine of closed-off theories the naturalistic philosophy of language requires retention of the classical concepts for observation in the context of the quantum theory. The resulting equivocation is unnecessary, if it is remembered that the Newtonian concepts need not be retained. It is sufficient to use a less precise vocabulary that Heisenberg calls "everyday" words used by physicists in order to describe the experimental set up, which is a macrophysical phenomenon. These vague words neither affirm nor deny any theory about the fundamental constitution of matter. And after the quantum theory was found to be experimentally adequate, the vagueness in these concepts was resolved by the equations constituting the statements of the quantum theory, because the quantum theory had become a semantical rule contributing to the complex meanings of these univocal terms. For this resolution of vagueness to occur it is not necessary for the Newtonian macrophysical laws to be made logical extensions of the quantum theory by logical reduction procedures, nor is it necessary for the Newtonian macrophysical laws to be replaced by a macrophysical extension of the quantum laws. It is sufficient merely that the scientist realize that quantum theory and not classical physics is the empirically acceptable microphysical theory.

Heisenberg's doctrine of closed-off theories is incorrect, and Einstein's semantical thesis expressed as an admonition to Heisenberg is correct, because the vocabulary used for observation after the quantum theory's acceptance is a univocal vocabulary with quantum concepts, and because the statements of quantum theory contribute to the meaning complex associated with the descriptive terms used for observation and with the Newtonian concepts completely excluded. The quantum theory decides what the scientist observes in the cloud chamber, because the description of the vapor trails as water droplets or in any other macrophysical terms is not a contradiction of the quantum theory and specifically of the indeterminacy relations. The macrophysical description does not affirm a Newtonian theory. The quantum theory is built into the univocal meanings associated with the observation description.

In his "Remarks on the Origin of the Relations of Uncertainty" in a memorial volume dedicated to him titled *The Uncertainty Principle and Foundations of Quantum Mechanics* (1977), which was in press at the time of his death in 1976, Heisenberg concludes the brief four-page article by noting that there have been attempts to replace the traditional language with its classical concepts by a new language which should be better adapted to the mathematical formalism of quantum theory. But he adds that the development of language is a historical process, and says that artificial languages such as Esperanto have never been very successful. He states that during the preceding fifty years, physicists have preferred to use the traditional language in describing their experiments with the precaution that the limitations given by the uncertainty relations should "always be kept in mind". He concludes that a "more precise" language has not been developed, and that in fact it is not needed, since there seems to be general agreement about the conclusions and predictions drawn from any given experiment in the field.

Heisenberg never repudiated his doctrine of closed-off theories. His statement that the contemporary physicist must keep quantum effects "in mind" when the physicist is describing macrophysical objects, even while not explicitly accounting for experimentally undetectable quantum effects, can be construed in terms of his equivocal semantics. But it can be reconstrued more elegantly in terms of a univocal semantical thesis, in which having to keep quantum effects in mind is *prima facie* evidence of a semantical change resulting from the development of the modern quantum theory. The semantical change consists in the fact that the concepts employed for macrophysical description contain component parts, that are describable by reference to the quantum theory. The univocal semantical thesis does not imply Hugh Everett's "many worlds" interpretation (which is in principle empirically untestable), nor does it require or imply any other reductionist development of a macrophysical quantum theory, i.e. a macrophysical theory which is deductively or reductively a logical extension of the microphysical quantum theory.

A semantical theory of equivocation is the result of the acceptance of a naturalistic philosophy of the semantics of language, and the assumption that meanings are simple, indivisible wholes. Logical Positivists and operationalists such as Nagel also maintain such semantical views, when they too wished to explain semantical change in scientific theory while at the same time to maintain semantical permanence in scientific observation.

But all such views are untenable, because they imply what can only be called "double think": they demand that the physicist indulge in a contrived cognitive duplicity with himself, a pretext of simultaneously both knowing and not knowing the modern quantum theory. However, concepts are not known like physical objects; they are knowledge. Scientists never did carry on the kind of cognitive duplicity that the equivocal semantical theories require, and philosophers no longer expect that they should.

The philosophy of science that Heisenberg practiced as a result of Einstein's influence, is historically more important than the one he expounded as a result of Bohr's influence. In the philosophy he practiced, he anticipated the contemporary Pragmatist philosophy of science by at least a quarter of a century. Because contemporary Pragmatism is based on an artifactual view of the semantics of language, it implies the interdependence of truth (or belief) and semantics. On the naturalistic view of semantics, the truth of empirical statements is entirely dependent on meanings which are determined independently by the natural processes of perception; while on the artifactual view truth and meaning are mutually determining in empirical statements believed to be true Thus when he used Einstein's admonition for developing the uncertainty relations, Heisenberg had made a prior commitment to the empirical truth of the equations expressing them, and he then used them for his theory-dependent observation of the Wilson cloud chamber vapor trails made by the free electron. Years later Feyerabend will call this "counterinduction", even though he did not recognize its occurrence in Heisenberg's practice, and would probably have produced a very different philosophy of quantum theory had he seen it.

Furthermore, given philosophical realism, and that on the realist view semantics describes ontology, then the artifactual philosophy of the semantics of language also implies the mutual determination of truth and ontology. This too was practiced by Heisenberg. When he interpreted the uncertainty relation realistically, as Einstein had interpreted the Lorentz transformation realistically contrary to Lorentz, Heisenberg used the uncertainty relations to describe the quantum theory's ontology, the wave-particle duality, and to defend the Copenhagen thesis that wave and particle are manifestations of the same individual entity. Years later Quine will call this practice "ontological relativity". This practice has the important implication of excluding from scientific criticism all prior ontological commitments, such as Einstein's commitment to determinism; prior ontological commitments did not function as criteria for scientific criticism in Heisenberg's judgment about duality. Similarly contemporary Pragmatists reject the view held by both the Romantics and Positivists, that prior ontological commitments may operate as criteria for scientific criticism. Pragmatists admit only empirical criteria for scientific criticism, and they let the statement of the tested and nonfalsified theory describe the ontology for the theory's domain.

The discussion in this chapter about Heisenberg's semantics of quantum mechanics and his doctrine of closed-off theories has assumed the viewpoint of the contemporary Pragmatist philosophy of science. Other philosophical perspectives have been assumed both by physicists and by philosophers of science for viewing the history of quantum mechanics. The physicist Max Jammer has written two books from the perspective of naturalistic philosophies of the semantics of language. In his *Conceptual Development of Quantum Mechanics* (1966) he assumes the perspective of Bohr's philosophy of observation. Then in his *Philosophy of Quantum Mechanics* (1974) he assumes explicitly the doctrine of operationalism and the Logical Positivist philosophy of the semantics of language, and he references Carnap, Braithwaite, and a few lesser lights. In the latter work he describes Heisenberg as an operationalist "deep in his heart", but adds that he was not a "pure" operationalist because of his use of Einstein's admonition in developing the indeterminacy relations. In neither book does Jammer find any fundamental philosophical significance in Heisenberg's use of Einstein's admonition. It is somewhat ironic that Jammer writes on the philosophy of quantum mechanics from a philosophical perspective that the development of quantum mechanics helped render anachronistic among contemporary professional philosophers of science many years before Jammer wrote these books. In his *Quantum Mechanics and Objectivity* the philosopher Patrick A. Heelan observes that Heisenberg's philosophy is very different from Bohr's. But his book, originally a doctoral dissertation written at the Roman Catholic University of Louvain, Belgium, offers a metaphysical orientation. Heelan describes Heisenberg as the most metaphysical of modern scientists, and he laments that the genre of philosophical writing on physics in recent years has been dominated by a British and American school which tends to place metaphysics somewhere between mysticism and crossword puzzles. Heelan examines Heisenberg's views in terms of an opposition between "empiricism" and "rationalism"; this distinction does not correspond to the opposition between the naturalistic and artifactual philosophies of the semantics of language. Such comparisons of Heisenberg's philosophy with the metaphysical views of various historic metaphysicians that Heisenberg references is of course possible. But a better understanding of Heisenberg is to be had by examining the views of those persons to whom he was consciously receptive in his pioneering work in microphysics, and specifically in his strenuous intellectual effort to forge a semantical interpretation for his empirically successful yet radically distinctive formalism. Those persons who exercised such influence on Heisenberg's thinking were not the historic metaphysicians, but rather were the two great physicists of his time to whom he had personal access, and whose professional opinions and approval he valued more than anyone's: Einstein and Bohr.

KARL POPPER AND FALLIBILIST CRITICISM

Karl Popper (1902-) was born in Vienna, Austria. He enrolled in the University of Vienna in 1918, where he studied physics, mathematics, and philosophy. In 1928 he received his Ph.D. for a dissertation titled *On the Problem of Method in the Psychology of Thinking*. He never returned to the subject of psychology again during his professional career, because he became convinced that methodology of science is exclusively a matter of logic and objective knowledge. Popper was personally acquainted with Rudolf Carnap and other members of the Vienna Circle, and although he had been invited to address the group at a meeting in which he set forth his philosophy of science, he was never a member of the Circle. In 1937 he went as a senior lecturer to Canterbury University College in Christchurch, New Zealand, and then in 1945 he was appointed to a readership at the London School of Economics in the University of London. In 1949 he was made professor of logic and scientific method at the London School, and was knighted in 1964.

Einstein's Influence and the Fallibilist Thesis of Criticism

In his intellectual autobiography in Schilpp's *The Philosophy of Karl Popper* (1974) Popper states that Einstein was the most important influence on his thinking. The influence was not a personal one, since Popper and Einstein did not actually meet until 1950; the influence was through Einstein's public lectures and published works. The year 1919 was the fateful year in Popper's intellectual life; he was interested in the views of several thinkers including Marx's theory of history, Freud's theory of psychoanalysis, and Alfred Adler's theory called "individual psychology". Popper relates in his "Science: Conjectures and Refutations" (1957) in *Conjectures and Refutations* (1963), that he had come into personal contact with Alfred Adler and cooperated with Adler in the latter's social work with children and young people in the working class districts of Vienna during the last years of the Austrian Empire and the consequent revolution. In the summer of 1919 Popper became dissatisfied with the views of Marx, Freud and Adler, because the persons who accepted and advocated these theories were strongly impressed by the theories' purported explanatory power, and because study of these theories had the effect of an intellectual conversion or revelation. Most objectionable to Popper was the fact that once the reader's eyes were opened to the theory, he found that the theory was verified everywhere it could be applied. Unbelievers were dismissed as persons who could not see the verifications. In Popper's view the apparent strength of these theories' explanatory power, is their principal weakness.

Popper saw in Einstein's theory a striking contrast to the situation he found in the views of Marx, Freud and Adler. Eddington's solar eclipse observations in 1919 brought the first important test to bear upon Einstein's relativity theory of gravitation. This test was distinctive, because in the test there was a risk involved in the theory's prediction. Had Eddington's observations showed that the predicted effect is definitely absent, then Einstein's theory would simply have been refuted. And the risk in Einstein's case was very great, since the predicted effect was different from that which was expected on the basis of Newton's theory, which had long demonstrated great success culminating with the discovery of the planet Neptune. In his autobiography Popper said that what impressed him most was Einstein's own clear statement that he should regard his theory of relativity as untenable, if it should fail certain tests. This was an attitude that was very different from the dogmatic attitude of the Marxians, the Freudians, or the Adlerians. Einstein was looking for crucial experiments where agreement with his predictions would by no means establish his theory; but where disagreement with his predictions, as Einstein was the first to say, would show his theory to be untenable. Thus in 1919 Popper concluded that the critical attitude, which does not look for verifications but rather looks for crucial tests that can refute the tested theory, is the correct attitude for science, even though the crucial tests can never establish the theory. This is Popper's fallibilist philosophy of scientific criticism.

Explanation, Information, and the Growth of Science

Popper's philosophy recognizes the dynamic character of science that cannot be recognized with the philosophy of the Positivists. His statements on the dynamics of science are found in appendices to the 1968 edition of his *Logic of Scientific Discovery*, in his "Truth, Rationality, and the Growth of Scientific Knowledge" in *Conjectures and Refutations*, and in "The Rationality of Scientific Revolutions" in *Problems of Scientific Revolutions* (ed. Harre, 1975), as well as elsewhere in his literary corpus. His fallibilist thesis is not only a philosophy of scientific criticism, it is also a philosophy of scientific explanation and of the growth of scientific knowledge. As a philosophy of scientific criticism, it says that the empirical test outcome can never establish or "verify" a scientific theory, but the test can refute or "falsify" the theory. However, even before a theory's empirical claims are actually tested, it is possible to determine whether or not it is a scientific explanation. It is not a scientific explanation if it is not empirically testable. Another way that Popper describes this condition is that what makes a theory scientific is its power to exclude the occurrence of some possible events, and he calls the singular statements that describe these excluded events "potential falsifiers". This way of speaking introduces his idea of various degrees of explanatory power; the more that a theory forbids or excludes and therefore the larger the class of potential falsifiers, then the more the theory tells us about the world. Popper calls the variability of degree of explanatory power the amount of information content of a theory or explanation. The idea of the amount of information content may be illustrated by reflection on the logical conjunction of two statements α and β. It is intuitively evident that the conjunction $\alpha\beta$ has no lesser amount of information content than do the component statements taken separately, and it usually has more information content than its components. This is because there are more potential falsifiers for the conjunction than for the component statements taken separately; the conjunction is false if either component is false. In some contexts Popper calls information content "empirical content", and he calls the falsifiability of the theory its "testability". All of these terms refer to a logical relation between a theory or a hypothesis and its class of potential falsifiers.

Popper relates the idea of information content to probability theory. He says that the amount of information content is inversely related to the degree of probability that may be associated with a hypothesis. This view can be illustrated also by the logical conjunction: if the probability value P(α) is associated with the statement α and the probability value P(ß) is associated with the statement ß, then by the probability calculus the probability P(αß) associated with the conjunction αß must be less than the probability values P(α) and P(ß). Therefore as the information content of a theory increases, the associated probability must decrease. Popper maintains that the whole problem of the probability of hypotheses as viewed by Carnap is misconceived, because on Carnap's idea of degree of confirmation, scientists should prefer statements having higher associated probabilities, while on Popper's view scientists should prefer theories with higher information content. Therefore in contrast to Carnap's idea of degree of confirmation Popper advances the idea of degree of "corroboration", although in some contexts Popper also uses the phrase "degree of confirmation" in a sense that is synonymous with his idea of degree of corroboration. On the corroboration thesis a scientific theory that has greater information content (because it is more universal, or because it is more accurate than an alternative theory) also has a higher degree of corroboration, if when it is tested it is not falsified. Like the idea of information content, the idea of corroboration is based on the idea of falsifiability, but a theory would not be said to have been corroborated until it had been tested and found to have no falsifying test outcome; the degree of corroboration actually attained does not depend only on the degree of falsifiability. A statement may be falsifiable to a high degree yet it may be only slightly corroborated or it may be falsified.

The measures for corroboration, $C(h,e)$, explanatory power, $E(h,e)$, and probability, $P(h,e)$, for hypothesis h and for basic statement e of evidence describing a test outcome, are related by certain equations. The inverse relation between the measures of corroboration and probability are related as follows:

$$C(h,e) = 1 - P(h,e),$$

and he is willing to admit a proposal by Kemery in *The Journal of Symbolic Logic* (1954) that the relation may also be expressed in terms of information science concepts as:

$$C(h,e) = 1 - \log P(h,e).$$

And Popper relates corroboration and explanatory power as:

$$C(h,e) = E(h,e)\ (1 + P(h)\ P(h,e)).$$

Popper states that the measure of the degree of corroboration, $C(h,e)$, may be interpreted as a measure of the rationality of belief in the statistical hypothesis, h, in the light of test outcomes, e, only if e consists of reports of the outcome of sincere attempts to refute the hypothesis by the severest test that can be devised, rather than attempts to verify h. But the corroboration does not measure the degree of rationality in our belief in the truth of h, since $C(h,e) = 0$ whenever h is logically true; rather it is the measure of accepting tentatively a problematic guess. On the other hand the measure $E(h,e)$ may be interpreted as the measure of the explanatory power of h with respect to e, even though e is not a report of any genuine and sincere attempts to refute h; $E(h,e)$ is a purely logical relation to the infinite class of potential falsifiers.

The idea of relatively greater or lesser degrees of information content and falsifiability, provides the basis for Popper's concepts of scientific progress, the growth of scientific knowledge, and the aim of science. He advances a "metascientific" criterion of progress, that enables the scientist and methodologist to know in advance of any empirical test, whether or not a new theory would be an improvement over existing theories, were the new theory able to pass crucial tests in which it is compared to older, existing alternatives. He calls this criterion the "potential satisfactoriness" of the theory, and it consists of the degree or amount of information content. Simply stated, his thesis is that the theory that tells us more is preferable to one that tells us less, and the theory that tells us more is also that which is most falsifiable. From this it follows that the aim of science is high empirical information content as well as successful performance in tests. It is the criterion of high information content that makes the growth of science rational. The aim of science is not high probability, and the rationality of science does not consist of constructing deductive axiomatic systems, since there is little merit in formalizing a theory beyond the requirements for testing it. Nor does the growth of science consist of the accumulation of observations; rather it consists of the repeated overthrow of scientific theories and their replacement by more satisfactory theories. The continued growth and progress of science is essential to the rational and empirical character of scientific knowledge. The growth is continuous, because criticism of theories, which are proposed solutions, in turn generate new problems. Problems occur when expectations are disappointed. Science starts from problems, not from observations, although unexpected observations give rise to new problems. Popper views science as progressing from old problems to new problems, to new problems of increased depth as it progresses from old theories to new theories of increased information content. Popper also views progress in science as approaching more and more closely to the truth, where truth is understood as a correspondence with the facts and as a regulative idea. Just as there are degrees of information content, so too there are degrees of approach to the truth that he calls "verisimilitude".

In his "Rationality of Scientific Revolutions" Popper therefore sets forth two criteria for the rationality of scientific revolutions, which are also two logical properties that enable the scientist to evaluate any new theory even before it is tested. The first criterion may be called a criterion of discontinuity: the new theory must conflict with the old one in the sense that it leads to conflicting results. Popper says that in this sense scientific progress is always revolutionary, and that Marx's refrain "revolution in permanence" is applicable to science. The second criterion may be called a criterion of continuity: the new theory must be able to explain fully the success of its predecessor in the sense that there are applications in which the old theory must appear to be a good approximation to the results of the new theory, while there are other cases where the new theory yields different and better results than the old one. Scientific revolutions are rational because unlike ideological revolutions, which are sociological, the former cannot simply break with tradition.

Against Psychologism, Induction, and Naturalistic Semantics

Popper's philosophy of knowledge is a critique of psychologism and a defense of the objectivity of knowledge. In the opening chapter of *Logic of Scientific Discovery,* which is titled "A Survey of Some Fundamental Problems", he devotes a section to the elimination of psychologism. This section follows the opening section on the problem of induction, which he views as a fallacy resulting from the psychologistic philosophy of knowledge. He sets forth his own theory of knowledge in the fifth chapter titled "The Problem of the Empirical Basis" and the opening section is a critique of the psychologistic view that perceptual experiences are the empirical basis for science. In his "Demarcation Between Science and Metaphysics" (1955) in *Conjectures and Refutations* he criticizes Carnap's theory of meaninglessness, which he describes as a "naturalistic theory of meaningfulness" of linguistic expressions. The linguistic expressions of particular relevance are those singular statements that are used for describing observations in science. All of these ideas are interrelated according to Popper: induction as the logic for making generalizations and hypotheses, psychologism which proposes perception as the empirical basis of observation in science, and the naturalistic theory of the semantics of language. Popper rejects all of them together. In his "Epistemology Without a Knowing Subject" (1967) and his "On the Theory of the Objective Mind" (1968) published as chapters three and four in his *Objective Knowledge* (1972), in Part I of *The Self and Its Brain* (1977), and also in an appendix to *The Open Universe* (1982), Popper sets forth his own philosophy of the three "worlds" of reality in which subjective psychology and objective knowledge are in different worlds.

The development of Popper's own philosophy of science began with the objective of demarcating empirical science and pseudoscience (e.g. Astrology, Marxism, Freudianism, and Adlerian psychology). His solution to the problem of demarcation is the criterion of empirical falsification, which he also uses to demarcate empirical science from metaphysics, and he contrasts this criterion with the criterion of meaningfulness that Carnap and other Logical Positivists used for distinguishing science from metaphysics. Carnap's criterion of meaningfulness is based on the naturalistic philosophy of language. Popper argues that the Positivists have never succeeded in distinguishing science from metaphysics or in distinguishing theory from observation, that metaphysics need not be meaningless even though it is not a science, and that Positivism has tended to exclude scientific theories as meaningless while failing to exclude metaphysics as meaningless. Popper maintains that there is no observation without theory, and that the observation terms occurring in observation language are "theory impregnated", such that observation terms are a type of theoretical term that Carnap calls disposition terms. The reason that the Positivists have not succeeded in distinguishing science from metaphysics, is that they cannot define meaningfulness, and they cannot define meaningfulness because they interpret the problem in a naturalistic way, as though it were a problem in natural science or in psychology. Popper maintains that the Positivists have confused the psychology of knowledge with the logic of knowledge, which is to say that they have adopted a psychologistic philosophy of knowledge. Popper rejects both behaviorism and psychologism, and maintains that the content of thought, the meanings of words, the semantics of language, are not determined either by the natural laws of the physical world or by the natural laws of psychology. The world of objective knowledge, which is governed by the laws of logic, is a third world that is autonomous from the world of objective physical nature and also from the world of subjective psychology. In *The Self and Its Brain* he argues against behaviorism and physicalist reductionism by the display of ambiguous drawings that he emphasizes may be interpreted in different ways by voluntary action, in order to demonstrate the existence of world 2, the world of the mind and of subjective mental experiences. And he argues against the psychologistic view by stating that the objects of world 3 are intersubjectively testable. Hence there are the three separate worlds which cannot be reduced to one another: world 1 is the world of objective physical nature, world 2 is the subjective world of psychological experience, and world 3 is the objective world of human artifacts or creations including knowledge. Popper emphasizes that while the three worlds interact through world 2, nevertheless the world of objective knowledge is autonomous of the world of subjective psychological experience including perceptual experiences. Advocates of psychologism and the naturalistic theory of the semantics of language fail to recognize the autonomy of world 3 from the other two worlds. In recent years the information scientist Bertram C. Brookes in his article "The Foundations of Information Science: Philosophical Aspects" in *The Journal of Information Science* (1980) proposed that the task of information science as a discipline can be defined as the exploration of the world of objective knowledge understood as Popper's world 3, and that this discipline is distinct from documentation and librarianship.

Popper's rejection of inductive logic is based on his thesis that world 3 is autonomous from worlds 1 and 2. He references Einstein's stated view that there is no logical path leading to the universal laws that scientists search for, and that these laws can only be reached by intuition. Popper accepts Hume's thesis that universal statements cannot be justified by the singular statements describing observations, and he rejects Wittgenstein's verifiability criterion of meaningfulness adopted by Carnap and the other Logical Positivists of the Vienna Circle. He also rejects the probabilistic inductive logic developed by Carnap and set forth in the latter's *The Logical Foundations of Probability*, and he wonders why anyone would ever write such a book. In Popper's view there is no logic of scientific discovery; there is only a psychology of scientific discovery. He explains that the title of his own book, *The Logic of Scientific Discovery* in not about the psychological processes involved in inventing new scientific theories, but rather is about the growth of scientific knowledge by the proposal and criticism of new theories, or as he also says, by conjectures and refutations.

Popper's philosophy of scientific knowledge is a sustained critique of Positivism, but it is not just a critical rejection; he has his own alternative philosophy of observation. The Positivists maintained that there is a clear distinction between theory and observation, such that one could separate the language of theory from the language of observation with each containing its own distinctive vocabulary and its own class of universal of statements. The universal statements containing only observation terms are produced by inductive generalization, while those containing theoretical terms are invented by the scientist's creative imagination. However, with the recognition that theory determines what is observed, the separation between theory language and observation language can no longer be sustained, and the ideas of theory and observation must be reconceptualized. And since the existence of an observation language was thought to be the empirical basis for science, the empirical basis for science also must be reconsidered.

The Positivists had attempted to base empirical science on "atomic statements", "protocol statements" and "judgments of perception" stated in the observation language. Popper rejects these ideas with his rejection of the naturalistic philosophy of meaning. Instead he proposes the idea of the "basic statement", which he defines as a singular statement which together with the universal statements of theory can serve as a premise in an empirical falsification of a theory. The basic statement is fundamentally different in concept from Carnap's protocol statement. The protocol statement is thought to be justified by perceptual experiences and thereby to constitute a foundation for science. But Popper maintains that this is a confusion between the subjective psychological aspect of knowledge and the objective logical aspect. Perceptual experiences are subjective and psychological; they can motivate a decision and hence an acceptance or a rejection of a statement, but a basic statement cannot be justified by them any more than they can be justified by thumping on a table. Basic statements are objective in the sense that they can be intersubjectively tested by repetition of the conditions that occasioned them. And they can be falsified, since they operate as premises from which other statements can be deduced, which in turn can be tested. As a result there can be no ultimate statements in science, as the Positivists believed; all statements in empirical science can be refuted by falsifying some of the conclusions that may be deduced from them.

But it is not necessary that a basic statement should be tested in order for it to be accepted; it is only necessary that the basic statement be testable. The function of basic statements is to test theories. Every test of a theory must stop at some basic statement, which the scientists have agreed to accept at least for the present time. To the extent that the basic statements are accepted on the basis of agreement, they are conventional. But the agreement is not arbitrary or capricious; the decision is made by reference to a theory and the problem that the theory is proposed to address. Theory dominates experimental work from its initial planning to its completion in the laboratory. Popper summarizes his views on the empirical basis of science by means of a memorable metaphor: There is nothing absolute about science; it does not rest upon solid bedrock, as it were. The bold structure of its theories rises as it were above a swamp like a building erected on piles, which in turn are driven down to whatever depth is found to be satisfactory to carry the structure for the time being.

Popper's reconceptualization of the empirical basis of science is also a reconceptualization of the concept of theory in science. Unlike the Positivists, Popper does not define the concept of scientific theory in terms of theoretical terms. Instead he views theories as universal statements, and there is no longer any distinction between empirical laws and theories, since there is no longer any distinction between theory language and observation language based on a distinction between theoretical terms and observation terms. All the universal statements in science are conjectures that are testable and falsifiable, and these conjectures are invented by the human mind; none of them are produced by inductive generalization. To give a causal explanation of an event means to deduce a statement which describes the event using as premises of the deduction one or more universal laws as theories together with singular basic statements that describe the initial conditions. Popper's ideas for such terms as "theory", "law", and "cause" are fundamentally different from the Positivists' ideas for these terms, because Popper's ideas are separated from the subject matter or ontologies described by the sciences.

Empirical science is not purely formal like mathematics or logic, but neither is it defined in terms of certain substantive concepts about reality as it is described by science today. Future science may have revised the substantive content of today's science and yet science will still be science as

Popper has defined it. As Popper says in reply to Kuhn's concept of science in "Normal Science and Its Dangers" in *Criticism and the Growth of Knowledge* (1970), science is "subjectless". Such could not be said of science by the Positivists, for whom the naturalistic philosophy of the semantics of language requires that certain substantive concepts permanently established by observation, must always be retained as definitive of the empirical character of science. The rejection of the naturalistic philosophy of the semantics of language implies the reconceptualization of such metascientific terms as "theory", "law", "explanation", and "cause" in a manner that disassociates these ideas from any particular ontology that the semantics of science may describe at any point in history. Empirical science becomes a sequence of alternative ontologies instead of a specific ontology. And with his criterion of increasing information content Popper believes that the sequence of ontologies is not a disconnected sequence, but rather one that reveals objective and rational scientific progress. Unfortunately Popper himself did not follow through on these ideas when he supported Einstein's criticism of the Copenhagen interpretation of quantum theory, and advanced his own "commonsense realism" ontology.

On Computers, Induction Machines, and Scientific Discovery

In his *Logical Foundations of Probability* and elsewhere Carnap proposed using a computer to make empirical generalizations with inductive logic. Throughout his career Popper has rejected the idea of inductive logic, but in *Realism and the Aim of Science* (1982) he admits to induction machines of a certain type. For such a machine he postulates a simple universe containing individuals and a limited number of properties that the individuals can have. This universe furthermore operates with a number of so-called "natural laws". Popper says that for this universe a machine can be created, so that in some reasonable period of time it will discover the laws that are valid in the postulated universe during the time period. If the laws of its universe are modified, the machine will show its capacity for finding a new set of laws. It would be capable of drawing up statistics about various distinguishable occurrences and of calculating averages. If the postulated universe is complicated further to include among its natural laws, the laws of succession, the general or conditional frequencies having a certain degree of stability, etc., then the machine can be enhanced to be able to formulate hypotheses, to test the hypotheses, and to eliminate those that should be eliminated. Such a machine can learn from experience.

But this inductive machine is limited to the universe that its architect has created for it. The architect of the universe decides what are to be individual events, and what constitutes a property or a relation. In general it is the architect of the machine who decides what the machine can recognize as a repetition. And even more fundamentally it is the architect of the machine who decides what kinds of questions the machine is to answer. All these considerations mean that the more important and difficult problems are already solved by the human designer, when he constructs the machine and the universe it can recognize. All the things that Positivists such as Carnap had thought to be simply given by nature, the meanings that according to the naturalistic theory of the semantics of language are delivered by the natural operation of human perception, are in Popper's view the product of the creative and imaginative powers of the human designer. These powers enjoy a freedom, that is permitted by the artifactual character of objective knowledge, and that is necessary for the creation of the hypotheses and theories that have characterized the growth of knowledge by science. The basis of this freedom is the nondeterministic relation between world 3 on the one hand and worlds 1 and 2 on the other. Carnap had admitted that an induction machine cannot create hypotheses, and that theories are inventions created by the human mind. But Popper does not admit to the Positivists' separation between empirical generalizations on the one hand and theories on the other; he maintains that there is no observation without theory. He also argues that no human or computer can predict the future growth of scientific knowledge by scientific methods without committing the fallacy of historicism. In his *Poverty of Historicism* (1975) as well as in *Realism and the Aim of Science* he maintains that historicism involves unconditional predictions, which are impossible. Prediction in science requires universal laws, which are always conditional.

As it happens, the computerized development of hypotheses and conjectures is precisely what information scientists attempt to accomplish by their artificial-intelligence computer systems, which Simon calls "discovery systems". These computer systems are instrumental to the scientist's development of hypotheses. They are not historicist, but are conditioned upon inputs that require the same kind of preparation or initial conditions that Popper says are needed for what he calls an induction machine, but the computer systems do not consist of induction procedures.

The Schism in Physics and Metaphysical Research Programmes

The term "schism" in the context of the philosophical discussions of the quantum theory, did not originate with Popper; it was introduced by Heisenberg. In his "Recent Changes in the Foundations of Exact Sciences" (1934) in *Philosophical Problems of Quantum Mechanics* Heisenberg notes a "peculiar schism", that he says is inevitable in the investigation of atomic processes. He is not referring to a sociological phenomenon in the physics profession or to an issue that must be resolved; he views the schism positively as a development in physics. As Heisenberg uses the term "schism", it refers to the different concepts used by classical physics and quantum physics and the different ontologies that they describe. On the one hand there is the need for macrophysical classical concepts of space and time, which are used in quantum physics for the description of experiments and of the apparatus of measurement in experiments. On the other hand there is the mathematical expression suitable for the representation of microphysical reality, the wave function in multidimensional configuration spaces that allow of no easily comprehensible interpretation. The dividing line between the classical and the quantum physics is the statistical relations. Heisenberg gives a summary of the Copenhagen interpretation of the statistical nature of the quantum theory in "The Copenhagen Interpretation of Quantum Theory" in his *Physics and Philosophy* (1958). His interpretation of the probability function is that it has both a subjective and an objective aspect. The subjective aspect makes statements about the observer's incomplete knowledge, while the objective aspect makes statements about what Heisenberg calls "tendencies" and "possibilities", and in this latter aspect he parenthetically refers to Aristotle's idea of *potentia*. The probability function in the quantum theory is subjective and represents incomplete knowledge, because the observer's measurements are always inaccurate. One reason that they are inaccurate is the ordinary errors of measurement that occur both in classical physics and in quantum physics. But the reason that is distinctive to quantum physics is the inaccuracy that is caused by a disturbance introduced by the measurement apparatus in the measurement process. Heisenberg illustrates this by means of an ideal experiment involving a gamma-ray microscope, that is used to observe an electron. In the act of observation at least one light quantum of the gamma ray must have passed the microscope, and must first have been deflected by the electron. Therefore the electron must have been impacted by the light quantum and must have changed its momentum. The uncertainty of this change is given by the uncertainty relations. When the probability function is written down, it includes both these inaccuracies, and there must be at least two such disturbing observations in an atomic experiment. The probability function also contains an objective element, but it is not like the description of motion in classical physics. The classical physicist would like to say that between the initial and the second observation the electron has described an unknown path in a cloud chamber. But Heisenberg says that between the two observations the electron has not described any path in space and time, since the electron has not been anywhere. The probability function does not represent a course of events in the course of time, but rather represents statistical possibilities or tendencies, which are actualized by the second act of observation. The transition from the possible to the actual takes place with the act of observation involving the interaction of the electron with the measuring device. Heisenberg states that the transition applies to the physical and not to the psychological act of observation, and he also states that certainly quantum theory does not contain "genuine subjective features" in the sense that it introduces the mind of the physicist as a part of the atomic event.

The objective aspect of the statistical quantum theory described in terms of the transition from the possible to the actual is due to the wave-particle duality, which Heisenberg illustrates by another ideal experimental set up involving monochromatic light passing through a screen with two holes or slits in it, and then registering on a photographic plate. Viewed as a wave phenomenon there are primary waves entering the slits, and there are secondary spherical waves starting from the slits, that interfere with each other to produce a pattern on the photographic plate. But the registration on the plate is a quantum process, a chemical reaction. If the quantum passes through either slit, the other one is irrelevant. But the existence of the other slit is in fact relevant, because the photographic plate registers an interference pattern. Therefore the statement that any light quantum must have gone through either one or the other slit is problematic. Heisenberg maintains that this problematic outcome shows that the concept of the probability function does not allow a description in space and time of what happens between the two observations. The description of what "happens" is restricted to the observation process in which there occurs the transition from the possible or *potentia* to the actual.

Popper's earlier views on quantum theory are set forth in his *Logic of Scientific Discovery* and his more mature statement is set forth in his *Postscript to the Logic of Scientific Discovery* (1982). The latter work is a collection of three volumes: *Realism and the Aim of Science, The Open Universe: An Argument for Indeterminism,* and *Quantum Theory and the Schism in Physics.* Popper brings to statistical quantum theory a prior ontological commitment, which he calls "commonsense realism". In Popper's view physics has historically developed out of one or another metaphysical view which he calls a "metaphysical research programme". A metaphysical research programme is a set of ideas that are currently untestable, and therefore are called metaphysical; in Popper's philosophy the demarcation between science and metaphysics is testability thus giving metaphysics a residual character relative to science. The metaphysical research programme supplies the physicist with both a metaphysical view or ontology about the general structure of the world and a metascientific view about such things as the criteria for a satisfactory scientific explanation based on the ontology contained in the metaphysical research programme. Science needs metaphysical research programmes, because they largely determine its problem situations. Popper cites Einstein's way of looking at the Lorentz transformation as an example of how a metaphysical research programme can supply a new way of looking at things, that may change science completely. Metaphysical research programmes change and are replaced as some parts become testable and are incorporated into science. The relation between the testable theory and the research programme is part of the history of the problem situations of the science, along with the problems arising from inconsistency among theories and empirical falsifications of theories.

Unlike Heisenberg, Popper views the schism in physics in more sociological terms and in terms of the issues that have given rise to the schism. And unlike Heisenberg, he does not view the current schism in physics favorably; in his opinion the acceptance of the Copenhagen interpretation and the rejection of what he calls the Faraday-Einstein-Schroedinger metaphysical research programme have left physics without any unifying picture of the world, without any theory of change, and without any general cosmology. The current schism in physics is a clash between two metaphysical research programmes, neither of which in his view seems to be doing its job. In *Quantum Theory and the Schism in Physics* he summarizes the current schism in terms of three issues: (1) indeterminism vs. determinism, (2) realism vs. instrumentalism, and

(3) objectivism vs. subjectivism. All three issues are closely related to one another and to the interpretation of the probability function in the statistical quantum theory. The schism has its orthodox group, and it has a variety of dissenters. On the dissenting side of the schism he locates the views of Einstein, de Broglie, Schroedinger and Bohm, which he characterizes together as determinist, realist and subjectivist. On the orthodox side of the schism he locates the Copenhagen school including Bohr, Heisenberg, Pauli and Born, which he characterizes together as indeterminist, instrumentalist and objectivist. He does not consider Heisenberg's views to be realist, and he effectively lumps Heisenberg together with Bohr, who was explicitly instrumentalist in his view of the formalism of quantum theory. Popper proposes a new, unifying metaphysical research programme, that he maintains offers a consistent ontology for both macrophysics and microphysics. In his autobiography he states that his views on quantum theory were greatly influenced by those of the physicist Alfred Lande, and he states in the *Postscript* that Lande anticipated his own interpretation of the quantum theory. Therefore, a brief examination of Lande's interpretation of the statistical quantum theory is in order before proceeding further in the discussion of Popper's own interpretation, the particle-propensity interpretation.

Lande's New Foundations of Quantum Physics

A brief biography of Alfred Lande (1888-1975) can be found in an obituary published in *Physics Today* (May, 1976). Lande was a German-born American physicist, who received a doctorate in physics in 1914 from the University of Munich, where he studied under Sommerfeld. In 1918 he co-authored a paper with Born, that refuted Bohr's model of coplanar electronic orbits. In 1931 he immigrated to the United States, where he taught theoretical physics at Ohio State University until his retirement in 1960. Lande originally advocated the Copenhagen interpretation of quantum theory, but publicly disassociated himself from it with the publication of his *Foundations of Quantum Theory* (1955). His most mature statement of his views is his *New Foundations of Quantum Mechanics* (1965), which includes ideas published in many of his previous papers.

As a physicist Lande had his own agenda: the solution of what he calls "The Quantum Riddle", which is the derivation of the laws of quantum mechanics from a nonquantal and nondeterministic basis without the *ad hoc* assumptions that he finds in the Copenhagen interpretation. In the process of developing his deductive explanation of quantum laws from three nonquantal postulates, he maintains that uncertainty is a physical principle for both classical and quantum physics, and he advances and defends a particle interpretation of both Heisenberg's uncertainty relations and Schroedinger's wave function. Both of these views were central to Popper's philosophy of science twenty years before Lande rejected the Copenhagen interpretation of quantum theory, and Lande references Popper's views in his own literary corpus. However, Lande never mentions Popper's propensity interpretation of probability theory and statistical quantum theory, and in fact maintains a contrary ontology with respect to the reality of the waves associated with the Schroedinger wave function.

In "Probability in Classical and Quantum Theory" in *Scientific Papers Presented to Max Born* (1953) as well as in his two later books, Lande argues that classical thermodynamics cannot be reduced to deterministic mechanics, and that it is futile to search for hidden causes behind any distribution that satisfies the rules of probability either in classical or quantum physics. To illustrate his thesis he describes an experiment in which ivory balls are dropped through a tube onto the center of a steel blade, resulting in an observed 50:50 average ratio of balls falling to the left or right. On the determinist view the 50:50 ratio is possible only if it is already contained in the initial conditions, which in turn either implies an infinite regress to still prior conditions, or is left unexplained. Lande rejects both these options. Instead he concludes that random distribution is a physical reality, and that determinism is a purely academic construction, because a program of giving a deterministic theory of statistically distributed events leads nowhere. Statistical theory can only reduce one probability distribution to another, and when there are ensembles of events conforming to error theory, these events are not reducible to deterministic mechanics.

In *New Foundations* he states that the belief in determinism is as much beyond the domain of physics as the belief in indeterminism; both ideas are metaphysical theses. Observation only shows that equal preparation, as far as equality can be achieved, always leads to unpredictably different results. Lande elevates this general insight to the physical principle of uncertainty. In contrast to ordinary experience, classical mechanics was deterministic, while on the other hand experience and quantum mechanics agree. Unpredictability understood as the acausality of individual events, must be seen as an irreducible feature of natural science. Statistical mechanics can describe predictable averages for unpredictable individual events. In quantum mechanics it is Heisenberg's great merit that he established quantitative limits for the uncertainty of prediction, but Lande also states that unpredictability of future events does not preclude the reconstruction of past individual cases using a deterministic theory.

Lande rejects Heisenberg's thesis that between two observations in atomic physics the electron is nowhere. In his discussions of uncertainty and measurement in *New Foundations* he admits that while in classical physics a measurement value can be attributed to the object immediately before, during, and after the measurement, in quantum physics there is an active, unpredictable, and unavoidable participation of the instrument or "meter" in producing the result, in which the microphysical object is thrown from its previous state into a new state. Therefore in quantum physics the measured value can be ascribed to the atomic object only immediately after the measurement is completed, and any subsequent measurement erases all traces of the first state and produces an entirely new situation. Nevertheless Lande maintains that it is always possible to reconstruct one and only one path between the two space-time positions according to the laws of classical mechanics *post factum*, even though the path cannot be predicted. He distinguishes between direct and indirect measurements; the former are coincidences in space and time, and are the basis for all other measurements, which are indirect measurements. Energy, momentum, and velocity are relevant examples of indirect measurements; velocity by definition requires measuring two adjacent positions at two adjacent times.

Lande rejects the Copenhagen thesis that effectively equates "indirectly observed" with "not observed", and then with "not observable", and finally with "nonexistent" and "meaningless". The Copenhagen school wrongly maintains that only direct measures count as observation. To say as they do, that position and momentum cannot be measured simultaneously is only a half truth. If one includes "directly", then it is trivial because momentum can never be measured directly. And without the word "directly" the statement is wrong, because the momentum value acquired within a given position increment can be determined by reconstruction of space-time data with the help of theory. The root of the difficulty with reconstructing values of indirect observables is the ambiguity of their definition, which always requires theory. Lande maintains that classical theory can be used to make the indirect measurements needed to describe the path of an electron. The controversy about the meaning of an atomic measurement is due to an erroneous connecting of the first measurement with a set of possible future measurements. When the wave function is used as a mathematical representation of just one physical state, there is no confusion; but when it is used to connect one measurement with a set of future possible measurements, misunderstanding occurs which results in different interpretations of the wave function, including the Copenhagen dualistic thesis that the wave function describes a physical state of matter which is spread out in space and time, and which suddenly contracts to one point when the particle is measured.

Lande rejects subjective interpretations, and states that quantum physics deals with records of instruments rather than any observer's consciousness, with physical objects rather than mental pictures, and with statistical distributions rather than lack of knowledge by human observers. Knowledge and conscious reading by observers are as irrelevant in atomic physics as they are in any other branch of physical science. Echoing Einstein's programmatic aim of all physics (but without referencing Einstein), Lande says that the object of natural science is to suppose that the real world exists without human advice and consent, and then to search for general regularities which may help to manipulate things and events. The significance of all that quantum theory stands for is to provide formulas, tables, and other rules of correlation between events, and in particular between probabilities of transition. To speak of the contraction of the wave packet upon an observation is as senseless in Lande's opinion as to speak of a sudden contraction of a statistical mortality table upon an individual fatality; a probability wave does not guide actual events any more than a mortality table guides actual mortalities, and it shrinks no more than a mortality table shrinks when an actual death occurs. The subjectivist confusion begins in Lande's view, when the material body used as a measuring instrument is regarded as a subject, and when it is then said that quantum theory has changed the relation between subject and object; this makes a great impression on those who mistakenly identify statistical distributions recorded by instruments with knowledge or lack of knowledge of observing subjects.

Lande advances a particle interpretation of the Heisenberg uncertainty relations and the Schroedinger wave function, and he criticizes the Copenhagen dualistic interpretation. A central part of his criticism is his alternative interpretation of the two-slit diffraction experiment, in which the diffraction pattern is construed by the Copenhagen school as an interference pattern, that must be taken as evidence for the wave nature of the electron, which in turn must also be construed as a particle before its entry into the slit and then again upon its impact on the photographic plate. Lande references the Stern-Gerlach experiment, the theory of William Duane (1923), and the work of Paul Ehrenfest and Paul S. Epstein (1924). He explains that Duane's quantum theory was not immediately recognized as a way out of the Copenhagen duality paradox, because his proposed statistical particle theory of diffraction pertains to X-rays, and was offered in support of the photon theory of light, and also because in 1923 diffraction of electrons was not yet discovered. Lande references a letter written to him by Born stating that Duane's 1923 paper on the particle theory of X-ray diffraction was well appreciated at the time of its publication, and that it is a riddle as to why its significance was overlooked when the diffraction of matter was discovered a few years later. Lande remarks that he could not find any hint of recognition in any of the works of Bohr, Born, de Broglie, Dirac, Einstein, Heisenberg, Pauli, or Schroedinger, that Duane's quantum rule is relevant to the alleged dilemma of matter diffraction and duality.

According to Duane's quantum rule for linear momentum, the incident matter particles do not spread out as continuous matter waves or manifest themselves as though they do. It is the crystal with its parallel lattice planes, which is already spread out in space, and which reacts as one rigid mechanical body to the incident particles, that produces the diffraction pattern. Duane's rule yields the same observed diffraction directly without appealing to any wave interlude. Therefore, the idea of a dualistic change from matter particles to waves and then back to particles is a quite unnecessary if not also a fantastic invention in Lande's opinion. According to his criteria for scientific criticism the scientific value of a theory is measured not only by its power to account for observed data, but also by criteria of simplicity, freedom from *ad hoc* assumptions, and reducibility to more general postulates. As a result of Duane's theory, quantum physics has discovered that even such wave-like phenomena as matter diffraction through crystals can be understood in a consistent unitary way as produced by matter particles alone, obeying the conservation laws of mechanics under special restrictions known as quantum rules, matter particles which react to bodies containing periodicies in time and space. Lande thus states that electrons always behave as particles, and never misbehave as waves; he calls Duane's quantum rule the "missing link" between wave-like appearances and particle reality. To the two recognized general quantum postulates, Planck's rule for energy exchange and Sommerfeld-Wilson's rule for angular momentum exchange, Lande adds Duane's quantum rule for linear impulse changes as the third postulate for quantum physics. Lande thus answers the problem of the two-slit diffraction experiment, the problem of which of the two slits did the particle pass through: he states that for its contribution to the diffraction pattern, it does not make any difference where exactly the diffraction takes place. The electron changes its momentum in reaction to the harmonic components of the matter distribution of the crystal screen with two slits as a whole. All that matters is the conservation of change and of total momentum in the reaction between electron and diffractor.

For these reasons Lande maintains that the Copenhagen school starts from "wrong physics", when they maintain that wave-like appearances of matter diffraction are due to the periodic wave action of the electron. The correct view is that the appearances are due to the periodic structure of the bodies in space (the crystal) and in time (the oscillators) via the three corresponding quantum rules for the momentum and energy activity of the periodic bodies. He calls his particle interpretation "practical realism", and offers reinterpretations of Heisenberg's and Schroedinger's equations. The Heisenberg uncertainty relations describe objective statistical dispersion. Heisenberg's claim, that simultaneous exact position and momentum measurement pairs is meaningless and nonexistent, is incorrect because it confuses lack of predictability (which is true) with lack of measurability (which is false). Unpredictable data can be reconstructed, including position and momentum measurement pairs that are more accurate than Planck's constant; and what can be measured, exists. The doctrine of the indeterminacy of existence is a "semantic artifice" rather than legitimate physics. Nor is denying that a particle always is somewhere warranted by diffraction experiments, because each particle reacts to a space-extended periodic component in the matter distribution of the diffractor. To say that the particle is nowhere is a "linguistic extravaganza" and not a philosophical innovation.

As for Schroedinger's equation, Lande says that it does not deal with matter waves, but with probability amplitudes; it is a probability table not essentially different from any mortality table. The real constituents of matter are discrete particles, which occasionally give the appearance of wave action, and the real constituent of light is a continuous electromagnetic field, which sometimes gives the appearance of photonic particles. The Schroedinger wave function is a probability curve describing betting odds for future events; it is not a real thing even when the curve looks wave-like. Lande uses the phraseology of Dr. Samuel Johnson (a critic of Bishop Berkeley's *esse est percipi*, who kicked a great stone and exclaimed "I refute him thus") saying that you can kick a stone, and you can kick an electron and even a water wave and an electromagnetic wave, and be hurt by them, thus proving their reality. But you cannot kick or be hurt by a wave-like curve representing probabilities of events. For Lande, physical interaction is the only correct ontological criterion for physical reality. He also takes exception to Born, his former colleague who had initially developed the statistical interpretation of the Schroedinger wave function as a probability amplitude for particles, but who later made what Lande calls "belated concessions" to the Copenhagen dualistic interpretation. He references Born's "Physical Reality" appearing in *Philosophical Quarterly* (1953) in which Born sets forth his own ontological criterion, the criterion of invariance. In this article Born is not explicitly opposing Lande, but rather is opposing the Idealist metaphysics and the Logical Positivist philosophy of phenomenalism.

Born explains his criterion of invariance as follows: Most measurements in physics are not concerned with things that interest us, but are concerned with some kind of projection which is defined in relation to a system of reference. In every physical theory there is a rule which connects the projections of the same object on different reference systems; the rule is called a law of transformation, and all transformations have the property of forming a "group", where the sequence of two consecutive transformations is a transformation of the same kind. Invariants are quantities having the same value for any system of reference, and therefore are independent of the transformations. The main advances in the conceptual structure of physics consist in the discovery that some quantity which was formerly regarded as the property of a thing is in fact only the property of a projection. The historical development of the theory of gravitation from pre-Newtonian physics to relativity theory is one example. Another example is the development of quantum physics. An observation or measurement in quantum physics does not refer to a natural phenomenon as such, but to its projection on a system of reference which is the whole apparatus used in the experiment. Using instruments the physicist can obtain certain restricted but well described information, which is independent of the observer and of his apparatus, namely the invariant features of a number of properly devised experiments. Bohr's complementarity principle means that the maximum knowledge of the quantum can only be obtained by a sufficient number of independent projections of the same physical entity. The final result of complimentary experiments is a set of invariants characteristic of the entity, and these invariants are called "charge", "rest mass", "spin", etc. In every instance, when we are able to determine these quantities, we decide we are dealing with a definite particle. The words "photon", "electron", etc. signify definite invariants, that can be constructed by combining a number of observations.

Born maintains that the idea of invariance is the clue to a rational concept of reality, not only in physics but in every aspect of the world. The power of the mind to neglect the differences of sense impressions and to be aware only of their invariant features is the most impressive fact of man's mental structure. He proposes translating the term "*gestalt*" not as "shape" or "form" but as "invariant", and to speak of invariants of perception instead of sense impressions as the elements of our mental world. In the closing paragraph of his article Born considers the reality of waves according to his ontological criterion of invariance. He says that we regard waves on a lake as real, though they are nothing material but are only a certain shape of the surface of the water. The justification for this view is that they can be characterized by certain invariant quantities like frequency and wave length, or as a spectrum of these. Born says that the same thing holds for light waves, and he asks rhetorically why the physicist should withhold the epithet "real" even if the waves represent in quantum theory only a distribution of probability.

In his *New Foundations* Lande replies to Born's rhetorical question from the viewpoint of his own criterion of interaction: Particles are real and Schroedinger waves are not real for the same reason that sick people are real things, while the wave-like curve which symbolizes the probability distribution during a fluctuating epidemic is not a real thing. Lande says that a given formalism can always be interpreted in a variety of ways. At the conclusion of his *New Foundations* he gives seven alternative interpretations of the Schroedinger wave function including Schroedinger's, de Broglie's, Bohm's, Heisenberg's subjective interpretation, Heisenberg's objective interpretation together with Bohr's instrumentalist interpretation, and Lande's own interpretation. He does not include Popper's propensity interpretation. He states that this list is indicative of the present confusion regarding the wave function, and paraphrases Mao Tse Tung saying that while it may be good politics to let a hundred flowers bloom and let a hundred schools contend, it is not good enough for science. He asserts that only his interpretation stands up to realistic criticism in accordance with "monolithic" quantum mechanics.

Popper's Particle-Propensity Interpretation of Quantum Theory

Popper explains the basis for the schism in physics as follows: On the one hand Einstein was a determinist, who believed that the statistical nature of quantum theory is due to the physicist's ignorance of the underlying deterministic laws, which have not yet been discovered. Therefore Einstein chose a subjective interpretation of probability based on the scientist's ignorance. On the other hand Heisenberg was an indeterminist, but his introduction of the observer's disturbance of the quantum phenomenon by the measurement apparatus resulted in the combination of both the objective and subjective interpretations of the probability function in the Copenhagen interpretation of the quantum theory, because the only objective interpretation of probability available at the time was the frequency interpretation. The frequency interpretation is applicable only to mass phenomena, while the quantum theory pertains to singular events. Therefore in order to describe the single quantum event, it seemed necessary to view probability as describing the scientist's ignorance resulting from the disturbance. For this reason the Copenhagen interpretation also relies on the subjective interpretation of probability. Popper's propensity hypothesis advances an objective interpretation of the probability calculus and of probabilistic theories in physics, and it is an objective interpretation that is applicable to singular events. Popper has arguments for probability interpretations that are exclusively objective, but any objective interpretation requires a realistic philosophy with an indeterministic ontology. Therefore he also advances arguments for realism and indeterminacy, as well as for objectivism.

He has several arguments against the subjective interpretation of probability and for the objective interpretation. Some quantum theorists such as Pauli introduce the idea of induction into discussions about the statistical nature of quantum theory. Popper rejects this application of inductivism for the same reasons that he rejects all applications of the idea of induction; induction is psychologistic and confuses world 2 with world 3. He also argues that the idea of explaining the statistical outcomes of experiments and predictions in terms of the ignorance of the physicist is absurd. Empirical science absolutely never explains anything in terms of the researcher's ignorance; it always explains phenomena in terms of other phenomena.

Popper's argument for realism is based on his fallibilist thesis of scientific criticism. Simply stated, he argues that the possibility of falsification is evidence of the existence of the real world that is independent of human knowledge. He furthermore argues that the fact that theories are conjectures does not imply that they do not describe the real world. Rational criticism results in better theories that have greater verisimilitude. Popper argues against instrumentalism, which he associates with both Bohr and Heisenberg. In "Three Views Concerning Human Understanding" in *Conjectures and Refutations* he references Heisenberg's thesis that physical theories such as Newton's are not falsified, but rather have had their applicability restricted by later theories such as relativity and quantum mechanics. This view is an aspect of Heisenberg's doctrine of closed-off theories, although Heisenberg did not set forth his doctrine of closed-off theories as an instrumentalist thesis. In a footnote in this paper Popper states that Heisenberg's instrumentalism is far from consistent, and that he has many anti-instrumentalist remarks to his credit, but that Heisenberg's view of quantum theory necessarily leads to an instrumentalist philosophy by neglecting falsification and stressing application. A mere instrument cannot be falsified, and the instrumentalist view may be used *ad hoc* to rescue a theory threatened by contradictions. Popper maintains that such am evasion was the reason that Bohr advanced his principle of complementarity, the renunciation of the attempt to interpret atomic theory as a description of anything; the self-consistent formalism need not be reconciled with its inconsistent applications, if it is left uninterpreted. On Popper's view the unfalsifiability thesis of the instrumentalist view makes instrumentalism incapable of explaining scientific criticism and scientific progress; only by reaching for refutations can science hope to learn and to advance.

Popper argues against determinism, and in this respect he takes exception to Einstein, although he says that he may have changed Einstein's mind about determinism in a conversation at Princeton in 1950. Popper distinguishes between metaphysical determinism, which is a thesis about the whole world, and scientific determinism, which is a thesis about the part of the world described by a scientific theory. He classifies Einstein as a metaphysical determinist, and reports that in his discussions with Einstein he referred to him by the name Parmenidies, because like the ancient philosopher Parmenidies, Einstein's metaphysical determinism implies that the future is entirely contained in the past, and that change is not real but is merely an appearance. Popper also argues against scientific determinism, and specifically he denies that Newtonian mechanics implies a deterministic ontology. He describes the theories of classical physics as *prima facie* deterministic, by which he means that the deterministic character is a property of the theory and not of the real world. He maintains that classical physics does not imply determinism any more than quantum physics does, because there is always an irreducible and stable statistical element in any predictions made with a *prima facie* deterministic theory; and it is always necessary to add to the deterministic theory a probability assumption to explain the statistical component in the prediction, because statistical conclusions require statistical premises. Popper quotes at length Lande's description of the experiment with the ivory balls and steel blade, which Lande uses to argue that statistical results require statistical assumptions about the initial conditions. Therefore Popper rejects attempts to explain the statistical outcomes subjectively by reference to lack of knowledge of the experimenter for the reasons given above, and he maintains that the law-like behavior of statistical sequences is for the determinist ultimately inexplicable.

Popper developed his propensity interpretation of probability in 1950 specifically to address the interpretation problem arising from statistical quantum theory, but it is also intended to be applicable to all physics. While it is but one of many interpretations for the probability calculus, it is the best for physics in Popper's view. Popper distinguishes three objective interpretations of the probability calculus: the classical interpretation, the frequency interpretation, and his propensity interpretation. The classical interpretation is that the probability measure P(α,ß) is the proportion of equally possible cases compatible with the event ß, that are also favorable to the event α. The frequency interpretation is that P(α,ß) is the relative frequency of the events α among the events ß. The propensity interpretation is a refinement of the classical interpretation. In the classical interpretation experimentation is not needed, because it deals with equally possible cases, such as the two sides of a coin or the six faces of a die.

The propensity interpretation substitutes weights for equally possible cases, and the weight are experimentally determined measures of the propensity or tendency of a possibility to realize itself upon repetition. Thus in the propensity interpretation the measure P(α,ß) is the propensity of α given experimental conditions ß; it is the sum of the weights of the possible cases that satisfy the condition ß which are also favorable to α, divided by the sum of the weights of the possible cases that satisfy ß. The propensity interpretation is closely related to the frequency interpretation; the latter is about frequencies in actual finite sequences of experiments, while the former is about virtual finite sequences. In the propensity interpretation probability statements are about some measure of a physical property of the whole repeatable experimental arrangement, a measure of a virtual frequency, and the probability distribution is taken to be a property of the single experiment. The fact that the probability distribution in the propensity interpretation is a property of a single experiment is the strategic characteristic of this interpretation for quantum theory. Previously in *Logic of Scientific Discovery* Popper had attempted to modify the frequency interpretation so that it could address single events by means of what he called "formally singular statements". He abandoned this idea, when he developed the propensity interpretation. Now he says that the frequency measurements function to test the conjectured virtual frequency, which is a conjecture like any other scientific hypothesis.

The propensity interpretation is consistent with Popper's particle interpretation of the quantum theory, that he had advanced years before in *Logic of Scientific Discovery*. According to Popper`s particle interpretation the Heisenberg uncertainty relations are statistical scatter relations that describe the lower limits of the dispersion of particles; they are not the upper limits of the accuracy of measurements, as Heisenberg maintains. The uncertainty relations apply only to the magnitudes that belong to the particle after the disturbing measurement has been made. The particle always has position and momentum, and both position and momentum up to the instant of measurement can be ascertained in principle with unlimited accuracy. It is not the impossibility of precise measurement, but the statistical scatter that makes it impossible to predict the path of the particle after the disturbing measurement operation. The scatter relations are statistical predictions about paths, and the paths must be measurable in order to test the statistical theory. For these reasons Popper rejects

Heisenberg's view, as Popper sees it, that the uncertainty relations express limits to our subjective knowledge instead of expressing objective statistical scatter relations, and that measurements are impossible because of the nonexistence of the entities measured. What is impossible is producing scatter-free, dispersion-free quantum states. The statistical laws add to our knowledge. They do not set limits to our knowledge; they set limits to the scatter relations and tell us that the scatter is an objective reality that cannot be suppressed.

The propensity interpretation solves the problem of the relationship between particles and their statistics, and between particles and waves. Popper calls the Copenhagen wave-particle dualistic interpretation the "great quantum muddle". The great quantum muddle results from the mistake of taking the probability distribution function as a physical property of the elements of the population. Popper believes that this mistake is historically due to the fact that the works of de Broglie and Schroedinger led physicists to view the wave as the structure of the particle, and thus to view the particle as a "wave packet" or a "wavicle". Popper maintains that the statistical wave function is a property characterizing a sample space and not a property of the elements of the sample space. The elements have the properties of a particle. The propensity interpretation achieves the application of probability theory to single cases, but it does not do this by speaking about single electrons or protons; it speaks about propensities, which are properties of each instance of the whole repeatable experimental situation involving a single particle.

Propensity statements in physics describe properties of the situation, and are testable if the situation is typical. Popper accepts Lande's explanation of the two-slit experiment, and he references what he calls the Duane-Lande space periodicity formula. The two-slit experiment is a space periodicity experiment, in which the particle interacts with the whole experimental situation including the crystal. More specifically from the viewpoint of the propensity interpretation, it is the whole experimental arrangement which determines the propensities. The possible results of any one experiment are different in the case of both slits being open from the case where only one is open; propensities are dependent on possibilities, such that the results will differ with different experimental arrangements, one slit or two. Thus in the two-slit arrangement the particle will pass through only one of the slits and in a sense will remain unaffected by the other slit. What the other slit influences are the propensities of the particle relative to the entire experimental arrangement and not relative to the particle itself: the propensities for reaching the one point or the other point on the screen with the two slits. The Schroedinger wave equation enables the physicist to determine the propensities, and it entails the Heisenberg scatter relations, which limit the possible predictions.

Popper states that Schroedinger had anticipated one of the most important aspects of the propensity interpretation, namely the objectivity and reality of the waves in configuration space. One of the features of Popper's propensity interpretation is his thesis that the propensities are real, just as forces are real, and he speaks of propensity fields, just as contemporary physicists are accustomed to speak of force fields. The propensities are dispositional relational properties of the experimental set up. The waves are propensities of the particles to take up certain states under the conditions of the experimental set up, and the propensity waves are therefore no less real than electromagnetic waves. Lande believes that if he admitted to the reality of the Schroedinger wave, then like Born he would have to make what he called "concessions" to the Copenhagen dualistic thesis. Therefore Lande maintains that the Schroedinger wave function interpreted as a probability wave is merely a statistical function that is no more real than a mortality table, which Lande did not view as real. But Popper uses Lande's criterion of interaction, and argues that because the probability waves can interact to produce interference, they must be real, and not merely be mathematical tables. Popper supports Lande's rejection of the Copenhagen dualism, but unlike Lande, Popper says that he prefers to speak of the particle and its associated propensity fields, instead of speaking of the particle and its associated mathematical probability function.

In the 1982 introduction to *Quantum Theory and The Schism in Physics* Popper proposes a crucial experiment for deciding between the Copenhagen interpretation and his propensity interpretation of the uncertainty relations. At issue is the subjective interpretation in the Copenhagen view: whether knowledge alone is sufficient to create uncertainty, or whether it is the physical situation that is responsible for the statistical scatter. The experimental situation involves two slits in two opposing screens, through which electrons pass, one through each screen. The sizes of the two slits are greatly different, while the degree of inaccuracy in the knowledge of the size of a slit is presumably small relative to the difference in the slit sizes. Popper expects different statistical scatters from each of the slits, while the accuracy with which the size of the slits is known, is the same for each slit.

On Crucial Experiments and Scientific Revolutions

Unlike the Positivist philosophy of science, which has been interred to its resting place in the history of philosophy, Popper's philosophy of science is still a living philosophy in the sense that it is still accepted and debated in the professional literature. Popper has addressed more than one generation of philosophers during his life time. Initially his philosophy was a critique of the Positivists, who viewed his philosophy as an unconventional novation, while today his philosophy is criticized by the contemporary Pragmatists, who view his philosophy as the conventional wisdom. The central issue in which Popper represents the conservative position is the problem of the decidability of scientific criticism including most notably the decidability of crucial experiments. The origin of the problem is the thesis shared by both Popper and the Pragmatists, and also enunciated by Einstein, that theory determines what is observed. To the Pragmatists this thesis implies that the description of the observed results from an experimental test cannot be understood in the same way by different scientists who maintain alternative theories in an experimental test that is crucial in the sense that it purportedly decides between the alternative theories. If theory determines what is observed, then scientists maintaining different theories do not observe the same thing, and the observed outcome from the crucial experiment cannot decide between the alternative theories. To Popper, the 1919 eclipse experiment, which was the historic crucial experiment deciding on behalf of Einstein's theory of relativity, demonstrates conclusively that crucial experiments are decisive.

It should be noted at the outset that even in his earliest writings Popper has maintained that falsification is never finally and permanently conclusive, because the singular basic statements that are potential falsifiers may be revised, thus occasioning the revision of a falsifying test outcome. The empirical test may be said to be conclusive only to the extent that interested scientists agree to accept certain basic statements. Popper states that in some cases it has taken scientists a long time before a falsification is accepted, and that it is usually not accepted until a falsified theory is replaced by the proposal of a new and more adequate theory. But Popper does not find this historical fact to be problematic, even though in his view it is responsible for having led the Pragmatists to accept irrationalism and relativism in methodology of science. In his introduction to *Realism and the Aim of Science* he gives several examples of successful falsifications, that furthermore have led to important scientific revolutions. If the development of Einstein's relativity theory can be said to be the formative influence in Popper's views, then the development of the quantum theory can be said to be the formative influence in the contemporary Pragmatist philosophy of science. The topic of crucial experiments has assumed its controversial status in the professional literature because of the Copenhagen interpretation of quantum theory. The Copenhagen interpretation denies that a crucial experiment can decide between the wave and particle interpretations of microphysics, because the electron has the properties of both wave and particle. The Pragmatist, W.V.O. Quine, invoked Duhem's philosophy of physical theory not only because of Duhem rejection of the decidability of crucial experiments, but also more fundamentally because of Duhem's thesis of the organic character of the semantics of theory language in physics. In his "Two Dogmas of Empiricism" Quine extended Duhem's thesis of the organic or wholistic character of the semantics of physical theory, to make it a general theory of the semantics of language as such, including the language used by physicists to describe observed experimental test outcomes. As a result of this extended thesis, which is now conventionally called the Duhem-Quine thesis, the wholistic character of the semantics of language explains why crucial experiments are undecidable not only in the wave-particle issue in quantum theory, but even more generally for all scientific criticism. Even where one of the alternatives is the Copenhagen dualistic interpretation, as in Lande's list of seven interpretations, the crucial experiment cannot effectively decide among them, according to the contemporary Pragmatist philosophy. The issue of crucial experiments has become a focal point in philosophy of science for the larger issues of the decidability of scientific criticism and of the nature of the semantics of language in general. The historic transition from the Positivists' naturalistic philosophy of the semantics of language to the contemporary artifactual philosophy of the semantics of language has resulted in two alternative artifactual philosophies of the semantics of language: The one is the organic or wholistic thesis advocated by the Pragmatists, which they use to attack the decidability of crucial experiments and of scientific criticism in general. The other is the logical or mechanistic thesis advocated by Popper, which he uses to defend the decidability of crucial experiments and the rationality of scientific criticism in general.

In his "Three Views Concerning Human Knowledge" (1956) reprinted in his *Conjectures and Refutations* Popper discusses Duhem's views on crucial experiments. He notes that Duhem shows that crucial experiments cannot establish a theory by refuting its alternatives, and emphasizes that Duhem does not say that theories cannot be refuted in crucial experiments. Popper maintains that crucial experiments can be used to decide between alternative theories, as occurs when a new theory in proposed as a superior alternative to an older theory. The new theory is tested by applying it to cases for which it yields results that are different from what is expected from the older theory. Such cases are "crucial" in the Baconian sense that they indicate the crossroads between two or more theories, but not in the Baconian sense that any theory can be established. Popper then turns to Duhem's thesis that in every test it is not only the theory under investigation that is tested, but also the whole system of assumptions made by the theory, such that it is never possible to be certain which of the assumptions is refuted by the test. Popper states that if the scientists consider each of the two theories in the crucial test together with all the background knowledge assumed by both theories, then the scientists decide between the two systems, which differ only over the two alternative theories in the test. Then Popper adds that scientists do not assert the refutation only of one of the theories by the test, but rather the theory together with the background assumptions. By this he does not mean that every statement in the theory and its assumed background is refuted, but only that there is at least one statement that is erroneous, and that it may be in either the theory or the assumed common background. Thus he also says that in future tests parts of the background knowledge may be rejected as responsible for the falsification of the theory in the current crucial test.

Popper then proposes to "characterize the theory under investigation" in the crucial test precisely as that part of the vast system of knowledge for which the scientist has an alternative in mind, and for which he has therefore designed the crucial test. This may be taken as Popper's basis for individuating theories: theory α is distinguished from theory ß, because α makes a claim or statement than is an alternative to that made by ß, and because α consists in the language that makes it an alternative to ß. Thus it may be said that Popper individuates theories by reference to the theories' semantical properties as manifested in the crucial test situation. However, Popper does not define theory language by reference to the crucial test situation as such; he often states that the background knowledge includes theories other than the tested one. In his philosophy, therefore, theory language is any testable general statement regardless of whether or not it is being tested, which is to say that he defines theory by reference to its syntactical properties and not by reference to its pragmatic properties. Furthermore Popper's concept of theory language may be contrasted with that of the Positivists, who believed that it is possible to define theory in terms of its semantical properties by means of their distinction between theoretical and observation terms; Popper rejects this distinction.

In his "Truth, Rationality, and the Growth of Knowledge" (1961) reprinted in *Conjectures and Refutations* Popper turns to Quine's use of Duhem's philosophy. Quine maintains a wholistic view of empirical testing, and in his "Two Dogmas of Empiricism" in *From A Logical Point of View* he states that our statements about the external world face the tribunal of experience not individually but as a corporate body. Popper replies that this wholistic view of tests, even if it were true, would not create a serious problem for the fallibilist philosopher of science. He repeats his thesis that the fact that scientists take a vast amount of background knowledge for granted, is not to say that the scientist must uncritically accept it; the background knowledge too may be challenged and tested. Even though all of the background assumptions may be challenged, it is quite impossible to challenge all of the assumptions at the same time. All criticism must be "piecemeal", which Popper says is only another way of saying that the fundamental maxim of every critical discussion is that one should "stick to the problem". The misguided attempt to question all background assumptions merely leads to the breakdown of a critical debate.

Furthermore even though the falsification of a theory does not reveal where the error is, nevertheless it is still possible to find the hypothesis that is responsible for the refutation, i.e. to find which hypothesis is responsible for the refuted prediction. The fact that such logical dependencies may be discovered is established by the existence of independence proofs for axiomatized systems; these are proofs that show that certain axioms of a system cannot be derived from the rest. Popper argues that the existence of such proofs shows that Quine's wholistic view of the global character of all empirical tests is untenable, and that it explains why even without axiomatized physical theories, the scientist may still have an inkling of what has gone wrong with the theory. In *Realism and the Aim of Science* Popper affirms as historical fact, that scientists are sometimes highly successful in attributing to a single hypothesis the responsibility for the falsification of a complex theory or of a system of theories, and he argues that this success remains to be explained if one adopts the wholistic view of empirical testing.

In 1962 Thomas Kuhn wrote *Structure of Scientific Revolutions* in which he used the wholistic thesis to interpret the history of science. In 1970 he defended his wholistic interpretation against critics in *Criticism and the Growth of Knowledge*. The leading critic in this same book was Popper, who contributed "Normal Science and its Dangers". In his earlier statements in defense of the decidability of crucial tests Popper did not explicitly address the basis of the wholistic view of testing, namely the thesis that the semantics of language is wholistic. The wholistic thesis of the semantics of language means that the meanings of terms are mutually determined in the context of the discourse in which they occur, such that alternative contexts consisting of alternative theories produce a semantic ambiguity or equivocation that is propagated through all of the related language. Therefore when considering the alternative theories investigated in a crucial test, all that constitutes the background assumptions is ambiguous. In other words there is really no common background, because one semantical interpretation is given to the language expressing the background assumptions by one of the theories in the test and another interpretation is given by the other theory in the test. Often the two alternative semantical interpretations are spoken of as two different languages, and there is said to arise a problem of translation from one to the other. This thesis is strategic to Kuhn's critique of the Positivists, because the lack of any common semantics for alternative theories that makes impossible a common background for crucial tests, also makes impossible a common observation language.

Kuhn maintains that the kind of scientific progress that Popper describes with its crucial experiments and falsifications can occur only within a linguistic framework, and he calls this type of scientific progress "normal science", which Kuhn opposes to another type which he calls "extraordinary science" or "revolutionary science". Revolutionary science is a transition from one language framework to another, where the term "framework" in the discussion refers to discourse having a univocal semantical interpretation and associated ontology. Popper rejects this theory of scientific revolution as irrational, when he criticizes Kuhn in *Criticism and the Growth of Knowledge*. While admitting that "normal science" in Kuhn's sense does exist, Popper argues that such normal science is dogmatic. He says that science is essentially critical; that it consists of bold conjectures controlled by criticism, and that it may be called revolutionary in this rational sense. He rejects Kuhn's relativism, the thesis that the linguistic framework cannot be critically discussed, and he calls this "the myth of the framework". Comparison of different frameworks is always possible on Popper's view, and so is critical discussion therefore. Even totally different languages are not untranslatable. And it would be simply false to say that the transition from Newton's theory of gravitation to Einstein's theory is an irrational leap, and that the two are not rationally comparable; the transition to Einstein's theory was genuine progress in comparison with Newton's. Popper concludes that the myth of the framework is in our time the central bulwark of irrationalism, and that it exaggerates a difficulty with communication and criticism into an impossibility. In place of criticism as is found in Popper's falsification thesis, Kuhn proposes turning for enlightenment concerning the aim of science to psychology and sociology. But Popper rejects this proposal, and states that compared with physics, sociology and psychology are riddled with fashion and with uncontrolled dogmatism. He believes that such a proposal is a backward regression that cannot solve the difficulty.

In his "The Rationality of Scientific Revolutions" in *Problems of Scientific Revolution* Popper distinguishes between the sociological and the logical or rational dimensions in the history of science, when he distinguishes ideological from scientific revolutions. By an ideology he means any nonscientific theory, creed, or view of the world that is attractive or interesting to people including scientists. He cites the Copernican and Darwinian revolutions as examples of scientific revolutions that gave rise to ideological revolutions, because each changed man's view of his place in the universe. But these were also scientific revolutions in so far as each overthrew a dominant scientific theory, the one a dominant astronomical theory, the other a dominant biological theory. He also cites Einstein's relativity theory as a revolution, a truly scientific revolution, that gave rise to operationalism and supported Positivism, even though Einstein later rejected these ideologies. And Popper also refers to the subjectivist interpretation of quantum theory as an ideology, although in 1982 he proposed a crucial experiment that he thought could decide against it.

The wholistic thesis of the semantics of language is used by Pragmatists to explain events that have been observed in the history of science: the impediment that language creates both to the development of new theories and to the communication of new theories within a profession. However, Popper relegates all semantical analysis to the status of a variation on the essentialist metaphysical thesis; in his autobiography in *Philosophy of Karl Popper* he admonishes the reader never to let himself be "goaded" into taking seriously problems about words and their meanings. He maintains that words "merely" play a technical or "pragmatic" role in the formulation of theories, just as the letters in written words play such a role in the formation of the words. It should be noted that contemporary Pragmatists are as anti-essentialist as Popper; one need only recall Quine's rhetorical ridicule that an essence is merely "a meaning wedded to a word". Popper's philosophy does not offer a theory of semantical description to reconcile the phenomenon of semantical change with his views on the decidability of criticism.

The Philosophy of Science

Popper`s philosophy of science addresses the basic questions currently discussed by the professional philosophers of science. His explicit rejection of the Positivist and the essentialist naturalistic philosophies of the semantics of language, represents a basic problem shift, a reconceptualization of science in philosophy of science. His theory of scientific explanation has been called the hypothetico-deductive thesis. In his chapter on theories in *Logic of Scientific Discovery* he states that to give a causal explanation of an event means to deduce a statement that describes the event, using as premises of the deduction one or more universal laws together with certain singular statements called initial conditions. Later in "Aim of Science" in *Ratio* (1957), which is reprinted both in *Objective Knowledge* and in *Realism and the Aim of Science*, he defines a causal explanation as a set of statements by which one describes a state of affairs to be explained, statements he calls the "*explicandum*", by deduction from a set of explanatory statements, which he calls the "*explicans*". The *explicans* must logically entail the *explicandum*, and it must not be known to be false. Furthermore, the *explicans* must be independently testable, so that it is not *ad hoc*. This means that the *explicandum* must not be the only evidence relevant to the *explicans*; the *explicandum* must have a variety of testable consequences, and especially consequences that are different from the *explicandum*.

The Logical Positivist concept of explanation is also described as hypothetico-deductive in the above sense. But there are also fundamental differences between Popper's and the Logical Positivists' views. Of central importance to Popper's concept of scientific explanation is the thesis that causal explanation need not describe certain things, or in other words that it need not have a certain semantics describing a certain ontology needed to supply science with foundations, such as the phenomenologist ontology. Popper's view therefore differs from the Positivist view that causal explanation must have a semantics with such ontological categories as sensations, elementary phenomena, or sense data. And it also differs from the Romantic view of causal explanation in social science, which requires a mentalistic ontology. In *Poverty of Historicism* Popper rejects the Romantic requirement of intuitive understanding of purpose and meaning produced by sympathetic imagination, a view which is also called the *verstehen* thesis of explanation in social science. He maintains that this requirement goes beyond causal explanation, and proposes his doctrine of the unity of method in both natural and social science, the method that he describes in *Logic of Scientific Discovery*. In Popper's philosophy of science "causal explanation" is defined in terms of the function that theories perform in realizing the aim of science; it is not defined in terms of some fundamental ontology. His view of causal explanation is the result of his rejection of the naturalistic philosophy of meaning; without the naturalistic theory of semantics there is no basis for requiring any particular ontology including the particular ontology's concept of causality, in order to be able to give a causal explanation. Rejection of the naturalistic thesis implies the rejection of all ontological criteria for causal explanation as well as the rejection of the distinction between observation language and theory language and of the idea of the existence of an ontological foundation for science. Thus Popper says that explanation is of the known by the unknown in the sense of conjectural, instead of by the known in the sense of the permanently established foundation. In this respect Popper is in the company of the contemporary Pragmatists; Quine for example calls the view that there are ontological criteria for causal explanation, the "genetic fallacy".

But in Popper's philosophy of science the rejection of ontological criteria for causal explanation became complicated in later years by his idea of metaphysical research programmes. The metaphysical research programme is not atemporal and eternal like the essentialists' or the Positivists' ontological foundations; it is part of the historical problem situation at a particular juncture in the history of a science, and it is also untestable at the point in time. Most notably in Popper's view, at the given point in the history of the science the metaphysical research program functions as an ontological criterion for what constitutes a satisfactory explanation. This complication arises from Popper's way of demarcating between science and metaphysics, which appeared many years before he introduced the idea of metaphysical research programmes into his philosophy of science, as he did in his later discussions of quantum theory. As early as 1955 in "Demarcation Between Science and Metaphysics" in *Conjectures and Refutations* he states that all physical theories say much more than the physicist can test, and that whether this "more" belongs to physics or should be eliminated as a metaphysical element is not easy to say. And in 1958 in "On the Status of Science and of Metaphysics" reprinted in the same book he says that one can discuss irrefutable metaphysical theories rationally in the sense that one can discuss their ability to solve the problems that they purport to solve, that is, in relation to their problem situation. This complication has its origin in the residual status of metaphysics in Popper's philosophy; metaphysics for him contains a great heterogeniety of types of knowledge, which have nothing in common but their irrefutable character and therefore their nonscientific status. Historically philosophers have not treated metaphysics in so residual a manner, but instead have offered positive characterizations of metaphysics, which have sometimes been called "transcendental metaphysics", and which are not typically viewed merely as protoscience; for example issues such as realism versus idealism are viewed as transcendental and as incapable of empirical resolution at any time. In the concluding paragraph of the concluding section of the concluding volume of the *Postscript*, Popper states that there may be a criterion of demarcation within metaphysics between what he calls "rationally worthless" metaphysical systems on the one hand, and metaphysical systems that are worthy of discussion and thought on the other hand. He does not characterize the basis for such a demarcation within metaphysics, but his motivation for recognizing the existence of protoscience metaphysics within residual metaphysics seems clearly to have been the result of the influence of Kuhn. In the 1982 "Introductory Comments" in *Quantum Theory and the Schism in Physics* Popper compares metaphysical research programmes to Kuhn's concept of paradigm, while stressing that metaphysical research programmes must be seen in terms of a situation that can be rationally reconsidered, and that scientific revolutions viewed as changes of paradigms are due to rational criticism. In this context he references his 1975 "Rationality of Scientific Revolutions", where he distinguishes between scientific and ideological revolutions, and then sets forth criteria for rational criticism of scientific revolutions like Einstein's even before any experimental testing is attempted.

Popper's rejection of the naturalistic theory of meaning is implied by his falsificationist theory of scientific criticism. Theories are conjectures that are created by the human imagination, and similarly the meanings associated with the theories' constituent terms must also be created artifacts distinguished as world 3 objects. The theories do not originate by any natural process such as induction, and similarly the constituent meanings do not originate by any natural process such as perception. The theories are not permanently established by verification or confirmation, and similarly the meanings are not permanently established by virtue of any foundational ontology. Theories are routinely falsified as a part of the progress of science, and similarly the ontologies described by the meanings are changed with the revision of theory. The paradigmatic case is the transition from Newton's mechanics to Einstein's relativity theory. Einstein's theory does not include Newton's as a special case, but rather contradicts and corrects Newton's theory, and therefore describes an alternative ontology. But in such cases the new theory offers a higher degree of information content as indicated by the relative sizes of the classes of potential falsifiers, such that even before empirical tests are attempted, it is possible to recognize that the new theory is preferable if it survives the test. Crucial experiments are methodologically and historically important decision procedures in the progress of science. In the case of the transition from Newton's theory to Einstein's theory a crucial experiment was performed in 1919, in which Einstein's theory made the more accurate prediction within the range in which the deviation between the two theories was experimentally distinguishable. Crucial experiments are not only effective for deciding between theories, but are characteristic of the growth of science toward greater information content and verisimilitude. Popper rejects the wholistic variation on the artifactual theory of meaning, which implies that crucial experiments are invalid because the alternative theories cannot share common background assumptions with univocal semantics, and which implies in general that scientific criticism is undecidable. Both in his 1982 introduction to *Realism and the Aim of Science* and as early as his *Logic of*

Scientific Discovery in 1934, Popper has maintained that the falsifying basic statements like all empirical statements cannot be verified, and that therefore it is impossible to prove conclusively that an empirical scientific theory is false. He also states that every falsification can be tested again for motivating an agreement among interested scientists about the test outcome. He maintains that there have historically been successful scientific revolutions, which were introduced by successful falsifications, and he rejects the view that falsification plays no role in the history of science. But he offers no theory of meaning description, that would enable him to reconcile the phenomenon of semantical change with his thesis of crucial experiments and the rational growth of science. Without a metatheory of semantical change, Popper cannot explain why communication problems are merely difficulties instead of impossibilities, because he cannot explain why they are difficult.

Popper's concept of scientific explanation and the rejection of the naturalistic theory of meaning implied by the fallibilist theory of scientific criticism, in turn implies a new concept of the aim of science, that is very different from the views of the Positivists. In his philosophical development two different types of statements of the aim of science may be distinguished, firstly the logical statement and then the later institutional statement. As early as 1934 in his discussion of the degrees of testability in the *Logic of Scientific Discovery* he states that theoretical science aims to obtain theories that are easily falsifiable, because the theories have a large information content, a large class of potential falsifiers. This concept of the aim of science is integral to Popper's view of the growth of scientific knowledge based on the idea of increasing empirical information content resulting from increasing falsifiability. Similarly in "Truth, Rationality, and the Growth of Knowledge" he states that the task of science is a search for interesting truth in the sense of truth that has a high degree of explanatory power or empirical information content. In his later statements Popper added to these ideas of the aim of science the role of the historical problem situation with his idea of the metaphysical research programme. In this his introductory chapter to *Realism and the Aim of Science* he describes science as a social institution, that results from human actions that are unforeseen and unintended. He states that science grows through the institutionalized cooperation and competition of scientists, who are not only motivated by their own subjective curiosity but also by their wish or aim to make a contribution to the growth of objective knowledge. The phrases "social" and "unforeseen and unintended" seem to refer to Popper's views on the nature of social science and to his rejection of all historical relativism. Popper defines social science as the study of the unintended consequences of social behavior. But what is unforeseen in the growth of science is the new theories that result from conjectural scientific research. The content of theories in future science is in principle unpredictable in Popper's view, and he rejects all historicisms that purport to predict history including the history of science.

The strategic relevance of Popper's reference to the institutional character of science in the context of objective knowledge becomes evident when contrasted with Kuhn's view that in the history of science the ontology of a prevailing theory assumes an institutional status, and it is very likely that Popper was led to think of the aim of science in institutional terms as a result of Kuhn's views. Kuhn's thesis that the prevailing theory or paradigm assumes institutional status, means that the ontology of the prevailing paradigm functions as the criterion for scientific criticism, and that therefore commonly recognized revolutionary developments in the history of science which introduce a new theory and ontology into a science, must be viewed as an institutional change with no larger framework providing continuity. In Popper's view this radical discontinuity is historical-relativist and irrational. In his "Rationality of Scientific Revolutions" he paraphrases Marx, saying that the growth of science is "revolution in permanence", but Popper intends this phrase to mean that there exists criteria for scientific criticism that are invariant through even the most revolutionary developments, that make scientific change rational and meaningfully progressive. Thus the force of Popper's statement that the growth of objective scientific knowledge is a social institution, is that the objective nature of science makes revolutionary scientific change a change within an enduring set of institutional value standards, instead of a breakdown of the institution. The criteria for scientific criticism that operate as the institutional values of the scientific community are in Popper's view independent of the semantics and ontology of the prevailing theory or paradigm. As he says, science is subjectless. In his 1982 "Introductory Comments" to *Quantum Theory and the Schism in Physics* Popper compares his idea of metaphysical research programmes to Kuhn's idea of paradigms, but nevertheless maintains that metaphysical research programmes can be rationally reconstructed and rationally criticized, even though they cannot yet be empirically tested.

Popper's rejection of the naturalistic theory of meaning had the interesting consequence of leading him to exclude the consideration of the topic of scientific discovery from the philosophy of science, which he viewed as exclusively a matter of logic and objective knowledge; he believes that the topic of scientific discovery is exclusively a psychological and subjective matter. The conjectures resulting from the discovery process belong to world 3, but the discovery process itself belongs to world 2, and events in world 2 cannot determine the contents of world 3. While this view offers very adequate recognition to the freedom in the creative discovery process, it also relegates a whole area of interest for philosophers to the empirical studies of the psychologists. This exclusion is perhaps due partly to Popper's identification of traditional discussions of discovery with the "logic" of induction; when he rejects inductive logic, he therefore rejects all logic from the discovery process. He later modified this view, when he explained what he would admit to be possible with an "induction machine". Considering the work done by contemporary information scientists, Popper's later statements are more plausible. Up to the present time at least, contemporary information scientists would find it difficult to deny that the system designer of a discovery model must decide on what Popper calls "a repetition" of an event, which is to say that the designer must firstly conceptualize the input to the system. The discovery systems are not unconditioned, but must draw from the current state of the science under investigation for their input language.

Comment and Conclusion

Popper`s philosophy was occasioned by Einstein's development of relativity theory, a milestone episode in the history of science that Popper took to be paradigmatic of scientific progress. And Popper's philosophy is also a milestone in the history of philosophy, because it represents a fundamental problem shift. While Carnap and other Positivists continued their efforts to establish theoretical science including Einstein's theory, on firm ontological foundations, Popper rejected the naturalistic theory of meaning that supposedly supplies such a foundation, and accepted the revision of scientific explanation as a matter of course. Positivist foundational problems, such as the problem of the meaningfulness of theoretical terms, became pseudo problems as a result of Popper's problem shift, while the problem addressed by Popper, the rational growth of science without foundations, has become central to philosophy of science.

Popper's philosophy was not occasioned by the development of the modern quantum theory, and he spent much of his professional career attempting to reconcile his philosophy and the modern quantum theory. It may be said that just as Carnap had attempted to reconcile Positivism and Einstein's relativity theory, so too Popper had attempted to reconcile his philosophy and the new quantum theory, except that Popper also presumed to revise the semantical interpretation of quantum theory. In the meanwhile the Pragmatist philosophers have taken up a role relative to quantum physics, accepting it as the paradigm of modern physics, that Popper had taken up relative to Einstein's relativity theory. As a result Popper's philosophy now represents the conservative position in the contemporary professional literature of philosophy, a position that more often casts him in the role of the defensive old guard than the aggressive *avant garde.*

One of the distinctive aspects of the historical development of quantum theory is the persistent plurality of semantical and ontological interpretations compatible with the same experimental measurements and mathematical formalism. This plurality has caused lengthy controversies among the physicists; Lande's list of seven alternative interpretations may be taken as indicative of this plurality. Popper's response to this situation in modern microphysics was to create still another interpretation for the quantum theory, his particle-propensity interpretation, because like Einstein, he rejects the schism in physics and believes that a uniform ontology for both microphysics and macrophysics in necessary. This strategy has its dangers; both philosophers and physicists have often found that physicists who philosophize are usually naive philosophers, and that philosophers who physicize are usually naive physicists. The Pragmatists reacted differently; for them the quantum theory is the paradigmatic episode in the history of science, and their more accepting attitude has occasioned another problem shift in philosophy of science. While some Pragmatists express reason to advocate one or another particular interpretation of quantum theory as distinctively interesting from the viewpoint of philosophy of science, that reason is not the particular philosopher's own ontological commitments; rather it proceeds from a belief in the importance of the particular interpretation for scientific discovery (discovery is the topic that Popper did not consider even to be a part of philosophy of science). The focus on the problem of scientific discovery has in turn occasioned the problem shift: philosophers have reconsidered the semantical and ontological pluralism represented by the different interpretations of quantum theory; they have concluded that the pluralism is an outcome of certain properties of language, and that it is therefore a strategic condition for continuing scientific growth (growth is the topic that Popper considered to be central to philosophy of science). In brief Popper's approach is to attempt to adjust the semantics and ontology of quantum physics to his philosophy, while the Pragmatists' approach has been to attempt to adjust philosophy of science to account for the phenomenon of semantical and ontological pluralism in science and to identify its function.

As it happens, Popper's rejection of the naturalistic theory of meaning had supplied philosophers with the point of departure for addressing this phenomenon of semantical pluralism, and they did so in ways that Popper did not accept. The philosophical view that affirms an artifactual character of the semantics of language admits to a wholistic variation, that introduces an unresolvable cultural and historical relativism into science, which in turn makes problematic the intersubjective objectivity and rationality that Popper considers to be necessary for the growth of science. The affirmation of this wholistic variation and its consequent linguistic relativism, is occasioned by the thesis that scientific change involves semantical change. Popper's philosophy does not address the problem of semantical change, because he identifies all attempts at semantical description or "meaning analysis" with essentialism. As a result contemporary philosophers of science have moved on to new problems that Popper is unprepared to address.

THOMAS KUHN ON REVOLUTION AND PAUL FEYERABEND ON ANARCHY

Thomas S. Kuhn (1922-) was born in Cincinnati, Ohio. He received a bachelor of science degree *summa cum laude* from Harvard University in 1943. His first exposure to history of science came as an assistant to J. B. Conant in a course designed to present science to nonscientists. He received his Ph.D. from Harvard in 1949, and has since taught history of science at Harvard University, at the University of California at Berkeley, and at Princeton University. In 1979 he was appointed M. Taylor Pyne Professor of History of Science at the Massachusetts Institute of Technology.

Paul K. Feyerabend (1924-) was born in Vienna, Austria. He was inducted into the Austrian army during World War II, and was wounded in a retreat from the advancing Russian army in 1945. After the war he studied theater at the Wiemar Institute, and then went to the University of Vienna, where he received a Ph.D. in philosophy in 1951. He then went to England to study under Wittgenstein, but Wittgenstein's untimely death sent him to Popper, whose views he had initially accepted but later rejected. He emigrated to the United States in 1959, and has since taught philosophy at the University of California at Berkeley, where he has further developed his philosophy of science, which is critical of Popper's.

Conant On Prejudice And The Dynamic View Of Science

James B. Conant (1883-1978) is the principal influence on the professional thinking of Kuhn. Kuhn dedicated his *Structure of Scientific Revolutions* to Conant, "Who Started It", and Conant often acknowledged Kuhn's contributions to the "Case Histories in Experimental Science" course that Conant started at Harvard University. Conant received his doctorate in chemistry at Harvard in 1916, and then taught chemistry at Harvard from 1919 until 1933, when he accepted an appointment to be the university's president. In 1953 he resigned his position at Harvard to accept an appointment as U.S. High Commissioner of the Federal Republic of Germany, and then later as U.S. Ambassador to Germany. In 1970 he wrote *My Several Lives: Memoirs Of A Social Inventor*, an autobiography describing these three phases of his professional life. Conant's views on the history and nature of science are set forth in a series of books. The earliest is his *On Understanding Science: An Historical Approach* (1947), which he later expanded into *Science And Common Sense* (1951). A year later he published *Modern Science And Modern Man* (1952), which contains "The Changing Scientific Scene: 1900-1950" in which he elaborates his "skeptical approach" to modern quantum theory. In 1964 he published *Two Modes Of Thought*, which contains several references to Kuhn's *Structure of Scientific Revolutions* in context supportive of Kuhn's thesis.

Conant advocates what he calls the "dynamic view" of science, and he contrasts it with the "static view", which he identifies with the Positivist philosophy and specifically with the philosophy set forth by Karl Pearson in the latter's *Grammar of Science*. The static view represents science as a systematic body of knowledge, while the dynamic view represents science as an ongoing and continuing activity. On the dynamic view the present state of knowledge is of importance chiefly as a basis for further research activity. Conant defines science as an interconnected series of concepts and conceptual schemes that have developed as a result of experimentation, and that are fruitful of further experimentation and observations. He explicitly rejects the Positivist view that science is a quest for certainty, and he emphasizes that science is a speculative enterprise and a quest that is successful only to the degree that it is continuous.

Conant also maintains what he calls his "skeptical" view. On this view microphysical theory does not actually describe reality, but rather is a "policy" that serves as a guide for fruitful future research activity. He maintains that the wave-particle duality thesis in the quantum theory has changed the attitude of physicists, such that science is now viewed in terms of "conceptual schemes", which arise from experiment and are fruitful of more experiments. The wave-particle duality is one such conceptual scheme, and it justifies his "skeptical" approach, because this conceptual scheme does not describe what light "really" is. Instead modern physics describes the properties of light and formulates them on the simplest possible principles. The history of science is a succession of conceptual schemes. Conant references the view of the Harvard Pragmatist philosopher, William James, who maintained that man's intellectual life consists almost wholly in the substitution of a conceptual order for the perceptual order from which experience originally comes. Different universes of thought arise as concepts and percepts interpenetrate and melt together, impregnate and fertilize each other. As a result the series of conceptual schemes in the history of science is one in which the conceptual schemes are of increasing adequacy to the perceptions in experimentation. Conant initially believed that natural sciences have an accumulative character that reveals progress, but following Kuhn's *Structure of Scientific Revolutions* (1962) Conant modified his view of the accumulative nature of science. He continues to find accumulative progress in the empirical-inductive generalizations in science and also in the practical arts, but he excludes accumulative progress from the theoretical-deductive method which admits to scientific revolutions.

Conant identifies the static view with the logical perspective, while he admits the psychological and the sociological perspectives in his dynamic view. The sociological perspective reveals that science is a living organization, which can exist because of close communication that enables new ideas to spread rapidly, and that enables discoveries to breed more discoveries. Scientists pool their information, and by so doing they start a process of cross-fertilization in the realm of ideas. As a social phenomenon, science is a recent invention starting with the scientific societies of the seventeenth and eighteenth centuries, and then evolving in the universities in the nineteenth century. Communication was initially through letters, then later through books, and now through journals. He maintains that historically one of the more important psychological aspects of the development of science is prejudice, a matter toward which he admits he himself has an ambivalent attitude. On the one hand the traditions of modern science, the instruments, the high degree of specialization, the crowd of witnesses that surround the scientist, all these things exert pressures that make impartiality in matters of science almost automatic. If the scientist deviates from the rigorous role of impartial experiment or observation, he does so at his peril. On the other hand Conant says that to put the scientist on a pedestal because he is an impartial inquirer is to misunderstand the historical situation. This misunderstanding results both from the dogmatic character of textbooks and from the view of Positivist philosophers such as Karl Pearson. Conant emphasizes the stumbling way in which even the ablest of the scientists of every generation have had to fight through thickets of erroneous observation, misleading generalization, inadequate formulations and unconscious prejudice. He notes that these problems are rarely appreciated by those who obtain their scientific knowledge from textbooks and by those who expound on "the" scientific method.

Conant exhibits his thesis in his description of the chemical revolution, in which the phlogiston theory of combustion was replaced by the theory of oxygen. He notes that for one hundred fifty years an anomaly to the phlogiston theory, the fact the a calx weighs more than its metal, was known to exist, but that the theory itself was never called into question until a better one was developed to take its place, namely Lavoisier's new conceptual scheme. In the meanwhile the phlogiston theory was an obstruction to the development of the new conceptual scheme, as scientists attempted to reconcile the anomaly to the phlogiston theory. Conant also notes that even after the new conceptual scheme was advanced to overthrow the phlogiston scheme, there continued to be debate, and that the proponents of the new conceptual scheme were no more shaken by a few alleged facts contrary to the new scheme, than were the advocates of the old scheme by facts anomalous to the earlier scheme. Lavoisier pursued his conceptual scheme in spite of embarrassing experimental findings, which only after his death were found to be erroneous findings. Conant's thesis in this examination of the chemical revolution is that both sides in the controversy had put aside experimental evidence that did not fit into their respective conceptual schemes. And in his view what is most significant is the frequent fact that subsequent history may show that such arbitrary dismissal of "the truth" is quite justified. He concludes that to suppose that a scientific theory stands or falls on the issue of one experiment is to misunderstand science entirely. Conant characterizes the first fifty years of the nineteenth century that culminated in the chemists' atomic theory of matter, as a period of "the conflict of prejudices". He notes that one who is not familiar with this episode in the history of science will be amazed to discover that all the relevant ideas and all the basic data for the atomic theory were at hand almost from the outset of the nineteenth century; an analysis of the arguments, pro and con, shows that certain preconceived ideas then current among scientists blocked its development. Still, Conant rejects the view that the scientific way of thinking requires the habit of facing reality quite unprejudiced by any earlier conceptions. In his *Science and Common Sense* he admits that prejudices are emotional and nonlogical reactions, yet he also maintains that every scientist must carry with him the scientific prejudices of his day the many vague, half-formulated assumptions which to him seem "common sense". Apparently as a result of his acceptance of prejudice as an inevitable fact in the dynamics of science, Conant unabashedly declares that his dynamic view of science is itself his "prejudice", and adds that he makes "no attempt to conceal it".

It may be said that one of the differences between Kuhn and Conant is that the latter regards prejudice as merely an inescapable fact in the history of science, while the former regards it as having a positive function that is inherent in the dynamics of science. In Kuhn's doctrine of "normal science", what Conant calls "prejudice", Kuhn calls by the less pejorative phrase "paradigm consensus". But unlike Conant, Kuhn does not view prejudice as merely an individual phenomenon with one scientist taking one prejudice and another taking an opposing prejudice. In Kuhn's view paradigm consensus is a sociological-semantical phenomenon, and this semantical perspective did not come from Conant. In spite of Conant's dynamic view including reference to William James about percepts being impregnated with concepts, Conant's view of the semantics of language is not dynamic; it is static. His static view of the semantics led him to his "skeptical approach", just as it led Bohr to his instrumental view of the formalisms of quantum physics, and for the same reason: without a theory of semantical change, neither Bohr nor Conant could admit a realistic interpretation to the wave-particle duality of the modern quantum theory. Kuhn had the advantage of another influence, that of Paul Feyerabend, who supplied him with an appreciation of semantical change. This semantical theory, which Kuhn integrated into a theory of scientific revolutions, is Feyerabend's doctrine of semantic incommensurability.

Nagel and Feyerabend on Meaning Variance

Semantic incommensurability is a special case of the more general semantic phenomenon that Feyerabend calls "meaning variance", the phrase that he uses to refer to semantic change. Accordingly it is instructive to consider firstly Feyerabend's thesis of meaning variance. This thesis is argued in his "Explanation, Reduction, and Empiricism" in *Minnesota Studies in the Philosophy of Science* (1962), where he opposes it to the contrary thesis that he calls meaning invariance, which he finds characteristic of the Logical Positivist philosophy and specifically of the views of Carl Hempel and Ernest Nagel. Together with Paul Oppenheim, Carl Hempel set forth the "nomological-deductive" thesis of scientific explanation in "Logic of Explanation" in *Philosophy of Science* (April, 1948), and a later statement by Hempel is given in chapters five and six of his *Philosophy of Natural Science* (1966). Nagel set forth his thesis of reduction in chapter eleven of his *Structure of Science* (1961). Hempel and Oppenheim emphasize the logical-deductive nature of scientific explanation, while Nagel addresses more explicitly the semantical aspect of theoretical explanation and reduction. Since the semantical aspect is at the center of Feyerabend's thesis of meaning variance, a brief consideration of Nagel's discussion of the reduction of theories is in order, to understand what Feyerabend is opposing. As it happens, Nagel might also be said to have a thesis of meaning variance, but his view of semantical change is not the same as Feyerabend's.

Initially the Logical Positivist interest in reduction was part of the Unity of Science program. When it became evident that this program is unmanageably ambitious, the reductionist program was limited to the characteristically Logical Positivist problem of relating theoretical terms in theories to an observation language. This type of reduction is accomplished by what Carnap called "reduction sentences" and by what Hempel called "bridge principles". Nagel is in the Logical Positivist tradition, but his treatment of logical reduction is somewhat less programmatic and more closely related to episodic developments in the history of science. He is more interested in those cases in the history of science, in which a relatively autonomous theory is absorbed by or "reduced to" some other more inclusive theory, a type of development that he believes is a recurrent feature of the history of modern science. In this type of episode the set of theoretical statements or experimental laws, as the case may be, that is reduced to another theory is called the "secondary science", while the theory to which the reduction is effected is called the "primary science". Reductionism is a type of explanation in science, and Nagel explicitly defines it as the explanation of a theory or of a set of experimental laws established in one area of inquiry by a theory formulated in some other domain. He is principally interested in those types of reduction in which concepts are required for describing phenomena in one area that were not formerly employed in the other area, even when the two areas were described with the same vocabulary. He refers to this type of reduction as a "heterogeneous" reduction, because it describes a qualitative dissimilarity between the phenomena in the domains of the two theories involved in the reduction. On the other hand a reduction without different vocabulary and describing a qualitative similarity is what he calls a "homogeneous" reduction. Nagel finds only the heterogeneous type to be problematic.

Nagel employs a theory of meaning in which a descriptive term may have as many meanings as there are explications. He illustrates his thesis in his examination of the reduction of thermodynamics to statistical mechanics and of the semantics of the term "temperature", as that term's meaning is affected by the successful reduction. Even before the reduction is made, there is much to be said about the semantics of the terms involved, because a term such as "temperature" has several meanings resulting from overtly performed instrumental operations. Nagel exemplifies the multiple meanings of the term "temperature" by noting that a person who understands temperature in terms of an ordinary mercury thermometer, would have difficulty understanding what is meant by a temperature of fifteen thousand degrees, if he also knew that no mercury thermometer could be used to measure such an extreme temperature. But if the person had studied physics, he would discover that the term "temperature" in physics has a broader application from a more embracing set of rules of usage describing other measurement procedures. Nagel references Bridgman's idea of "operationalist definitions", and states that such rules of usage are explications aimed at specifying the meanings of descriptive expressions such as "temperature" in terms of other observable ones, which in any given context must be traced to certain descriptive expressions that are selected to be observable primitive expressions. It is noteworthy that in Nagel's theory of semantical specification as in Bridgman's, each such specification describing an alternative measurement procedure constitutes a cognitively distinct meaning of the observation term. Yet these multiple meanings are not unrelated equivocations, since the diverse measurement procedures will yield the same measurement values where more than one is deemed applicable. Thus the term is *empirically* unambiguous while at the same time it is *cognitively* equivocal. Nagel extends Bridgman's semantical thesis for observation terms to theoretical terms. He gives as examples of theoretical explications of "temperature", the explication in the science of heat with the help of statements describing the Cournot cycle of heat transformation, and therefore in terms of such theoretical primitives as "perfect nonconductors", "infinite heat reservoirs" and "infinitely slow volume expansions". Nagel emphasizes that while the term "temperature" is explicated in the science of heat in terms of both theoretical and observational primitives, it is not the case that the term understood in the sense of the first explication is cognitively synonymous with "temperature" construed in the sense of the second. This is one way in which the thesis of multiple meanings serves the Logical Positivist well: the Positivist does not want the meanings of observation terms to be contaminated with the meanings of theoretical terms, and so it is important to him that the set of meanings supplied by the various theoretical explications and the set supplied by the observational explications be separate and distinct. The thesis that multiple explications do not result in cognitive synonymy but rather in empirically unambigious cognitive equivocation, thus enables him to say that even when a revolutionary new theory is developed, it will produce a new set of theoretical explications but will not revise the set of observational explications. In this way there is room for meaning variance in the theoretical meanings, and yet there is also room for meaning invariance in the observational meanings. It is interesting that Nagel's approach is different from Carnap's, because the latter distinguishes theoretical terms as having "incomplete" semantics, such that theoretical terms could change their meanings by becoming more complete even in a heterogeneous reduction. Carnap did not employ any thesis of empirically unambigious equivocation like Nagel; Nagel is more faithful to Bridgman.

Nagel next considers the formal conditions for a reduction. In the reduction of thermodynamics to statistical mechanics the Boyle-Charles' law is made a logical consequence of the principles of mechanics, when these principles are supplemented by a hypothesis about the molecular constitution of a gas, a statistical assumption about the motions of molecules, and a postulate concerning the experimental notion of temperature with the mean kinetic energy of the molecules. Nagel sets forth two formal conditions for the reduction: the condition of connectability and the condition of derivability. The first condition requires that assumptions be introduced which postulate suitable relations between what is signified by a descriptive term (e.g. "temperature") in the secondary science, and traits represented by theoretical terms already present in the primary science (e.g. the kinetic energy of molecules). The second condition, the condition of derivability, requires that together with the above mentioned assumptions all the laws of the secondary science including those containing the connected terms, must be logically derivable from the theoretical premises and their associated "coordinating definitions" in the primary science. The coordinating definitions or "correspondence rules", as Nagel also calls them, are the same as what Carnap called "reduction sentences", and what Hempel calls "bridge principles". By whatever name, these are the sentences that connect theoretical terms occurring in a theory with the observation terms in the empirical statements the theory explains deductively. Both the primary and secondary theories involved in a reduction are presumed to have whatever coordinating definitions that they need before the reduction is effected. When both of these conditions are satisfied, the reduction can be effected, and the experimental and theoretical laws of the secondary science are made logical consequences of the theoretical assumptions including the coordinating definitions of the primary science.

After his discussion of the formal conditions, Nagel extends his semantical thesis of multiple meanings to reduction. After the reduction of thermodynamics to statistical mechanics is accomplished, the term "temperature" can be explicated in terms of the mean kinetic energy of molecules, and it thereby acquires still another meaning. This is the outcome of satisfying the condition of connectability. He explicitly denies that the connection made by the assumptions employed in the reduction are "logical" connections between established meanings of expressions, because the assumptions would then assert that there is either a synonymy or a one-way entailment in the relation to a theoretical expression in the primary science. Nagel maintains that the connecting assumptions are initially conventions that merely assign the additional meaning, and which later become empirical statements because further development of the theory makes it possible to calculate the temperature of the gas in some indirect fashion from experimental data other than the temperature value obtained by actually measuring the temperature of the gas. He rejects as "unwitting double talk" the objection to his thesis, that the reduction occurs because of a redefinition of the term "temperature". He maintains that the term "temperature" cannot be cognitively synonymous with the phrase "mean kinetic energy of molecules". He says that the terms in each of the two sciences have meanings unambiguously fixed by codified rules of usage or by established procedures appropriate to each discipline, and that these established meanings are not lost or changed as a result of the reduction.

Feyerabend is critical of the views of Hempel and Nagel, and he takes a fundamentally different view, fundamental because Feyerabend advances his "pragmatic theory of observation" in opposition to the Positivist naturalistic view of observation. This point of departure places Feyerabend in the same company as Einstein, Popper and Hanson, all of whom reject the Positivist separation of theory and observation. On the Positivist view observation statements are the products of natural processes, that supply the observation language with its meanings. Feyerabend on the other hand affirms an artifactual theory of meaning, when in "Explanation, Reduction, and Empiricism" he bases his pragmatic theory of observation on the distinction between nature and convention. In his view this distinction implies, contrary to the Positivist view, that the observational status of a statement must be separated from its meaning. Thus Feyerabend says that an observation sentence is distinguished from other sentences of a theory not by its meaning content but by the cause of its production, by which he means that its production conforms to certain behavioral patterns. His pragmatic theory of observation gives Feyerabend an alternative to any reductionist thesis such as Nagel's. He maintains that when a transition is made from one theory to another theory of wider scope, which Nagel calls the secondary and primary sciences respectively, what actually happens is something much more radical than the incorporation of an unchanged theory into the context of the primary theory, unchanged, that is, with respect to the meanings of the secondary theory's main descriptive terms as well as to the meanings of the terms of its observation language. What happens instead of a reduction is the complete replacement of the ontology and perhaps the formalism of the secondary science by the ontology and the formalism of the primary science, and the corresponding change in the meanings of the descriptive elements of the formalism of the secondary theory, providing that these elements of the formalism of the secondary theory are still used.

Feyerabend states that contrary to the Positivist reductionist thesis, the replacement affects not only the theoretical terms of the secondary science, but also at least some of the observational terms occurring in its test statements. He opposes the Positivist thesis that a comprehensive theory merely orders facts, and maintains that a general theory has a deeper influence on thinking. This deeper influence is the semantical influence of the context of the primary theory on the empirical statements and vocabulary of the secondary theory. The consequence of the distinction between nature and convention, that separates observability and meaning, is what Feyerabend calls the "contextual theory of meaning". This theory of meaning description implies a wholistic approach in his view, because he says that the contextual determination of meaning is not confined to a single scientific theory or even to a single language. Thus the unit of language involved in the test of a specific theory is not just the theory taken together with its own consequences, but rather is a whole class of mutually incompatible and factually adequate theories. This class is the context by which meanings are to be made clear. Feyerabend's rejection of the Positivist naturalistic causal theory of meaning and his proposal of his conventionalist contextual theory of meaning, lead him to attack two basic assumptions that he finds in Nagel's theory of reduction and explanation. These two assumptions are (1) deducibility and (2) meaning invariance. Meaning variance is one of the reasons that deducibility is impossible, but in addition to meaning variance, there are purely quantitative reasons why deducibility is impossible. In his treatment Nagel gave the reduction of Galileo's physics to Newton's physics as an example of a homogeneous reduction, one in which there is no meaning change resulting from the reduction. But Feyerabend says that there is a quantitative deviation between the Galilean and the Newtonian physics, an inconsistency due to the fact that one and the same set of observational data is compatible with very different and mutually inconsistent theories. This inconsistency, that makes deduction logically impossible, has two reasons. Firstly universal theories always make claims about phenomena that are beyond those that have actually been observed or that might be available at any particular time; it is this characteristic that makes them universal. Secondly the truth of any observation statement, such as a statement reporting a measurement reading, can be asserted only within a certain margin of error. The first reason allows for theories that differ in domains where experimental results are not yet available. The second reason allows for such differences even in those domains where observations have been made, provided that the differences are restricted to the margin of error in the observations.

The principal reason that deducibility is impossible in explanation and reduction of general theories is the inconsistency produced by the meaning variance, the semantical change resulting from the change of context. To illustrate this Feyerabend considers the purported reduction of Aristotle's theory of motion to Newton's theory. In this case Newton's theory offers the same quantitative measurements as Aristotle's, and there is no quantitative inconsistency. The reduction is achieved in the apparently simple manner of equating the concept of impetus in the Aristotelian theory with the concept of momentum in Newton's theory. On Newton's approach the meanings of the descriptive terms in the impetus theory are fixed by the procedures and assumptions of the theory. But Feyerabend maintains that the concept of impetus as fixed by the usage established in the Aristotelian theory of motion cannot be defined in a reasonable way in Newton's theory, because the usage involves laws that are inconsistent with Newtonian physics. Thus contrary to Nagel, the concept of impetus is not explicable in terms of the theoretical primitives of the primary science in a reduction, even if equating impetus with momentum is proposed as a physical hypothesis instead of an analytical one. Such a physical hypothesis merely says that wherever momentum is present, then impetus will also be present, and the measurements will be the same in both cases.

Feyerabend also finds meaning variance in the purported reduction of phenomenological thermodynamics to the kinematic theory of gases, the case considered in detail by Nagel. He describes Nagel's view as the claim that the terms of the statements which have been derived from the kinetic theory with the help of correlating hypotheses will have the same meanings that they originally had within the phenomenological theory, and he states that Nagel repeatedly emphasizes that these meanings are each fixed by its own procedures, that is by the procedures of the phenomenological theory, whether or not the theory has been or will be reduced to some other discipline. Thus the term "temperature" as fixed by the established usages of phenomenological thermodynamics, as Nagel says, is such that its application to concrete situations entails the strict nonstatistical law. Feyerabend states that the kinematic theory does not offer such a concept. There does not exist any dynamical concept in the phenomenological law, while on the statistical account fluctuations between two levels of temperature is allowed. He therefore says that the thermodynamic concept and the kinetic statistical concept of temperature are "incommensurable", and that replacement rather than incorporation or derivation characterizes the transition from a less general theory to a more general one. Feyerabend notes that both he and Nagel say that incorporation into the context of the statistical theory changes the meanings of the main descriptive terms of the phenomenological theory, but he adds that this is "double talk" by Nagel, because the law that has been reduced is no longer the same law. He says Nagel's view of change of meanings is somehow supposed to leave untouched the meanings of the main descriptive terms of the discipline to be reduced. There is a sense in which Nagel's view involves double talk. This double talk is not an inconsistency in Nagel's thesis, but rather is a logical consequence of his semantical thesis, the view that the terms in science are equivocal and have multiple meanings. But Feyerabend prefers to reject any such equivocation that would permit semantical continuity through the reduction, and instead prefers to retain the univocation in the terms at any point in time, and to affirm a change from one univocal meaning to another, even at the expense of a semantical discontinuity in the empirical explications. Consideration of the nature of this semantical discontinuity introduces the roles of inconsistency and especially "incommensurability".

The Sapir-Whorf Hypothesis

In his "Explanation, Reduction, and Empiricism" Feyerabend describes two ways in which theories can be related to each other such that meaning variance may occur. Those two ways are inconsistency and incommensurability. Given two historically successive theories denoted **T** and **T'**, the theory **T** will differ from the theory **T'**, either (1) if **T** is inconsistent with **T'** in the domain of deduced empirical laws where **T** and **T'** overlap, or (2) if the set of empirical laws that follow from theory **T'** will be incommensurable with those following from **T**. When the relation is that of inconsistency, the two theories are commensurable, which is to say semantically comparable. Feyerabend references Popper saying that the new and superior theory **T'** implies laws that are different from and superior to those implied by theory **T**. In this case the laws deduced from theory **T'** correct and replace those deduced from **T**, just as occurred in the case of Newton's theory correcting and replacing Kepler's and Galileo's laws. When theories **T** and **T'** are incommensurable, however, they do not have any comparable observational consequences. It is not even possible to say that the empirical laws that are deduced from one are superior or inferior to those that are deduced from the other. This semantic incommensurability is admitted by Feyerabend's pragmatic theory of observation. On this theory of meaning nature does not determine the content of thought and therefore does not guarantee consistency or even comparability of meaning. Instead the content of thought is a human artifact not unlike any work of art, and there may result differences between people's thinking that are so fundamentally different, that they may admit no basis for comparison or common denominator; they may be incommensurable.

In his "On the 'Meaning' of Scientific Terms" (1965), reprinted in

Realism, Rationalism, and Scientific Method, Feyerabend describes a theory and its predecessor to be incommensurable, if prior to the time the theory is proposed, there exists no more general concept having an extension that includes the extensions of the concepts of the two theories. He considers Einstein's relativity theory to be incommensurable with Newtonian celestial mechanics, because prior to Einstein the Reimann metric did not include time, and he says that this change in the transition to Einstein's theory was drastic enough to exclude common elements between the two theories. He also considers quantum theory to be incommensurable with classical physics, because prior to its advent the conservation laws were not applied to virtual states. Later Feyerabend further elaborates on his concept of semantic incommensurability by drawing upon the Sapir-Whorf hypothesis and specifically upon Whorf's thesis of linguistic relativity. Both Kuhn and Feyerabend briefly reference Whorf in their works published in the 1960's, and Feyerabend's elaboration of his thesis of semantic incommensurability is to be found in his *Against Method* published in 1975. But before turning to this work, a summary of the Sapir-Whorf hypothesis is in order.

Benjamin Lee Whorf (1897-1941) was a cultural anthropologist and linguist by avocation, who received a BA degree in chemical engineering in 1918, and spent his career with an insurance company eventually becoming Assistant Secretary, an officer of the corporation. He became interested in linguistics in 1924 and was almost completely self-educated in linguistics except for some nondegree courses that he took from Edward Sapir, a cultural anthropologist and linguist at Yale University. Sapir encouraged Whorf to study the language of the Hopi American Indians, and he financed Whorf's field studies. These studies occasioned Whorf's formulation of the Sapir-Whorf hypothesis, the thesis of linguistic relativity for which Whorf is now best known. This thesis is still controversial, and is in conflict with such absolutist views as Chomsky's thesis of innate linguistic universals. Whorf wrote many articles, but few of those that he submitted to academic journals were accepted and published in his life time in spite of the intrinsic merit of the papers. A posthumous anthology of his writings titled *Language, Thought and Reality* was published in 1956 (ed. Carroll, MIT Press). It may be said that there is an earlier and a later, expression of Whorf's thesis. The earlier statement made in the 1930's is his thesis of "cryptotypes" or "covert categories", while the more mature statement is the explicit statement of linguistic relativity made in "Science and Linguistics" in 1940. Whorf exemplifies the idea of the cryptotype with grammatical categories for gender. Gender may be manifested either by overt or by covert indicators. They are overtly manifested by morphemes, that are formal markers that occur in such languages as Latin or German. They are covertly manifested in English by what Whorf calls their "reactance", their association with definite linguistic configurations such as lexical selection, word order that is also class order, or in general by some kind of patterning. More precisely: overt categories are those having a formal mark that is present in every sentence containing a member of the category, while covert categories are all others, even those that are marked nonphonetically but only in certain types of sentences. And he defines his idea of reactance as a special type of "rapport", an idea that is roughly equivalent to the general idea of structure in language. Rapport is the linkage between the elements of language, that enables these elements to have semantical effect, and it is governed by what Whorf calls "an invisible central exchange". This central exchange of linkage bonds is what gives rise to the covert categories, or "cryptotypes" as he also calls them, that are submerged, subtle and elusive meanings corresponding to no actual word, yet which have a functionally important role in the grammar of a language. Words of a covert category are not distinguished by a formal mark but rather by a semantical class, by an idea that gives the grammatical class its unity, which is manifested by common reactance. Semantically the covert category is what Whorf calls a deep persuasion of a principle behind some phenomenon, like the ideas of inanimation, substance, force, or causation.

The thesis that language structure controls thought, which Whorf sets forth in his theory of covert categories, is central to his theory of linguistic relativity. He locates his development of linguistic relativity in the history of cultural anthropology in the lineage of Franz Boas and Edward Sapir. Boas had shown that a language can be analyzed *sui generis*, that is without forcing upon the language the categories of the "classical" tradition. Then in 1921 in his book *Language* Sapir inaugurated the linguistic approach to thinking, demonstrating the importance of linguistics to cultural anthropology. According to Whorf comparative linguistics now reveals that the background linguistic system, the grammar of each language, is not merely a sentence producing instrument for voicing ideas but rather is the shaper of ideas. And this is the essence of his thesis of linguistic relativity. The human mind cuts up nature, organizes it into concepts, and ascribes significance, because men are parties to an agreement that holds throughout the speech community, and that is codified in their language. Not all observers are led by the same physical evidence to the same picture of the universe, unless their linguistic backgrounds are similar or in some way can be "calibrated". For Whorf's term "calibrated" one is tempted to substitute Feyerabend's term "commensurated", except that Feyerabend does not believe that semantically incommensurable theories can ever be commensurated.

Whorf further elaborates on his linguistic relativity thesis in his "Language, Mind and Reality" (1942). In the context of a discussion of the Mantric Art of India he distinguishes two great levels: the realm or level of meaning or lexication, and the higher and controlling level of patterning of sentence structure that guides words which occur at the lexical level and that is more important than words. Lexication, the partitioning of the whole manifold of experience and the assigning of the parts to words, makes the parts stand out in artificial and semifictitious isolation. This process of lexication is controlled by the patterning function of sentence structure and thus by the organizing at a higher level, where the combinatory scheme occurs. These patterns are not individual sentences, but rather are schemes of sentences and designs of sentence structure. The patterns are manifested by using the mathematical or grammatical formulas into which words, values or quantities may be substituted. Each language does this partitioning and patterning in its own way, and each has its own characteristic form principles, that make consciousness a mere puppet, whose linguistic maneuverings are held in unsensed and unbreakable bonds of pattern. These passages suggest similarities between Whorf's view and Feyerabend's contextual theory of meaning, save for the fact that Feyerabend does not restrict the term "meaning" to a lexical function.

As it happens, Whorf explicitly states in several of his later articles that his thesis of linguistic relativity applies to empirical science. He views it as applicable not only because science including mathematics consists of language, but also because an awareness of the effect of language on the foundations of thought will facilitate what he describes as science's next great march into the unknown. He expresses regret that philosophers and mathematicians do not even have apprenticeship training in linguistics, and he states the opinion that further development in logic will proceed with the investigation of the structures of diverse languages. Like later philosophers, Whorf views the various specialized sciences as different languages, because he finds that there exist communication problems among the researchers in the different specialties, just as there are such problems among the speakers of different natural languages. He maintains that these communication problems do not simply breed confusion about details that the expert translator could resolve. The problems are much more perplexing, since the language of science is a "sublanguage", which incorporates certain points of view and certain patterned resistances to widely divergent points of view. These resistances not only isolate artificially the particular sciences from one another, but they also operate to restrain the scientific spirit from taking the next great step in its development, a step which entails viewpoints unprecedented in science and involving a complete severance from tradition. This great episode will unify the diverse sciences, and will be based on the discovery of the aspect of language consisting of patterned relations. The approach to reality through mathematics as used in science today is merely one special case of this. Whorf proposed that there is a premonition in language of an unknown and vaster world, which is quite different from the world as it is currently understood through the structure of the Indo-European languages which insist on substantives. The apparent necessity of substances is purely a result of the "Ayrian grammar". The logic of Aristotle is provincial, because it is based on the ideology of substantives, while modern physics with its emphasis on fields casts doubt on this ideology. Whorf prognosticates the emergence of a new type of language for science which is even more universal than that presently used, because it will be a transcendental logic of relations of pure patternment.

Feyerabend on Semantic Incommensurability

Feyerabend's later and more comprehensive statement of his incommensurability thesis is set forth in chapter seventeen and in a brief appendix in his *Against Method.* The centrality of the incommensurability thesis to his philosophy is indicated by the fact that this chapter and its immediately following appendix pertaining to the incommensurability thesis, take up approximately seventy pages of this three-hundred page book. Later in his *Science and a Free Society* (1978) he emphasizes that his intent in the discussion of incommensurability is to understand the changes that take place when a new world view enters the scene, and that this requires examining it from the perspective of the concerned parties, and not as it appears or is projected on to a later ideology years afterwards. The significance of incommensurability is that the concerned parties experiencing it cannot subject the new idea to what they regard as their own rationality, and must allow "reason" which is accessible to them to be violated. He views this analysis "from the inside" to be of the utmost practical importance, because it is what occurs in a scientific revolution, and every researcher should be prepared for such events, which would otherwise catch the researcher by surprise.

In the opening sentence of chapter seventeen of *Against Method* Feyerabend says that he has much sympathy with the clearly and elegantly formulated view of Whorf, and he gives a brief summary of Whorf's principle of linguistic relativity. In the appendix following the chapter he notes that Whorf's principle admits to two alternative interpretations. On one interpretation it means that observers using widely different languages will posit different facts in the same physical circumstances in the same physical world. On the other interpretation it means merely that observers using widely different languages will arrange similar facts in different ways. The former interpretation is the one that Feyerabend says he uses for his own incommensurability thesis, and he justifies this interpretation on the basis of the great influence that Whorf ascribes to grammatical categories and especially to the hidden "rapport system" of language. The covert classifications that result from this hidden rapport system or "central exchange", create patterned resistances to widely divergent points of view. Feyerabend says that if these resistances oppose not just the truth of the resisted alternative views, but the presumption that an alternative has been presented, then we have in instance of incommensurability. This is the closest that Feyerabend comes to a definition of incommensurability, because as he says, it is hardly ever possible to give explicit definition of it, since it depends on covert classifications and involves major conceptual changes.

The body of Feyerabend's chapter discussing incommensurability is organized into three theses, which are summarized at the end. His first thesis is that there are in fact frameworks of thought which are incommensurable, and he emphasizes that this is an anthropological thesis. He maintains Whorf's principle of linguistic relativity applies to scientific theories such as Aristotle's theory of motion, the theory of relativity, the quantum theory and classical and modern cosmology, because they are sufficiently "deep" and have developed in sufficiently complex ways that they may be viewed as incommensurable and as widely divergent natural languages. And he therefore also maintains that philosophy of science is an anthropology of science, and not a logic of science as both the Positivists and Popper had maintained. In the examination of the incommensurable theories, where facts asserted by each cannot be compared side by side even in memory, it is necessary to take the approach of the field linguist and learn the new theory from scratch. The irrationality of the transition to the new theory is overcome by the determined production of nonsense until the material produced is rich enough to permit recognition of new universal principles. The initial madness turns to sanity provided that it is sufficiently rich and sufficiently regular to function as the basis of a new world view. There is no translation involved; instead there is a learning process. This is how Feyerabend sees the transition from classical mechanics to quantum mechanics, and from Newtonian mechanics to relativity theory. His second thesis is that incommensurability has an analogue in the psychology of perception, and that the development of perception and thought in the individual passes through stages that are mutually incommensurable. This is contrary to the Positivist philosophy of observation, and Feyerabend references Piaget's work with perceptual development in children.

His third thesis is that scientific theories may be incommensurable even when they apparently treat of the same subject matter and the same problem. On a realistic interpretation, as opposed to an instrumentalist interpretation, incommensurable theories do not treat the same subject matter. A new theory such as relativity theory in physics does not treat the same problem that is treated by its predecessor, Newtonian mechanics, when the former replaced the latter. The new theory does not "solve" problems confronting the old theory, but rather it "dissolves" them and removes them from the domain of legitimate inquiry, because the new incommensurable theory has an ontology that replaces that of the older theory. When the faulty ontology of the older theory is comprehensive, as in the Newtonian physics, then every description inside the domain must be changed; it must be replaced by a different statement in the new theory or it may be replaced by no statement at all. The new ontologies of relativity theory and quantum theory do not just deny the existence of classical states of affairs, they do not even permit us to formulate statements expressing such states of affairs. Crucial experiments are therefore impossible, because one theory cannot establish or refute another theory incommensurable with the former. Each incommensurable theory has its own facts, and it can be refuted only by reference to its own kind of experience, that is to say, by discovering its internal contradictions. Their contents cannot be compared.

Aside from internal inconsistency, the only basis for preference for one of several mutually incommensurable theories is some subjective basis, such as the scientist's metaphysical prejudices, his religious convictions, or his personal judgments of taste. With these semantical considerations, return now to the discussion of Kuhn.

Kuhn on the Copernican Revolution

Kuhn published his influential and popular *Structure of Scientific Revolutions* in 1962. This highly interpretative philosophical history of science was preceded by his *Copernican Revolution: Planetary Astronomy in the Development of Western Thought* in 1957. The earlier work is less philosophically interpretative; it was written before Kuhn met Feyerabend, and it is interesting because it reveals the distinctive aspects of the influence of Conant. In the years between the two books Kuhn was a faculty member at the University of California at Berkeley, where he made the acquaintance of Feyerabend, who was also a faculty member. The latter book reveals the influence of Feyerabend as well as the continuing influence of Conant. The *Copernican Revolution* contains ideas that reappear in the *Structure of Scientific Revolutions*. One idea is that a central feature of scientific revolutions is that old theories are replaced by new and incompatible ones. In the latter book this thesis is elaborated in semantical terms, and it is the basis for his describing scientific revolutions as "noncumulative" episodes in the history of science. However, in the 1957 book his semantical view is that scientific observations are indifferent to the conceptual schemes that constitute theories, that observations must be distinguished from interpretations of the data that go beyond the data, such that two astronomers can agree perfectly about the results of observation and yet disagree emphatically about issues such as the reality of the apparent motion of the stars. He states that observations in themselves have no direct cosmological consequences. However, later in 1962 he maintains instead that observations depend on the particular theory held by the scientist.

A second idea reappearing in the 1962 book is his thesis that the "logic" of science does not completely control the development of science. The logic that he has in mind is a stereotype of Popper's view, that the occurrence of just one single observation which is incompatible with a theory, dictates that the scientist reject the theory as wrong and abandon it for some other one to replace the wrong one. Kuhn believes that the incompatibility between theory and observation is the ultimate source for the occurrence of scientific revolutions, but he also maintains that historically the process is never so simple. Scientists do not surrender their beliefs so easily. What was to Copernicus a stretching and patching to solve the problem of the planets for the two-sphere theory, was to his predecessors a natural process of adaptation and extension. Kuhn therefore finds in the history of science what he calls "the problem of scientific belief". That problem is: why do scientists hold to theories despite discrepancies, and then having held to them in these circumstances, why do they later give them up? The significance that Kuhn gives to this phenomenon reveals the influence of Conant. The problem of "scientific belief" is the same as what Conant meant by the phenomenon of "prejudice". Typically historians and philosophers of science did not consider this phenomenon as having any contributing role in the development of science, because it is contrary to the received concept of the programmatic aim of science. And in 1957 Kuhn was as ambivalent in his attitude toward the problem of scientific belief as Conant was toward the phenomenon of prejudice in science.

In the 1957 book Kuhn locates part of the reason for the problem of scientific belief in the scientist's education, a reason that he also calls "the bandwagon effect". This reason is carried forward into the 1962 book, where it has a very important place. In the 1957 book, however, he considers it to be of secondary importance. The other and more important part of the reason in the 1957 book is the interdependence of other areas of the culture with the scientific specialty. The astronomer in the time of Copernicus could not upset the two-sphere universe without overturning physics and religion as well. Fundamental concepts in the pre-Copernican astronomy had become strands for a much larger fabric of thought, and the nonastronomical strands in turn bound the thinking of the astronomers. The Copernican revolution occurred because Copernicus was a dedicated specialist, who valued mathematical and celestial detail more than the values reinforced by the nonastronomical views that were dependent on the prevailing two-sphere theory. This purely technical focus of Copernicus enabled him to ignore the nonastronomical consequences of his innovation, consequences which would lead his contemporaries of less restricted vision to reject his innovation as absurd. In his 1962 book, however, Kuhn does not make the consequences to the nonspecialist an aspect of his general theory of scientific revolutions. Instead he maintains that scientists persist in their belief in theories with observational discrepancies for reasons that are entirely internal to the specialty.

Kuhn on the Structure of Scientific Revolutions

The *Structure of Scientific Revolutions* is a small monograph of less than one-hundred seventy-five pages written in a colloquial style without mathematical equations or the symbolic logic expressions which philosophers of science at the time had made seem obligatory. It was also a very timely presentation of the ascending Pragmatist philosophy of science illustrated with a plethora of apparently illustrative examples from the history of science, which seemed to document the book's thesis. Although many tenants of his 1962 book were previously published by Kuhn in his "The Essential Tension" in 1959, later reprinted in a book of the same name in 1977, the 1962 book was probably the most popular book pertaining to philosophy and history of science published in the 1960's, if not also for many years afterwards. It is reported to have sold more than half a million copies in the first twenty years following its publication, and it was widely read outside the relatively small circle of professional philosophers and historians of science.

In "Reflections on My Critics" in *Criticism and the Growth of Knowledge* (ed. Lakatos and Musgrave, 1970) Kuhn offers some personal insights. He states that in his work as an historian of science, he discovered that much scientific behavior including that of the greatest scientists, persistently violated accepted methodological canons, and that he wondered why these apparent failures to conform to the canons did not at all seem to inhibit the success of the scientific enterprise. The accepted methodological canons that Kuhn has in mind are not only those of the Positivists but also Popper's fallibilist thesis. He states that his altered view of the nature of science transforms what had previously seemed aberrant behavior into an essential part of an explanation for science's success, and that his criterion for emphasizing any particular aspect of scientific behavior is not simply that it occurs, or merely that it occurs frequently, but rather that it fits a theory of scientific knowledge, a theory which he says may have normative as well as descriptive value. The seemingly aberrant behavior is what he had previously called the problem of scientific belief, the practice of ignoring anomalies.

The thesis of the book integrates the views of Conant and Feyerabend into a coherent description of the historical development of the modern natural sciences. Kuhn portrays the developmental procession as an alternation between two phases, which he calls "normal science" and "revolutionary science", with each phase containing the seeds for the emergence of the other. In the normal science phase the phenomenon that Conant called "prejudice" and that in 1957 Kuhn called the "problem of scientific belief", reappears as "paradigm consensus" in his 1962 book, and it assumes a positive function without the ambivalence that it formerly had in Kuhn's and Conant's minds. In an article remarkably titled "The Function of Dogma in Scientific Research" in *Scientific Change* (ed. Crombie, 1963) Kuhn maintains that advance from one exclusive paradigm to another rather than the continuing competition between recognized classics, is a functional as well as a factual characteristic of mature scientific development. In the revolutionary science phase the old paradigm around which a consensus had been formed is replaced by a new one, which is incommensurable with the old one. Feyerabend's thesis of semantic and ontological incommensurability gives new meaning to the already conventional metaphor "scientific revolutions".

Kuhn's thesis is not just an eclectic combination of the influences of Conant and Feyerabend. His concepts of normal and revolutionary science are aspects of his sociological thesis, in which the concept of science as a social institution is fundamental. To sociologists and cultural anthropologists the concept of social institution means a set of beliefs and values shared among the members of a group or community, and internalized by each individual member of the community. The shared beliefs control the individual's understanding of the world in which he lives, and the shared value system regulates his voluntary behavior including his interaction with others. It is in these sociological terms that Kuhn advances his startling new concept of the aim of science. In the normal science phase the consensus paradigm by virtue of its consensus status assumes institutional status in its scientific specialty, and the aim of normal science is the further articulation of the paradigm by a "puzzle-solving" type of research uncritical of the paradigm. The paradigm is the scientist's view of the domain of his science, and the institutional valuation that consensus associates with the paradigm makes conformity with it the criterion for scientific criticism. Thus what Kuhn previously called the "problem of scientific belief" is no longer problematic; the belief status of the paradigm is explained by its institutional status. This status effectively makes it what Conant called a "creed".

Research producing scientific change in the normal science phase is controlled by belief in the consensus paradigm, and the resulting scientific change is always a change within the institutional framework defined by the paradigm.

In striking contrast the revolutionary science phase is not a change within the institutional framework defined by the paradigm, but rather is a change to another paradigm. It is therefore an institutional change in the sense of a change of institutions. Following Feyerabend, Kuhn maintains that the new and old paradigms involved in such an institutional change are semantically and ontologically incommensurable, such that there can be no shared higher framework to control the revolutionary transition. Scientific revolutions are therefore no less revolutionary in the literal sense than are political revolutions, because in neither case are there laws to govern them. In view of this sociological thesis Kuhn's own dynamic view of science may be described as a sequence of five phases, which follows closely the sequence of several of the chapter headings in his book:

(1) Consensus Phase. Mature sciences are distinguished by "normal science", a type of research that is firmly based in one or several past scientific achievements, and that the members of the scientific specialty view as supplying the foundations for research. Unlike early science there are normally no competing schools and perpetual quarrels over foundations in a mature science. The achievements that guide normal science research are called paradigms, which consist of accepted examples that provide models from which spring particular traditions of scientific research. A paradigm is an object for further articulation and specification under new and more stringent conditions, and it includes not only articulate rules and theory, but also the tacit knowledge and pre-articulate skills acquired by the scientist. No part of the aim of normal science is to call forth new sorts of phenomena or to invent new theories. This conformism proceeds both from a professional education that is an indoctrination in the prevailing paradigm set forth in the student's current textbooks and laboratory exercises, and from a consensus belief shared by the members of the scientific specialty that the paradigm seems sufficiently promising as a guide for future research, that acceptance of it is both an obligatory and a justified act of faith. Conformity to the paradigm assumes a recognizable function, which is to focus the group's attention upon a small range of relatively esoteric problems, to investigate these problems in a depth and detail that would not be possible if quarrels over fundamentals were tolerated, and to restrict the

research resources of the profession to solvable problems, where the solutions are "solvable" precisely because they agree with the paradigm and are interpretable in its terms.

(2) Anomaly Phase. Normal science is a cumulative enterprise having as its aim the steady extension of the scope and accuracy of scientific knowledge represented by the prevailing paradigm. Successful normal science does not find any novelties. But anomalies occur as the extension of the paradigm proceeds over a period of time. In fact the paradigm is the source of the concepts needed for recognizing the new fact and for giving it its anomalous status. The normal reaction to an anomaly is a modification of the articulate rules and theories associated with the consensus paradigm, so that the anomalous fact can be assimilated. Success in such modification is a noteworthy achievement for a normal science researcher. Isolated anomalies that are not assimilated are normally set aside under the assumption that eventually they will be reconciled, and normal science research continues with the consensus paradigm. Scientists are not easily distracted by anomalies from continued exploration of the promise of a generally still satisfactory paradigm. Kuhn rejects Popper's falsificationist philosophy, stating that if every failure to fit were ground for theory rejection, all theories ought to be rejected at all times.

(3) Crisis Phase. So long as the consensus paradigm is relatively successful, no alternatives to it are advanced. But eventually the anomalies become more numerous and more serious, and also the modifications necessary to assimilate those anomalies that can be assimilated, produce a certain amount of paradigm destruction. In due course some members of the profession lose faith and begin to propose alternatives. The construction of alternative theories is always possible, because there is an arbitrary aspect to language that permits many theories to be imposed on the same collection of data. When the consensus underlying the prevailing paradigm begins to erode enough that some members begin to exploit this arbitrary element and to create new theories, the profession has entered the phase of crisis. Crises are the crossing of the threshold into extraordinary or revolutionary science.

(4) Revolutionary Phase. Kuhn postulates what he calls a "genetic parallel" between political and scientific revolutions. Just as political revolutions are inaugurated by a growing sense that existing institutions have ceased adequately to meet the problems posed by an environment that they have in part created, so too scientific revolutions are inaugurated by a

growing sense that an existing paradigm has ceased to function adequately in the exploration of the aspect of nature to which the paradigm itself had previously led the way. Political revolutions aim to change political institutions in ways that those institutions themselves prohibit. Their success therefore necessitates the partial relinquishment of one set of institutions in favor of another, and in the interim society is not fully governed by institutions at all. As alternatives are formulated, society is divided into competing camps, those who support the old institutions and those who support the new. Once this polarization has occurred, political recourse fails; there is no suprainstitutional framework for adjudication of differences. Kuhn says that like the choice between competing political institutions, that between competing paradigms is a choice between incompatible modes of community life. In a scientific revolution the semantical and ontological incommensurability between rival paradigms excludes the possibility of any common framework for communication or reconciliation.

Kuhn does not describe incommensurability in terms of Whorf's linguistic relativity thesis, as did Feyerabend thirteen years later. Instead Kuhn invokes Hanson's thesis of *gestalt* switch, and references Hanson's *Patterns of Discovery* published four years earlier. He compares the change of paradigm to the visual *gestalt* switch. A certain *gestalt* is needed for the physics student to see the world as seen by the scientist, when for example the latter sees the electron vapor trail in the cloud chamber and the *gestalt* which is learned by the student is provided by the prevailing normal science paradigm. When at times of revolution the normal science tradition changes, then the scientist's perception of his environment must be reeducated; he must see with a new *gestalt.* This change of paradigm is not achieved by deliberation and interpretation, but rather by a sudden and unstructured event like the *gestalt* switch. While the members are individually experiencing the *gestalt* switch, the profession is divided and confused, and there is a communication "breakdown" between members having different *gestalten* and paradigms.

(5) Resolution Phase. Kuhn does not believe that issues in scientific revolutions are resolved by crucial experiments or by any other kind of empirical testing. In normal science testing is never a test of the paradigm, but rather it is a test of a puzzle-solving attempt to extend the paradigm, and involves a comparison of a single paradigm with nature. Failure of the test is not a failure of the paradigm, but rather is a failure of the scientist. In revolutionary science tests occur as part of the competition between two

rival paradigms for the allegiance of the scientific community. However, these tests do not have a compellingly deciding function. There can be no scientifically or empirically neutral system of language or concepts for these tests, since the paradigms are incommensurable, and those who maintain the old paradigm must experience a "conversion" to the new *gestalt*. Tests serve only to persuade that the new paradigm is the more promising guide for future normal science research. The actual decision about the future performance of the new paradigm is based on faith. As early supporters of the new paradigm show success, others follow until there is a new normal science consensus paradigm. The procession has come full circle to a new consensus paradigm.

In the final chapter of *Structure of Scientific Revolutions* Kuhn discusses the concept of scientific progress that is consistent with his theory of the historical development of science. He maintains that the semantics of the term "progress" is determined by reference to the research work of normal science, and specifically by the puzzle-solving type of work in normal science in the absence of competing schools. Progress occurs in extraordinary science by the transition to a new consensus paradigm, because in the judgment of the specialized scientific community the new paradigm promises to resolve outstanding problems that had occasioned the crisis and transition, and to preserve the community's problem-solving ability to treat the assembled data with growing precision and detail, even though the ability to solve problems cannot be a basis for paradigm choice.

Kuhn's Criticism of Popper's Falsificationist Philosophy

Kuhn's philosophy of science is not only a post-Positivist philosophy critical of Positivism, it is also a post-Popperian philosophy that is critical of Popper's falsificationist theory of scientific criticism and scientific progress. The difference between Kuhn and Popper is explicable in large part by the differences in the episodes in the history of science, that have had a formative influence on their respective thinking. Popper's philosophy of science was principally influenced by the episode in which the physics profession made the transition from Newton's theory of gravitation to Einstein's relativity theory. On the other hand Kuhn's philosophy was principally influenced by the earlier episode in which astronomers made the transition from Ptolemy's geocentric theory to Copernicus' heliocentric theory. The noteworthy difference between these two episodes is that the transition to Einstein's theory is often viewed as involving a crucial test, the celebrated eclipse test of 1919, while the transition to Copernicus' theory, like the transition to Lavoisier's oxygen theory of combustion discussed by Conant, is not associated with any crucial tests but involved various nonempirical considerations. Popper views these nonempirical considerations as external impediments to progress in science, while Kuhn views them as internal and integral to the development of science. Kuhn's explicit criticism of Popper is given in "Logic of Discovery or Psychology of Research?" in *Criticism and the Growth of Knowledge*. In this paper Kuhn begins by describing the similarities between his views and Popper's, that also separate both their views from those of the Positivists. He notes that both he and Popper are concerned with the dynamic processes by which scientific knowledge is developed, instead of the logical structure of the products of scientific research, and that therefore both of them look to the history of science. He furthermore notes that both of them draw many of the same conclusions from the history of science particularly about which fields are sciences and which are not, that both are realists, and that both reject the Positivist idea of a neutral or theory-independent observation language.

Then Kuhn turns to the contrasts between his views and Popper's. He maintains that even though he and Popper draw the same conclusions about which fields are sciences and which are not, they arrive at their shared conclusions by very different ways, that may be contrasted as different *gestalts* of the same situations. Popper maintains that scientists test theories and attempt to falsify them with a critical attitude, while Kuhn maintains his thesis of normal science according to which a theory is not tested critically, but instead functions as a premise for puzzle-solving research with currently accepted theory supplying the rules of the game. Kuhn says that the type of tests that Popper discusses, such as the eclipse test of Einstein's theory of relativity in 1919, is rare in science, and he identifies this rare type of research as extraordinary or revolutionary science. He says that Popper has mistakenly characterized the entire scientific enterprise in terms that apply only to its occasional revolutionary parts, and that he is turning Popper on his head, when Popper demarcates scientific from nonscientific fields, because in his view it is the abandonment of critical discourse rather than its adoption, that makes the transformation of a field into a science. Once a field has made that transition, critical discourse recurs only at moments of crisis, when the basis of the field is again in jeopardy. Therefore Popper's and Kuhn's lines of demarcation coincide only in their outcomes and not in their criteria; for their respective criteria they reference different aspects of scientific activity.

Then Kuhn goes on to say that even during revolutionary phases of science, the choice between paradigms is not a choice in which critical testing can play a decisive role. Kuhn references Popper's "Truth, Rationality, and the Growth of Knowledge" in *Conjectures and Refutations*, where Popper states that the Ptolemaic theory was replaced before it had been tested. In this article Popper maintains that such instances reveal that crucial tests are decisively important, so that scientists have reason to believe that the new theory replacing the old one is better and nearer to the truth. But Kuhn argues that not only had these theories not been put to the test before they were replaced, but furthermore none of them were replaced before it had ceased adequately to support a puzzle-solving tradition. Kuhn notes that both he and Popper agree that no theory can be conclusively falsified, that all experiments can be challenged either as to their relevance or to their accuracy, and that every theory can be modified by a variety of *ad hoc* adjustments without ceasing to be the same theory. But he argues that in Popper's philosophy recognition of such things operates merely as a qualification of his philosophy, even though these things occur in the history of science. Kuhn cites as an example that the state of astronomy was a scandal in the early sixteenth century, but most astronomers nevertheless thought that normal adjustments to a basically Ptolemaic model would be sufficient to set the situation aright. In this sense the Ptolemaic theory had not failed any test. However a few astronomers including Copernicus thought that the difficulties must lie in the basic Ptolemaic approach itself rather than in the particular versions of Ptolemaic theory.

Kuhn says that Popper's error is the belief that logical criteria can dictate the falsification of a theory and determine theory choice during revolutions. Logical falsification presumes that a theory can be cast or recast such that all events are either corroborating, falsifying or irrelevant instances. But this cannot be done unless the theory is fully articulated and its terms sufficiently defined, so that it is possible to determine their applicability in every possible case. Kuhn says that no theory can in practice satisfy such a requirement, and that he had introduced the term "paradigm" to underscore the dependence of scientific research on concrete examples, that supply what would otherwise be gaps in the specification of the content and application of scientific theories. Kuhn illustrates the semantical and pragmatical considerations captured by the term "paradigm" with a discussion of swans and the stereotype theory that says "all swans are white". Kuhn says that after a scientist has made his investigation and has found no instances of nonwhite swans, making the generalization explicit adds little or nothing to what is already known from the investigation. And if later one finds a black bird that otherwise appears to be a swan, then one's behavior will be the same whether or not one has made the explicit generalization that all swans are white. With or without the explicit generalization a decision must be made with respect to the possibility of black swans. Observation cannot force a falsifying decision. Only if one had previously committed oneself to a full definition of "swan", one that will specify its applicability to every conceivable object, could one be logically forced to rescind one's generalization. And Kuhn says that there is no good reason for such a commitment to any such explicit generalization; it is an unnecessary risk. Similarly in science the scientist who is confronted with the unexpected, must always do more research in order to articulate his theory further in the area that has just become problematic. He may reject his theory in favor of another, and may do so for good reason, but no exclusively logical criterion can dictate the conclusion that the theory has been falsified, or that it has not been falsified. Just as the investigator of swans need not make the decision as to whether whiteness is a defining characteristic of swans, until he can investigate further the apparently anomalous case of the black but otherwise swan-looking bird, so too the scientist has the same freedom to choose, and is not logically compelled to conclude that current theory has been falsified by apparently anomalous instances and test outcomes. Kuhn says that further empirical investigation is needed to answer such questions as how do scientists actually make the choice between competing theories, and how scientific progress is to be understood. He says that the type of answer to these questions must in the final analysis by psychological or sociological. He agrees with Popper's rejection of answers given in terms of the scientists' psychological idiosyncrasies, but he advocates investigation of the common elements induced by education of the licensed membership of the scientific group.

Kuhn, Normal Science, and the American Gothic Sociologists

M. Jourdain, the *parvenu* in Moliere's comedy, *Le Bourgeois Gentilhomme*, had aspired to write prose, and was delightedly surprised when he was told that he had been speaking prose for more than forty years without knowing anything about it. Moliere's play has its analogue in contemporary American sociology, save for the stark absence of anything comical. Among the more mature scientific professions sociology is often viewed as a pretentious *parvenu* with a literature of platitudes expressed in jargon. And American academic sociologists have longed to demonstrate the manifest scientific progress that the more mature scientific professions have exhibited often in their histories. Consequently like M. Jordain, the American sociologists were delightedly surprised when they were told by Kuhn, that they have been theorizing about the conditions for scientific progress for years without knowing anything about it. Sociologists did not have to be told how to practice Kuhn's doctrine of enforced consensus; it is an accepted practice in their profession. They had only to be told that it is a new philosophy of science that produces progress. Specifically he told them that his sociological thesis of normal science describes the conditions for the transition of social sciences from "preparadigm" status to "mature" status. In several places in his writings Kuhn maintains that the social sciences are immature sciences, because they do not have consensus paradigms that enable them to pursue the puzzle-solving type of research that characterizes normal science. In his "Postscript" he states that the transition to maturity deserves fuller discussion from those who are concerned with the development of contemporary social science. Not coincidentally none were more concerned with such a transition than American academic sociologists. And as the custodians and practitioners of the theory of consensus and conformity, none have thought themselves more professionally suited for such discussion.

A paradigmatic example of Kuhn's influence is represented by Hagstrom's *The Scientific Community* (1965). This book written by a sociologist and referenced later by Kuhn in support of his own views, is a study of how the forces of socialization by professional education and of social control by colleagues within a scientific community, operate to produce conformity to scientific norms and values. Just as Kuhn attributed institutional status to the prevailing paradigm, so too, Hagstrom identifies the norms and values of science with currently accepted substantive views, and he therefore says that substantive disputes in a scientific community are a type of "social disorganization". "Disorganization" is as pejorative a term in sociology as "depression" is in economics. Hagstrom identifies his theory as a functionalist theory, and in functionalist sociological theory social disorganization is viewed as pathologically symptomatic of institutional disintegration. He mentions two types of social-control sanctions that operate in the scientific community to produce the requisite conformity to the norms and values. They are firstly refusal to publish papers in the professional journals, and secondly denial of opportunity for occupational advancement. Hagstrom acknowledges Kuhn's influence in his preface, and he references and quotes Kuhn in several places in the book, particularly where Kuhn discusses professional education in mature sciences. And Kuhn in turn later references Hagstrom's book in "Second Thoughts" and in the "Postscript" in support of his theses.

Kuhn's influence on sociologists was manifested in other ways also. A short time after Kuhn's 1962 book there appeared a new sociological journal, *Sociological Methods and Research.* In a statement of policy reprinted in every issue for many years the editors state that the journal is devoted to sociology as a "cumulative" empirical science, and they describe the journal as one that is highly focused on the assessment of the scientific status of sociology. One of the distinctive characteristics of normal science in Kuhn's theory is that it is cumulative, such that it can demonstrate progress.

In "Editorial Policies and Practices among Leading Journals in Four Scientific Fields" in the *Sociological Quarterly* (1978) Janice M. Beyer reports her findings from a survey of the editors of several academic journals. These interesting findings reveal three significant differences between the editorial policies of the journals of the physics profession and those of the sociological profession. They are: (1) The acceptance rate for papers submitted to sociological journals is thirteen percent, while the rate for physics journals is sixty-five percent; (2) the percent of accepted papers requiring extensive revision and then resubmitted to referees is forty-three percent for sociological journals and twenty-two percent for physics journals; and (3) the percent of accepted papers requiring no revision is ten percent for sociological journals and forty-six percent for physics journals. The scientist who is not a sociologist may reasonably wonder either whether sociologists are really as professionally ill-prepared to contribute to a professional scientific literature as these findings would indicate, or whether there is something Orwellian in this enforced practice of extensive revision of purportedly scientific findings as a condition for publication. The latter would seem to be the case, since Beyer explains her findings in terms of Kuhn's thesis of normal science, and attributes the reported differences in editorial practices to differences in paradigm development. She states that sciences having highly developed paradigms use "universalist" criteria for scientific criticism, which she defines as the belief that scientific judgments should be based on considerations of scientific merit, where "merit" means conformity with a consensus paradigm. Understood in this manner, universalism is just an imposed uniformity that is indifferent to the distinction between contrary evidence and the contrary opinions of author and referees.

Ironically the outcome of the self-conscious attempt to make sociology a "mature" science practicing normal science with a consensus paradigm was something quite different than what Kuhn's philosophy had described. Kuhn's philosophy described a consensus paradigm that is empirical, so that it can produce anomalies which initially are ignored, but which eventually accumulate and spawn revolutionary alternative theories. What actually happened in sociology, however, was that the sociologists imposed social controls upon the members of their profession, in order to enforce conformity not to an empirical theory, but to a philosophy of science. The philosophy of science that the sociologists enforce upon their membership is the Romanticist philosophy introduced into American sociology by Talcott Parsons. This philosophy, which Parsons brought from the University of Heidelberg in Germany where he was influenced by the views of Max Weber, was to supply the philosophical foundations for his "functionalist" sociology, or at least for his own peculiar variation on functionalism. Even though his functionalist sociology has since waned, Parson's Romantic philosophy of science continues to dominate American sociology.

Not only did the sociologists get things mixed up, when they adopted a philosophy instead of an empirical theory for their consensus paradigm, they furthermore got things backwards, when they made Romanticism their consensus philosophy of science. While the natural sciences rejected Positivism and then moved forward to the post-Positivist philosophy of contemporary Pragmatism, sociologists rejected Positivism and then moved backward to the pre-Positivist philosophy of Romanticism. The contemporary Pragmatist philosophy of science differs from both Positivism and Romanticism in a very fundamental way, because both of these latter include ontological considerations in their criteria for scientific criticism. They differ between one another only about which types of ontology they will accept: the Positivists reject all "mentalism" in social and behavioral science, while the Romantics require it. The contemporary Pragmatists on the other hand subordinate all ontological questions and commitments to the empirical adequacy of the scientific law or theory. As a result, the philosophy of science that the sociologists imposed on their membership is anachronistic and quite at variance with the philosophy that Kuhn uses for his philosophical interpretation of the history and dynamics of science.

The followers of Parsons accepted Weber's *verstehen* concept of social science explanation, whereby empathetic plausibility is the principal criterion for scientific criticism. Whatever one may think of Kuhn's solution to the problem of scientific belief and the thesis of the consensus paradigm that constitutes his solution to this problem, the issue of freely ignoring empirical anomalies in normal science becomes moot, when there can be no empirical anomalies. The *verstehen* criterion reduces scientific criticism to what one or another particular critic finds intuitively acceptable, empathetically plausible, or otherwise comfortably familiar, however covert or idiosyncratic, and makes it into quarrels about intuitions; empirically adequate work is rejected out of hand, if it "doesn't make sense" according to the intuition of the particular critic. This intuitional criterion may be contrasted with empirical criticism in modern physics. When modern physicists were confronted firstly with Einstein's relativity theory and then with quantum theory, their profession in each case decided to accept the new physics, because it is more empirically adequate in spite of the fact that it is anything but intuitively familiar or platitudinous. This is not possible even today in American academic sociology, because the American sociological profession accepts and enforces consensus about Weber's strong version of the Romantic philosophy of science, and consequently they can make no distinction between contrary empirical evidence and contrary intuitive opinion. Parsons had never referenced Kuhn, and probably had not even read him; he had his own agenda for sociology long before Kuhn. The enforced consensus about Parson's sociology may be explained in part by the appointment of Parsons to the presidency of the American Sociological Association (ASA). In his *The Coming Crisis of Western Sociology* (1970) the sociologist Alvin W. Gouldner observed that Parsons used this position to influence the appointments to other executive positions in the ASA including most notably both the ASA's Publications Committee and the position of editor of its *American Sociological Review*. Gouldner said that there existed a "continuity-convergence ideology" that produced a blanketing mood of consensus that smothers intellectual criticism and innovation.

No conspiracy theory involving Parsons, however, could adequately explain the sociologists' willingness to adopt his distinctive "functionalist" sociology and its associated German Romantic philosophy of science. The doctrinairism of the American sociological profession and its receptivity to Parson's Romanticism is also explained in part by the thesis of the functionalist sociological doctrine itself and in part by the timing of the appearance of Kuhn's sociological thesis of normal science. The thesis of the functionalist doctrine is that social controls producing conformity to a consensus of views and values, explain the existence of social order in any group. And this in turn implies that failure to conform is dysfunctional in a pejorative sense of being disorderly even to the extent of threatening complete disintegration of the group. Advocates of Parsons' functionalist sociology could not easily escape the inclination to apply these concepts to their own profession with Parsonian functionalism itself serving as the consensus view. Kuhn's theory of the development of empirical science is a logical extension of the Parsonian functionalist sociology: in the scientific community social conformity to a consensus paradigm functions not only to produce social order, with Kuhn's philosophy it also functions to produce social order and scientific progress.

Secondly Kuhn's theory made its appearance at an opportune time. Lundberg's initially popular Positivist program for American sociology had waned, because it never got beyond the stage of a programmatic proposal, and years earlier Parsons had launched his distinctive functionalist sociology from the prestigious platform provided by his faculty position as chairman of the sociology department at Harvard University. When Kuhn's sociological thesis of progress in science appeared, the *parvenu* scientific profession seeking acceptance among the empirical sciences was pre-disposed to impose a progress-producing consensus paradigm. The outcome of this combination of Parsonian Romanticism and Kuhnian "normal science" has been a chimerical science, which is about as normal as the gothic caricature of science depicted by Shelley's character, Victor Frankenstein - a Romantic grotesque deserving the epitaph "American Gothic" sociology.

As it happens, American Gothic sociology seems to have become the appalling spectre to prospective sociology students and to sociology students' prospective employers. In *Science and Engineering Doctorates* the National Science Foundation (NSF) reports statistics revealing a thirty-nine percent decline since 1976 in the number of doctoral degrees in sociology earned annually in the United States. This compares with a nearly seven percent growth in doctorates for all sciences during the same period. The NSF also reports that the median age of receipt of the doctorate in social science is between thirty-two and thirty-three years. And since the post-World War II "baby-boom" years of rising aggregate number of births did not end until 1964, it is clear that American academic sociology has been in decline during a period in which the pool of potential students has been rising. Therefore sociology's decline is not merely a demographic phenomenon circumstantial to the history of the profession; it is a pathological condition intrinsic to the American sociological profession's institutional values, normative standards, and research practices.

Ironically at the same time that the sociologists were accepting Kuhn's theory of scientific progress uncritically, the professional philosophers of science were attacking it with a veritable fusillade of criticism, most of which was directed at Kuhn's doctrine of normal science.

Popper's Criticism of "Normal Science"

Popper criticizes the aim of normal science as viewed by Kuhn, and he rejects the historical relativism that he finds in Kuhn's thesis. His criticism in reply to Kuhn is set forth in "Normal Science and its Dangers" in *Critcism and the Growth of Knowledge*. Popper notes that he and Kuhn agree that the normal work of the scientist presupposes a theory that supplies the scientist with a generally accepted problem situation for his work. Interestingly he also states that he has always said that some dogmatism is necessary, because giving in to criticism too soon may preclude finding out where the real power of a theory lies. And he says that while he has been only dimly aware of the distinction that Kuhn makes between normal and revolutionary science, he admits that normal science in Kuhn's sense does exist. But Popper maintains that the normal scientist in Kuhn's sense is a scientist who has been badly taught, since he does not think critically, a problem that Popper says he finds in quantum theory today. Popper expresses the opinion that uncritical normal science is dangerous both to science and to our civilization. He also takes exception to Kuhn's view that normal science as Kuhn conceives it is actually normal in the history of science. Kuhn's thesis of a single dominant theory may fit astronomy, but it does not fit the theory of matter or the biological sciences. Popper questions Kuhn's historical accuracy.

But Popper is principally concerned with Kuhn's historical relativism and with the thesis that philosophers of science should look to sociology and psychology of science instead of attempting a logical analysis, as Popper did in his own work. He argues that Kuhn's historical relativist thesis of the dynamics of science is not a sociological or a psychological one but rather a logical one, and he furthermore maintains that it is a mistaken one. He says that Kuhn's view that scientists must agree on fundamentals and on the framework of those fundamentals, in order to discourse rationally and critically, is what he calls "The Myth of the Framework". Popper admits that at any moment we are prisoners caught in the framework of our theories, expectations, past experiences, and language. But he adds that we are prisoners only in a Pickwickian sense, because if we try, we can break out of our framework into a better and roomier one. He emphasizes that his central point is that a critical discussion and a comparison of the various frameworks is always possible. He denies that different frameworks are like mutually untranslatable languages. In Popper's view the Myth of the Framework is the principal bulwark of irrationalism, and it merely exaggerates a difficulty into an impossibility. There are difficulties in discussion between people brought up in different frameworks, but Popper says that nothing is more fruitful than such discussions. An intellectual revolution may look like a religious conversion; a new insight may strike one like a flash of lightning. But this does not mean that one cannot evaluate former views critically and rationally in the light of new ones. It is simply false to say that the transition from Newton to Einstein is an irrational leap, and that the two theories of gravitation are not rationally comparable. In science we can say that we have made genuine progress, and that we know more than we did before such transitions occurred. Therefore, Popper says that all of Kuhn's own arguments go back to the thesis that the scientist is logically forced to accept a framework, since no rational discussion is possible between frameworks. This is not a historical, sociological, or psychological argument, but is a logical one, and it is a mistaken one. Popper says that science is "subjectless" in the sense that it is not bound to any framework.

Popper reaffirms his own thesis that the aim of science is to find theories, which in the light of critical discussion get nearer to the truth and have increased the truth content. Popper rejects Kuhn's proposal of turning to psychology and sociology for enlightenment about the aims of science and about the nature of scientific progress. He rejects all psychologistic and sociologistic tendencies, and furthermore says that in comparison to physics, psychology and sociology are riddled with fashions and with uncontrolled dogmas. He concludes by answering Kuhn's question, "Logic of Discovery or Psychology of Research?", with the reply that while Logic of Discovery has little to learn from the Psychology of Research, the latter has much to learn from the former.

Feyerabend on Theory Proliferation vs Consensus Paradigm

Feyerabend also criticizes Kuhn, and says that the doctrine of normal science is an ideology that Kuhn propagandizes among social scientists. His principal methodological criticism of Kuhn's philosophy is that Kuhn's theory cannot explain the transition from a monistic normal science to a pluralistic revolutionary science, since the impossibility of a semantically neutral observation language makes a plurality of alternative theories a precondition for the transition to be brought about. Feyerabend's criticism of Kuhn is given in his "Consolations for the Specialist" in *Criticism and the Growth of Knowledge.* Firstly he notes that he and Kuhn had discussed their views in 1960 and 1961, while both were at the University of California at Berkeley; and he says that while he recognizes the problems that interest Kuhn, notably the omnipresence of anomalies, he is unable to agree with Kuhn's theory of science, which he also calls an ideology. Feyerabend maintains that Kuhn's ideology can give comfort only to the most narrow-minded and the most conceited kind of specialist, that it tends to inhibit the advancement of knowledge, and that it is responsible for such inhibiting tendencies in modern psychology and sociology. He elaborates on his view that Kuhn's theory is an ideology. He states that Kuhn's presentation contains an ambiguity between the descriptive and the prescriptive mode of presentation, and that as a result more than one social scientist has pointed out to him that after reading Kuhn's book, he at last knows how to turn his field into a "science". Feyerabend reports that the recipe that these social scientists have taken from Kuhn consists of such practices as restricting criticism, reducing the number of comprehensive theories to one, creating a normal science that has this one theory as its paradigm, preventing students from speculating along different lines, and making more restless colleagues conform and do "serious work". He then asks whether or not Kuhn's following among sociologists is an intended effect, whether or not it is Kuhn's intention to provide a historicoscientific justification for sociologists' need to identify with some group. In criticism of Kuhn, Feyerabend concludes that it is actually Kuhn's intention to provide an ambiguity between the descriptive and the prescriptive modes of presentation, and that Kuhn wishes to exploit the propagandistic potentialities in this ambiguity. He says that Kuhn wants on the one hand to give solid, objective historical support to value judgments, which he and others regard as arbitrary and subjective, while on the other hand Kuhn also wants to leave himself a safe line of retreat. When those who dislike Kuhn's implied derivation of values from facts object, Kuhn's line of retreat consists of telling them that no such derivation can be made, and that the presentation is purely descriptive.

Feyerabend next turns his criticism to Kuhn's thesis as a descriptive account of science. The central thesis of his criticism of Kuhn is that the latter's theory of science leaves unanswered the problem of how the transition from the monistic normal-science period to a pluralistic revolutionary period is brought about. Feyerabend notes that both he and Kuhn admit to what he calls the methodological "principle of tenacity", which he defines as the scientist's selection from a number of theories, that one which in the particular scientist's view promises to lead to the most fruitful results, and then sticking to the selected theory even if the anomalies it suffers are considerable. He then asks how this principle can be defended, and how it is possible to change allegiance to paradigms in a manner consistent with it. He answers that the principle of tenacity is reasonable, because theories are capable of development and may eventually be able to accommodate the anomalies that their original versions were incapable of explaining. This is because relevant evidence depends not only upon the theory, but also upon other subjects, which are conventionally called "auxiliary sciences". Such auxiliary sciences function as additional premises in the derivation of testable consequences, and these premises "infect" the observation language in which the testable consequences are expressed, thereby providing the very concepts in terms of which experimental results are expressed. But it happens that theories and their auxiliary sciences often develop out of phase, with the result that apparently refuting instances may turn out not to indicate that a new theory is doomed to failure, but instead may indicate only that it does not fit in at present with the rest of science. Therefore scientists can tenaciously develop methods which permit them to retain their theories in the face of plain and unambiguously refuting facts, even if testable explanations for the clash with facts are not immediately forthcoming. The significance of the principle of tenacity, the practice whereby scientists no longer use recalcitrant facts for removing a theory, is that a plurality of alternative theories can coexist in a science at any given time. This pluralism is strategic to Feyerabend, because in his view the fact that theory determines observation implies that theories are not compared with nature, but must be compared with other theories. Alternative theories function to accentuate the differences between one another, such that the principle of tenacity itself may eventually urge the elimination of a theory. Hence, if a change of paradigms is the function of normal science then one must be prepared to introduce alternatives to a given theory. Feyerabend notes that in fact Kuhn himself has described in detail the magnifying effect which alternatives have upon anomalies, and has explained how revolutions are brought about by such magnifications. Feyerabend therefore proposes a second methodological principle, the "principle of proliferation", and he asks rhetorically, why not start proliferating theories at once, and why allow a purely normal science, as Kuhn conceives it, ever to come into existence?

Feyerabend then switches from a purely methodological perspective to a historical one, and replies to his own rhetorical question about theory proliferation vs normal science consensus. Using his methodological principles of tenacity and proliferation to examine the history of science, he maintains that normal science is a "big myth". He argues that even though there are scientists who practice puzzle-solving normal science, there is no temporally separated periods of monistic normal science and pluralistic revolutionary science. He supports a view initially proposed by Imre Lakatos, a professor of logic at the University of London, that the practices of tenacity and proliferation do not belong to successive periods in the history of science, but rather are always copresent. The interplay between tenacity and proliferation is an essential feature of the actual, historical development of science. It is not the puzzle-solving activity that is responsible for the growth of knowledge, but the active interplay of a plurality of tenaciously held views. It is the continuing intervention of new ideas and the attempts to secure for them a worthy place in the competition, that leads to the overthrow of old and familiar paradigms. Feyerabend furthermore maintains that revolutions are basically matters of appearance, and that during a revolution there is actually no profound structural change such as a transition from normal to extraordinary science as described by Kuhn. Thus, instead of advocating conformity to a monolithic consensus paradigm, as Kuhn does, Feyerabend issues what he calls a "plea for hedonism", by which he means the continuing practice of the theory-proliferating principle of tenacity.

Feyerabend took occasion to comment more favorably on Kuhn's philosophy, and to relate Kuhn's views to his own, where they manifest similarities. One aspect of Kuhn's philosophy that Feyerabend considers to be important is the concept of paradigm. Feyerabend says that Kuhn expanded on Wittgenstein's criticism of the Logical Positivist emphasis on rules and formal aspects of language, and that Kuhn made this criticism more concrete. He also says that by introducing the notion of paradigm, Kuhn stated above all a problem. Kuhn explained that science depends on circumstances that are not described in the usual accounts, that do not occur in science textbooks, and that have to be identified in a roundabout way. However, most of Kuhn's followers, especially in the social sciences, did not recognize the idea as a statement of a problem, but regarded Kuhn's account as a presentation of a new and clear fact. Feyerabend maintains that by using the term "paradigm", which is awaiting explication by research, as if explication had already been completed, they started a new and most deplorable trend of loquacious illiteracy. Feyerabend finds three note-worthy aspects in Kuhn's treatment of the relations between different paradigms. Firstly different paradigms use sets of concepts that cannot be brought into the usual logical relations of inclusion, exclusion, or overlap, and that incommensurability is the natural consequence of identifying theories with paradigms or, as Feyerabend calls them, traditions. Secondly different paradigms make researchers see things differently, such that researchers in different paradigms not only have different concepts, but also have different perceptions. Thirdly paradigms have different methods including intellectual as well as physical instruments for practicing research and evaluation results. He says that it was a great advance to replace the idea of theory with the idea of paradigm, which includes dynamic aspects of science. He notes that his earlier work had principally been concerned only with the first of the three mentioned aspects, and then only with theories.

Shapere's Criticism of Kuhn's Concept of Paradigm

Dudley Shapere, University of Chicago philosopher of science, argues that Kuhn's concept of paradigm is so vague as to be of questionable explanatory value, and he also rejects the relativism he finds in the concept of incommensurability. He wrote a critical review of Kuhn's *Structure of Scientific Revolutions* in the *Philosophical Review* (July, 1964), and shortly later he wrote a critique of the philosophies of both Kuhn and Feyerabend in "Meaning and Scientific Change" in *Mind and Cosmos* (ed. R.G. Colodny, 1966). In addition to the criticisms by Popper and Feyerabend, this is the criticism that Kuhn took most seriously of the many that his controversial book provoked in the literature. Unlike the criticisms of Popper and Feyerabend, which were principally directed at Kuhn's new concept of the aim of science, Shapere's criticism is directed at Kuhn's semantical views, and particularly at Kuhn's thesis of pre-articulate meaning set forth in the concept of paradigm.

Shapere finds particularly perplexing Kuhn's thesis that paradigms cannot be formulated adequately or articulated completely. He objects that if all that can be said about paradigms and scientific development, can and must be said only in terms of what are mere abstractions from paradigms, as Kuhn maintains, then it is difficult to see what is gained by appealing to the notion of a paradigm. He notes that in most of the cases that Kuhn discusses, it is the articulated theory that is doing the job that Kuhn assigns to the paradigm, yet in Kuhn's thesis the theory is not the same as the paradigm. Shapere says that Kuhn discusses the theory in these cases, because it is as near as he can get in words to the inexplicable paradigm. He therefore asks how can historians know that they agree in their identification of the paradigms in historical episodes, and so determine that the same paradigm persists through a long sequence of such episodes. Where, he asks, does one draw the line between different paradigms and different articulations of the same paradigms? On the one hand it is too easy to identify a paradigm, and on the other hand it is not easy to determine in a particular case what is supposed to have been the paradigm in that case. The inarticulate status of the paradigm makes individuation of the paradigm problematic. Shapere concludes that in Kuhn's theory anything that allows science to accomplish anything at all can be part of or otherwise somehow involved with a paradigm, with the result that the explanatory value of this concept of paradigm is suspect. He maintains that this idea of shared paradigms which are purportedly behind historically observed common factors that guide scientific research for a period of years, appears to be guaranteed not so much by a close examination of actual historical cases, as by the breadth of definition of this term "paradigm". He furthermore questions whether such paradigms even exist, since the existence of similarities among theories does not imply the existence of a common paradigm of which the similar theories are incomplete articulations. Shapere thus rejects what he calls the "mystique" of the single paradigm.

In addition to criticizing Kuhn's concept of paradigm, Shapere also criticizes the thesis of incommensurability that Kuhn took from Feyerabend. He maintains that Kuhn offers no clear analysis of meaning, and therefore no clear analysis of meaning change. The principal problem that he finds with the incommensurability thesis advocated both by Kuhn and by Feyerabend is that it destroys the possibility of comparing theories on any grounds whatsoever. He asks: if the incommensurable paradigms differ in all respects including the facts and the problem itself, then how can they disagree? Why do scientists accept one of them as better than the other? Neither Kuhn nor Feyerabend in his view succeeds in providing any extratheoretical basis for comparing and for judging theories and paradigms. The result he says is a historical relativism. Shapere proposes a resolution. He notes that the thesis of incommensurability requires that two expressions or sets of expressions must either have precisely the same meaning or else they must be utterly and completely different. He proposes what he calls a "middle ground" by altering this rigid notion of meaning. He proposes that meanings may be similar, such that they may be comparable in some respects even as they are different in other respects, and thus may be said to have *degrees* of likeness and difference.

Kuhn Replies

In "Reflections on My Critics" in *Criticism and the Growth of Knowledge* Kuhn replies to his critics. In these replies he distances himself from the sociologists while still affirming the sociological character of his theory of science, he minimizes the differences between himself and Popper while still affirming the uncritical attitude in normal science, and he differentiates his views on incommensurability from those of Feyerabend and also explains how a multiplicity of theories emerge in the crisis period without Feyerabend's principle of proliferation.

Firstly Kuhn distances himself from the sociologists. He states that he agrees with Popper; he says the received theories of sociology and psychology are "weak reeds from which to weave a philosophy of science", and he adds that his own work no more relies on current sociological theory than does Popper's. But he still maintains that his theory of science is intrinsically sociological, because whatever scientific progress may be, it is necessary to account for it by examining the nature of the scientific group, discovering what it values and what it disdains. Scientists must make decisions: they must decide what statements to make unfalsifiable by *fiat* and which ones will not be considered unfalsifiable; they must using probabilistic theory decide upon some probability threshold below which statistical evidence will be held to be inconsistent with theory; they must decide when a research programme is progressive in spite of anomalies, and when it has become degenerative because of them. He states that answers to such questions require a sociological type of analysis, because they are ideological commitments that scientists must share, if their enterprise is to be successful. So, the unit of investigation is not the individual scientist, but rather is the nonpathological, normal scientific group. He adds that while group behavior is affected decisively by the shared commitments, individuals will choose differently, because of their distinctive personalities, educations, and prior patterns of professional research, and these individual considerations are the province of individual psychology. And he says that he agrees with Popper in rejecting any role for individual psychology in philosophy of science.

Kuhn also addresses what Feyerabend called the ambiguity of presentation, the ambiguity between the descriptive and the prescriptive. He replies that his book should be read in both ways, because a theory of science that explains how and why science works, must necessarily have implications for the way in which scientists should behave, if their enterprise is to flourish. He states that if some social scientists have gotten the idea that they can improve the status of their field by firstly legislating agreement on fundamentals and then turning to puzzle solving, they have misunderstood him. Kuhn states that maturity comes to those who know how to wait, because a field gains maturity when it has achieved a theory and technique that satisfy four conditions that he sets forth (and it might be noted parenthetically that the practices recommended in Kuhn's four conditions are quite different than what is to be found in contemporary sociology). Those four conditions are as follows: (1) Popper's demarcation criterion must apply, such that concrete predictions emerge from the practice of the field. (2) Predictive success must be consistently achieved for some subclass of the phenomena considered by the field. (3) The predictive technique must have roots in the theory, which explains their limited success, and which suggests means for their improvement in both scope and precision. (4) Finally the improvement in predictive technique must be a challenging task demanding high talent and dedication. The statement of these four conditions leads to Kuhn's defense of his normal science thesis. He states that these conditions are tantamount to a good scientific theory, and he maintains that with such a theory in hand the time for criticism and theory proliferation has past. The scientists' objectives, then, are to extend the range and precision of the match between existing experiment and theory, and to eliminate conflicts both between the different theories employed in their work and between the ways in which a single theory is used in different applications. These are the types of puzzles that constitute the principal activity of normal science. And Kuhn says that the difference between him and Popper on this issue of criticism is only one of emphasis.

Kuhn then takes up the topic of semantic incommensurability, that he used to explain the communication breakdown occurring during revolutionary science, and he also discusses the topics of irrationality in theory choice and of historical relativism, that his critics find implied in the incommensurability thesis. Firstly he notes that his thesis is that the communication problem is not one of complete breakdown, and that partial communication occurs. Nevertheless Kuhn maintains a version of the incommensurability thesis. He says that a point-by-point comparison of two successive theories demands a language into which at least the empirical consequences of both theories can be translated without loss or change, and he denies that there exists such a theory-independent, semantically neutral observation language to enable such a comparison. He states that Popper's basic statements function as if they have this neutral character. He joins Feyerabend in stating that there is no neutral observation language, because in translating from one theory to another, the constituent words change their meanings or conditions of applicability in subtle ways. But Kuhn states that to him "incommensurable" does not mean "incomparable", and in this respect he departs from Feyerabend's incommensurability thesis. In his view the fact that translation exists, suggests that recourse is available to scientists who hold incommensurable theories. His explanation for the fact that translation communication is only partial and that translation is difficult, is given in terms of his concept of paradigm. The paradigm functions as an example that enables the scientist to recognize similar cases without having to articulate or to characterize the similarity relations explicitly in a generalization. He states that the practice of normal science depends on a learned ability to group objects and situations into similarity classes, which are "primitive" in the sense that the grouping of objects is done without supplying an answer to the question, "similar with respect to what?" In scientific revolutions some of the similarity relations change, such that objects that were grouped in a set are regrouped into different subsets than before. The example given by Kuhn of grouped objects is the sun, the moon and the stars, that were regrouped in the transition from the Ptolemaic to the Copernican celestial theory. And it may be noted that Feyerabend does not consider the transition to the Copernican celestial theory to be a case of semantic incommensurability.

Partial communication occurs, because in such a redistribution of similarity sets, two men whose discourse had previously proceeded with full understanding, may suddenly find themselves responding to the same stimulus with incompatible descriptions or generalizations. At this point there is an apparent shift in Kuhn's argument. He seems not to be saying that there can be comparability due to the existence of common primitive elements, that are regrouped. On the one hand he explains the difficulty of communication in terms of the tacit nature of the similarity relations, that are never made explicit or articulate, and that have undergone the change. But on the other hand the underlying continuity is not the primitive objects, but rather is the external stimuli in the stimulus-sensation portion of the human neural processing apparatus, as well as the neural processing apparatus itself, and the programming of the processing due to the shared cultural history of the two partially communicating scientists. Therefore scientists experiencing communication breakdown can discover by continued discourse the areas where their disagreement occurs, and what the other person would see and say, when presented with a stimulus to which his visual and verbal response would be different. Because of his theses of partial communication and of incommensurability-with-comparability, Kuhn can escape his critics' claims that his views of theory choice are irrational and that he is a historical relativist. He still maintains that there is an element of conversion in theory choice, because in the absence of a semantically neutral observation language the choice of a new theory is a decision to adopt a difficult native language, and to deploy it in a correspondingly different world. In a debate over theory choice neither party has access to an argument, which is compelling like logical or mathematical proofs. But their recourse to persuasion is for "good reasons", such as accuracy, scope, simplicity, or fruitfulness. These good reasons are the group's shared values, but not all scientists in the community apply these values in the same way. Consequently there will be variability that occasions revolutions. This is Kuhn's answer to Feyerabend's principal criticism. No principle of theory proliferation need be invoked to explain the transition to crisis and revolution, because unanimity of values will nonetheless produce the multiplicity of views, that bring on the transition from normal to revolutionary science. Variability in the application of uniform values produces variability in theories during normal science.

Feyerabend on Scientific Anarchy

In *Science and a Free Society* Feyerabend says in a section containing some autobiographical notes, that von Weizsacker (a former student of Heisenberg) has "prime responsibility" for Feyerabend's change to his anarchistic view. They met in Hamburg in 1965 and discussed the foundations of quantum theory. Feyerabend complained that alternatives to quantum theory had been omitted, but Weizsacker showed how quantum mechanics arose from concrete research. Feyerabend relates that it then became clear to him that general methodological rules imposed without regard to circumstances are a hindrance rather than a help, and that a person must be given complete freedom with no restrictions by any norms or demands regardless of how plausible they may seem to logicians and philosophers. Feyerabend concluded that such norms and demands must be checked by research, and not by appeal to ideas of rationality.

In *Against Method* (1975), Feyerabend's first book, he expounds his philosophy in terms of this political metaphor, "scientific anarchy", which he fully intends to be intellectually more radical than Kuhn's metaphor, "scientific revolution". Feyerabend's metaphor includes his principles of tenacity and theory proliferation to which he adds an antimethodological practice which he calls "counterinduction", a concept of scientific development that is opposed to the Logical Positivist critical method of confirmation and to Popper's critical method of corroboration. Counter-induction is opposed to all concepts of scientific rationality and methodology in which criticism is intended to eliminate some scientific theories as incorrect. Feyerabend advocates scientific anarchy, because he denies that there is any method or concept of rationality that is adequate to the history of successful science in any sense of the term. He is against all methodologies, because there is no methodological rule that has not been violated, and these violations are necessary for science. The only rule that he admits is that "anything goes". There is no institutional aim of science in his view, but instead each scientist may formulate his own individual aim of science, and "progress" may mean anything that one may wish.

In Feyerabend's view scientific knowledge is an ever increasing ocean of mutually incompatible and even incommensurable theories with each theory forcing the others into greater articulation. In this view counterinduction aims to introduce and to elaborate hypotheses, which are inconsistent with well established theories and with well established facts. This perpetual pluralism is possible, because even the most worthy theory has many anomalies where it does not fit the facts, while at the same time all factual statements contain theoretical assumptions. Not only is every factual description dependent on some theory, but there are also facts that cannot be unearthed except with the help of alternatives to the theory to be tested. These facts are unavailable so long as such alternative theories are excluded. In Feyerabend's view the practice of scientific research must not contain any rules requiring either consistency with so-called confirmed theories or with the choice between falsified and nonfalsified theories. The ocean of anomalies that always surrounds every theory is concealed by *ad hoc* hypotheses and by *ad hoc* approximations that are not the result of limited measurement accuracy, but which are adjustments to the theory to make it fit complicated cases.

Feyerabend illustrates counterinduction in the history of science with an examination of Galileo's defense of the Copernican theory against Aristotelian critics. In *Science and a Free Society* Feyerabend says that his views on Galileo expressed in *Against Method* are influenced by Philipp Frank. The relevant Aristotelian criticism is the "tower argument" according to which a stone dropped from a high tower would not fall vertically to the ground, if the earth were in motion as Copernicus' theory says it is, because the movement of the earth during the time of free fall would make the object fall at an angle away from the direction of the earth's movement. Feyerabend calls the observation of vertical fall of the stone a "natural interpretation" of the observation statement describing the motion of a falling stone, because the observational sensations are so firmly associated with the linguistic expression of the observation statement, and he says that it is very difficult to detect error in natural interpretations without alternative statements. In his examination of Galileo's reply to the tower argument Feyerabend maintains that Galileo used the Copernican theory to supply an alternative observational interpretation, and that Galileo's reply was a reinterpretation of the Aristotelian natural interpretation. In this manner Galileo appealed to the "real" motion of the falling stone, by which Galileo meant the stone's movement relative to absolute space. Galileo distinguished between Copernican and Aristotelian motion, and characterized them as "real" and apparent motions respectively, arguing that they are not the same.

Galileo's reply to the tower argument is an example of counter-induction. When a theory such as the Copernican theory is contradicted by facts, the counterinductive response is to turn around the situation and to use the theory as a "detection device" in Feyerabend's words. This procedure consists firstly of affirming the truth of the theory, and then of inquiring what changes in the facts will remove the contradiction between fact and theory. In this way hidden ideological components in the observation language expressing the facts are disclosed counterinductively. Once these ideological components are disclosed, the next step is to create a new observation language for the new theory. This is what Galileo did, and he used some propaganda to disguise that fact that he had invented the new observation language himself. His propaganda consisted in arguing that the human senses notice only relative motion, while they fail to notice motion had in common by such objects as falling stones and the earth, and he also used the *ad hoc* hypothesis that the earth is in permanent motion. Galileo believed in the truth of the Copernican theory, and he looked for the kinds of facts that supported that theory. One kind of fact, which is mentioned above, is that resulting from his reinterpretation of observed experience, such as the falling stone. His principle of the relativity of motion changed the conceptual component in observed fact. A second kind that Feyerabend identifies is the fact resulting from Galileo's invention and use of the telescope. Feyerabend says that Galileo did not know enough optical theory to enable the telescopic phenomena to function as independent evidence for the Copernican theory. Use of the telescope for celestial observation was also problematic to the Aristotelians, and what Galileo did was to use the agreement between the Copernican theory and the telescopic observation to argue on behalf of both of these views. The use of telescopic phenomena as evidence for the Copernican theory had to await the further development of the auxiliary science of optics.

Neither the telescopic phenomena nor the new idea of relative motion were acceptable to common sense at the time or to the Aristotelians, and the two associated ideas both seemed false. Yet these seemingly false and unacceptable phenomena were distorted by Galileo, and converted into strong support for Copernicus. Galileo replaced old facts with a new type of experience, which he simply invented for the purpose of supporting Copernicus, and he let apparently refuted theories support one another, in order to create a new world view. Feyerabend maintains that Galileo's arguments violate basic rules of scientific method, which were invented by Aristotle and canonized by Logical Positivists, such as Carnap and Popper. (Feyerabend occasionally calls Popper a Positivist.) And he states that Galileo succeeded precisely because he did not follow these rules. Had Galileo followed these methodological rules, he would have failed. Feyerabend's general thesis is that every methodological rule is associated with cosmological assumptions, so that using that rule implies that the cosmology in which it originates is correct. The rule that the Copernican theory must be tested is not unreasonable, but requiring that it be tested by confronting it with the *status quo* is not reasonable. What is reasonable is the purportedly "irrational" practice of waiting and ignoring large masses of critical observations and measurements, because the Copernican theory is an entirely new world view. It is necessary to retain the new cosmology, until it has been supplemented with the necessary auxiliary sciences, so that the language in which observations are expressed may be revised. Feyerabend finds what he illustrates with Galileo to be no less applicable today. He says that today's rational sciences survived, because irrational "prejudices" were permitted to have their way, and that it is advisable to let one's inclinations go against reason in any circumstances. Propaganda is of the essence. Science is more sloppy and irrational than its methodological image. Anarchistic deviations from rationality are necessary for progress. The image of twentieth century science is created by technological successes together with a "fairy tale" of how these technological miracles were accomplished. The fairy tale is that science is not an ideology, but rather is an objective measure for all ideologies. Science is an ideology, and successful science is very much a result of good luck and false beliefs. Feyerabend's thesis of scientific anarchy moves him far along in the direction of historical relativism. But the centrality of historical relativism in his philosophy of science is not fully evident without examination of the lengthy evolution of his philosophy of quantum theory and realism.

Feyerabend on Quantum Theory

From the time of his writing his dissertation in 1951, Feyerabend's philosophy of science has centered around the reconciliation of metaphysical realism with modern microphysics. The development of his thought on this matter might be viewed as a case of the moth and the flame. Initially he was critical of the Copenhagen interpretation, and particularly of Bohr's instrumentalist view of the quantum theory's formalism and of Bohr's complementarity thesis. Feyerabend received his views on metaphysical realism from Popper, but Feyerabend did not agree with Popper's attempt to supply the current quantum-theoretic formalism with the propensity interpretation. Instead Feyerabend defended the possibility of an altogether new microphysical theory. In the 1960's Feyerabend became involved in a long debate with Russell Hanson. As a result he reconsidered the merits of the current quantum theory, and the likelihood of its duality thesis and its quantum postulate being carried forward into a future microphysics. Then instead of continuing to advocate the revision of the current quantum theory into a microphysics that would be compatible with Popper's universalist realism, Feyerabend revised his concept of realism in a manner that no longer requires the universalism that Popper demands. Generalizing on Bohr's thesis of the relational character of quantum states when describing experimental findings with classical-colloquial concepts, Feyerabend formulated his nonuniversalist realism, which is a regional and historical relativist realism.

Feyerabend sets forth his statement of Popper's universalist realist philosophy in his "Attempt At A Realistic Interpretation of Experience" in *Proceedings of the Aristotelian Society* (1958). This paper is an abbreviated statement of his doctoral dissertation written in 1951 at the University of Vienna. The thesis of this paper, which he calls "Thesis I", is that the semantical interpretation of an observation language is determined by the theories that we use to explain what we observe, and that the interpretation changes as soon as those theories change. But he also states in this paper that one of the consequences of Thesis I is that we must distinguish between appearances or phenomena on the one hand and the things appearing on the other hand. In Feyerabend's view this distinction is fundamental to realism. On Thesis I the things appearing are those that are referred to by the observational sentences in a certain interpretation given by a realistic explanatory theory. In both this paper and in his "Complementarity" in *Proceedings of the Aristotelian Society* he criticizes the complementarity thesis of Bohr's interpretation of the modern quantum theory. In all discussions of the quantum theory Feyerabend always takes Bohr's statements and views to be authoritative and representative of the Copenhagen interpretation. In these earlier papers he acknowledges the influence of Bohm and of Popper upon his thinking. He notes that Bohr's idea of complementarity is based partly upon empirical investigations in physics and partly upon philosophical analyses, and he accordingly distinguishes between the experimental "fact of duality" and the philosophical thesis of complementarity. The fact of duality is the result of experimental findings. Experiments displaying interference effects can be explained by wave concepts, but they contradict explanations in terms of particle concepts. Conversely experiments displaying absorption and emission can be explained by particle concepts, but they contradict explanation in terms of wave concepts. Feyerabend maintains that there is no system of physical concepts, that can explain all these experimental facts about light and matter, which is to say, there is no universal theory of light and matter. He states that for a physicist who views duality as aspects of the same objective entity, that fact of duality proves that the theories available at the moment are inadequate, and such a physicist will search for a new theory and conceptual scheme, which satisfies two requirements: (1) the new theory must be empirically adequate, and (2) it must be universal. Such a theory conforms to what Feyerabend calls the "classical ideal", which is to say that it conforms to Thesis I, because it does not just describe appearances under certain experimental conditions, but rather it describes what light is and what matter is, the things appearing, in reality.

Feyerabend got this concept of realism from Popper. In "Complementarity" (1958) he references Popper's "The Aim of Science" published in *Ratio* (1957), and he says it as an excellent characterization of the classical ideal of scientific explanation and its connection with realism. In this article Popper affirms that explanations in science are given in terms of universal laws of nature, which are conceived as conjectural descriptions of the structural properties of nature, that is of the world itself. He explains that by "universal" he means that scientific laws and theories must make assertions about all spatiotemporal regions of the world. Popper also speaks of different levels of universality, which he exemplifies by the greater universality of Newton's laws relative to Kepler's and Galileo's laws. But Popper rejects a reductionist relation between Newton's and Galileo's physics. He states that whenever a new empirical theory of higher level of universality successfully explains an older theory, it does so by correcting the older theory. He adds that the idea of independent evidence can hardly be understood without the idea of discovery, of progressing to deeper layers of explanation without the idea that there is something to be discovered and to be discussed critically. Here "deeper layers of explanation" means explanation by means of more universal laws and theories, as exemplified by Newton's laws, which are deeper relative to Galileo's or Kepler's laws. This is the universalist realism that Feyerabend maintained, until he embraced relativism.

Feyerabend characterizes Bohr's philosophical thesis of complementarity as the exact opposite of the classical ideal of scientific explanation, and he says that the difference between the classical ideal and complementarity is an instance of the age-old issue between realism and Positivism. Bohr's complementarity thesis is an instance of Positivism, because Bohr maintains that the account of all evidence must be expressed in classical terms, and that it is not possible to dispense with what Bohr called the "forms of perception". Some philosophers such as Heisenberg consider Bohr's "forms of perception" to be neo-Kantian, and Feyerabend notes that Positivists do not customarily consider phenomena to have any forms, and Feyerabend therefore describes Bohr as a Positivist of a "higher order". He also states that Bohr's instrumentalist view of current quantum theory, which Bohr calls a "natural generalization of classical physics", is merely the result of retaining classical concepts. Both the retention of classical concepts and the instrumentalist view of quantum theory are contrary to Thesis I. He therefore says that complementarity is a statement of the fact of duality and is the way in which the classical concepts appear within the predictive schemes, which replace classical laws on the atomic level. He references passages contrary to Thesis I, in which Bohr states that the difficulties of atomic theory cannot be evaded by replacing the concepts of classical physics by new nonclassical conceptual forms. At the same time while Feyerabend views complementarity to be the result of retaining classical concepts, he does not simply deny the fact of duality, or that duality will be eliminated merely by philosophical reflection with the aid of

Thesis I.

With his distinction between the fact of duality on the one hand and the statement of complementarity expressing the fact of duality with classical concepts on the other hand, Feyerabend considers two approaches to a realistic microphysics. The first approach is to reinterpret the formalism of the modern quantum theory, which is a mathematical statement of the fact of duality. He admits that if the quantum theory is viewed as a predictive theory like celestial mechanics, then a realistic interpretation does not seem to be possible. But he adds that if the quantum theory is viewed as a theory containing new concepts for the description of nature, then a realistic interpretation "of a rather unusual kind" is definitely possible. This amounts to a proposal to construe the contemporary quantum theory with its duality thesis in accordance with Thesis I. Such a reinterpretation will not retain classical concepts, and will express the fact of duality without expressing complementarity. Feyerabend also says that the quantum theory used in this manner to form new concepts about the nature of physical systems, may permit some features of the macrophysical level to be derived from quantum mechanics, and therefore make duality compatible with the universality condition for realism.

But this first approach does not seem to be Feyerabend's preferred way to interpret microphysics realistically, and he says explicitly that the possibility of a realistic microphysics does not depend on supplying a realistic interpretation for the current quantum theory with its duality thesis. His second and preferred approach is to develop an entirely new microphysical theory. This new theory would satisfy two conditions: Firstly it would be universal, and secondly it would be empirically adequate. As a universal theory it will have a unified conceptual apparatus, which when applied to the domain of validity of classical physics, will be just as comprehensive as the classical apparatus. In other words the microphysical theory will be of a higher level of universality, such that it will also be a macrophysical theory, yet different from classical physics. Feyerabend explicitly compares the relation between the new universal microphysical physics and the classical physics, to the relation between the relativity theory of gravitation and the Newtonian theory of gravitation. The empirical adequacy criterion will be satisfied, when this realistic, universal macrophysical theory contains the current elementary quantum theory as an approximation. It may therefore contradict quantum mechanics without violating the universality criterion for realism. Feyerabend affirms that for a realist, the solution of the problem of duality need not be found in alternative interpretations of the current quantum theory, which he says is in all probability nothing but a predictive scheme anyway. Instead it can be found in the attempt to derive a completely new universal theory, which need not contain the duality thesis or complementarity. This new microphysical theory will supply new concepts for interpreting the observed fact of duality.

For ten years following these 1958 papers Feyerabend wrote a series of articles defending and advocating attempts to develop a new microphysics without duality. In these papers he contrasts his view that there can be a realistic microphysics without duality, with Bohr's view that all future microphysics must contain the duality thesis. In "Niels Bohr's Interpretation of the Quantum Theory" in *Current Issues in the Philosophy of Science* (1959) he discusses what he calls the "dogmatic elements" in Bohr's approach. He objects that Bohr treats duality as an unalterable experimental fact that must be included in any future microphysical theory; on his Thesis I description of experiments is not unalterable. Feyerabend argues that the only condition that need be satisfied by a future microphysical theory, is that it be compatible with experimental findings to a certain degree of approximation and within a certain degree of accuracy, that is required for the dogmatic elements of Bohr's approach. In this and other papers written during this period Feyerabend sets forth his interpretation of Bohr's philosophy, according to which all state descriptions of quantum mechanical systems are relations between the system and measuring devices in action, that is to say, between microscopic system and macroscopic apparatus. This relational character of quantum state descriptions results from the need to restrict the application of any set of concepts to a certain experimental domain because of the wave-particle duality. Bohr's relational view is contrasted with both the classical view and with Heisenberg's view of measurement in quantum theory. Feyerabend says that both classical physics and Heisenberg's view are variations on an "interactionist" view. In classical physics the interaction between the apparatus and the system can be explained in terms of the theory used to describe the system. And on Heisenberg's view the measurement of a quantum mechanical system involves an interaction that disturbs the system in unpredictable ways.

Feyerabend says that Bohr's relational view enabled Bohr to reply to the argument by Einstein, Podolsky and Rosen (EPR), who defended the thesis that quantum mechanical systems have definite classical states instead of indefinite states described by the indeterminacy relations. This argument postulates two systems which are separated to such an extent that no interaction can occur between them, and therefore measurement disturbance in one cannot affect the other. Bohr made his thesis of indefiniteness of state descriptions compatible with the EPR argument by assuming that states are relations between systems and devices rather than properties of the systems. The point is that while a property of the system cannot be changed except by interaction with the measurement device, a relation can be changed without such interaction. Position and momentum are therefore viewed by Bohr as relations rather than as properties of the quantum-mechanical system. Bohr attempts to express this by his distinctive use of the term "phenomenon", which he uses to refer to the observations obtained under specific circumstances including an account of the experimental arrangement. Therefore phenomena cannot be subdivided, and dynamical variables cannot be separated from the conditions of their application. Physical attributes no longer apply to the object per se, but apply to the whole experimental arrangement with different assertions (wave or particle descriptions) being made in different circumstances. Bohr relativized the dynamical variables in the quantum theory to the circumstances of the experimental situation, and years later following Bohr, Feyerabend would relativize all reality to the circumstances of the knower's situation.

But in 1962 in "Problems of Microphysics" in *Frontiers of Science and Philosophy* (ed. Colodny) Feyerabend was still defending the possibility of a universal and therefore realistic microphysical theory without duality. He says that between 1935 and 1950 the Copenhagen interpretation had become a "creed", and that the objections of a few opponents such as Einstein and Schroedinger were taken less and less seriously. But he notes that more recently there has occurred the development of a countermovement, which demands that the assumptions of the Copenhagen interpretation be given up and be replaced by a different philosophy. These "revolutionaries", as Feyerabend calls them, have shown not only that the empirical adequacy of the complementarity thesis is in doubt but also that even empirical success is not sufficient reason to say that there can be no valid alternative to complementarity. He insists that future researchers need not and indeed should not be intimidated by the restrictions that some "high priests" of complementarity would impose. One revolutionary in particular that Feyerabend has in mind is the physicist,

David Bohm. Initially Bohm had accepted the Copenhagen interpretation, but later he advanced an alternative thesis in his "Quantum Theory in Terms of Hidden Variables" in *Physical Review* (1951), and in more detail in his book, *Causality and Chance in Modern Physics* (1957). His "hidden-variable" thesis postulates the existence of a subquantum domain at a much smaller and presently experimentally inaccessible (therefore hidden) level of magnitude than the quantum domain that is described by modern quantum theory.

In "Professor Bohm's Philosophy of Nature", a review of Bohm's book in *British Journal for Philosophy of Science* (1961), Feyerabend says that complementarity can be interpreted in either of two ways. The way he finds acceptable is that in which it functions to provide an intuitive picture for wave mechanics, and as a "heuristic principle" guiding future research. He says that this first way is undogmatic, since it admits the possibility of alternatives including preferable alternatives, even though no satisfactory alternative exists presently. The second and unacceptable view is that of Bohr, who maintained complementarity as a basic philosophical principle incapable of refutation, and to which future microphysical theory must conform. In his review of Bohm, Feyerabend says that Bohm argues against Bohr's dogmatic view by affirming a role for speculation in modern empirical physics. In a discussion of the role of speculation in "Problems of Microphysics" Feyerabend rejects demands by Hanson that Bohm's theory must be set forth as an algebraically detailed and experimentally acceptable theory. He admits that such criticism is appealing to the great majority of physicists. But he maintains that such criticism puts the cart before the horse. The discussion among physicists of alternatives to the current theory plays a most important role in the development of physics, and a complicated physical theory cannot be invented in its full formal splendor without some preparation. This thesis was later elaborated upon by Feyerabend in his discussion of theoretical pluralism and counterinduction. At this stage of his thinking he advocates these ideas in order to encourage the development of a new microphysical theory not containing duality.

Norwood Russell Hanson, a professional philosopher of science, was an influential critic of Feyerabend's philosophy of quantum physics. In an article memorializing Hanson's death in 1967, and appearing in *Boston Studies in the Philosophy of Science,* Vol. III (ed. Cohen and Wartofsky, 1967) Feyerabend says that he changed his views about the Copenhagen interpretation, because of a series of debates with Hanson, and that by 1966 he had become persuaded of Hanson's view. Hanson brought a different agenda to the philosophy of microphysics than did Feyerabend. Hanson was not driven to defend the possibility of a universalist-realist microphysics, but rather was attempting to explain how the quantum theory as well as other theories are discovered. More specifically he focused on the role of semantics of observation and theory language in the discovery process. The evolution of their two agendas brought Feyerabend and Hanson into conflict. Integral to Hanson's agenda was the belief that the duality thesis will be contained in any future microphysical theory. This belief, which Hanson held with strong conviction, was due to the personal influence of P.A.M. Dirac, the physicist who developed the field quantum theory in 1928. On the other hand Feyerabend's agenda at that time was that a universalist-realistic microphysical theory is possible, precisely because the duality thesis need not be contained in any future microphysics, since according to Thesis I the observed experimental fact of duality can be revised by a new microphysical theory. Hanson's principal statement of his philosophy of science is set forth in his *Patterns of Discovery* (1958). In this work he recognizes the interdependence of observation and theory in a manner similar to Feyerabend's Thesis I, and Hanson describes observation as "theory-laden". In the "Introduction" to his *Realism, Rationalism and Scientific Method* (1981) Feyerabend comments that his Thesis I is not exactly the same as Hanson's doctrine that observation is theory-laden, because unlike Hanson, Hesse and others, he maintains that observation terms are fully theoretical and have no purely observational core. Feyerabend thus rejects Hanson's thesis of phenomenal seeing. Nonetheless Hanson was no more sympathetic than Feyerabend to Bohr's view that the concepts of classical physics must be used for observation in all of physics.

Hanson criticizes Feyerabend by maintaining that duality is stated by the quantum theory formalism itself, and that duality is not merely a philosophical thesis appended to the formalism, which might be replaced by an alternative interpretation not expressing duality. Hanson finds the duality thesis stated by the de Broglie-Einstein relations and also by the Dirac operator calculus which enables any wave-mechanical description to be transformed into an equivalent matrix-mechanical one. Feyerabend seems not actually to have maintained the position that Hanson criticizes, even in the first of his two approaches to a realistic microphysics given in "Complementarity" (1958). However, Hanson repeats this line of attack nearly ten years later in "Physical Implications of Quantum Physics" in The

Encyclopedia of Philosophy (ed. Edwards, 1967), where he characterizes Feyerabend as maintaining that the metaphysical views in the Copenhagen interpretation be abandoned as indefensible, and that the minimal scientific content consisting of algebraic transformations and factual data is quite compatible with some interpretation markedly different from the Copenhagen one. Perhaps this is just the way in which Hanson viewed Feyerabend's call for a new microphysics without duality, even though Feyerabend was very clear in stating that his second approach is not just an alternative interpretation of the elementary quantum theory, but rather is an entirely new microphysical theory related to elementary quantum theory as Einstein's relativity theory is to classical physics.

Nonetheless, the thrust of Feyerabend's attack is against Bohr's thesis that classical concepts in the complementarity description of the fact of duality must occur in microphysics including any future microphysics. In "Comments on Feyerabend's 'Niels Bohr's Interpretation of the Quantum Theory'..." (1959) Hanson states what he considers to be the minimal essentials of the Copenhagen interpretation: Firstly he maintains that past and present microphysical experience make it probable but in no sense necessary, that any future microphysical theory will incorporate the quantum postulate and the duality principle. Secondly he notes that there presently exists no coherent, currently workable and fully articulated conception of a microphysical theory, which can do without the quantum postulate and the duality principle. He maintains that Feyerabend is correct to score the strident statements of Bohr and Rosenfeld, when they violate the history of physics by suggesting that any future microphysics will of necessity guarantee things like complementarity. But he adds that Bohr's metaphysics is not an indispensable part of the Copenhagen interpretation, and he therefore distinguishes the "Copenhagen interpretation" from the "Bohr interpretation". He states that if the Bohr interpretation is cut away, then what remains is a "liberalized Copenhagen interpretation", which is entirely defensible. And he maintains that there are good contingent arguments in support of the expectation that any future microphysics will incorporate the quantum postulate and the duality principle, and emphasizes that presently there exists no working alternative to the current quantum theory notwithstanding all its awkward features. As a review of Hanson's philosophy of quantum theory will show, Feyerabend's response to Hanson's criticisms did not result in a "liberalized Copenhagen interpretation". What Feyerabend produced is an elevation of the Bohr interpretation to a generalized and quite radical relativistic philosophy of knowledge. It seems unlikely that Feyerabend understood what Hanson wanted to cut away.

Feyerabend on Relativism, Historicism, and Realism

The consequential outcome of the lengthy debate between Hanson and Feyerabend results less from their discussion about current quantum theory than from their discussion about the future of microphysics, if not also the future of Feyerabend's philosophy. Feyerabend found himself in the position of having to wait for some future physicist to produce a future scientific revolution in future microphysics, that would obligingly comply with his current philosophical specifications; and it may have occurred to Feyerabend that he might have to wait a very long time, even assuming that future physics were ever to accommodate him at all. In any event he was led to reconsider quite radically his agenda for a realistic microphysics. Instead of philosophizing to accommodate future physics to his universalist-realist agenda, he decided to philosophize to accommodate realism to the current quantum theory. Therefore he accepted Hanson's conviction that any future microphysics will very likely contain duality. But Feyerabend construed this to mean that duality must be expressed by complementarity, and in making his accommodation he did not cut away the Bohr interpretation and proceed with a "liberalized" Copenhagen interpretation, as Hanson had advocated. Instead Feyerabend drew upon Bohr's thesis of the relational nature of quantum states, which Feyerabend saw as contradicting *universalist* realism, and he then generalized on Bohr's relational thesis to affirm a nonuniversalist, *relativized* realism. Just as either the wave or particle manifestations of microphysical reality are conditioned upon respectively either one or another experimental arrangement, so more generally scientific knowledge is conditioned upon the historical situation and regional circumstances of the scientist. And even more generally all truth and knowledge including the particular Western tradition known as science, must be viewed in this historicist perspective.

It may be noted that Feyerabend had apparently been sympathetic to relativism even before his views on quantum theory had been influenced by Hanson. In 1962 he proposed his thesis of semantic incommensurability at the same time that Kuhn had used the idea to describe scientific revolutions. When critics pointed out the historical relativism implied in Kuhn's use of the incommensurability thesis, Kuhn began to modify the concept so as to evade the relativistic implications. But Feyerabend made no such concession, when he defended use of the idea. In "Consolations for the Specialist" (1971) he defended the relativistic implications of Kuhn's use of incommensurability, saying that the choice between incommensurable cosmologies is a matter of taste. In 1978 in his *Science in a Free Society* Feyerabend references Bohr's relational interpretation of the quantum theory, which Bohr had devised in response to the criticism of Einstein, Podolsky and Rosen, as an example of an incommensurable theory relative to classical physics. In this context he says that the change from one world view described by a theory to another world view described by another theory that is incommensurable with the first, is a change in universal principles, such that one no longer speaks of an objective world that remains unaffected by one's epistemic activities, except when moving within a particular world view. In this 1978 work Feyerabend continues to invoke universal principles. Bohr's relational thesis is referenced merely as an example of incommensurability, and seems not yet to have become integral to Feyerabend's cultural relativism.

Later in his "Introduction" to his *Realism, Rationalism and Scientific Method* (1981) Feyerabend states that quantum theory offers good reason to resist the universal application of his Thesis I and its realistic metaphysics. Logically to reject Thesis I is to reject common sense, and to announce that objectivity is a metaphysical mistake. But what physicists have actually done in effect is to reject the universal application of Thesis I, while still retaining in quantum theory some fundamental properties of common sense. In all but Bohm's hidden-variables quantum theory, a universal, realistic interpretation of the quantum theory has been replaced by a partial instrumentalism. He explains that the transition to a partial instrumentalism contains two elements, that are not always clearly separated. The first element is the existence of multiple metaphysical traditions. One tradition usually associated with common sense arguments in physics is the fact that there actually are relatively isolated objects in the world, and that physicists are capable of describing them. But there are also other metaphysical traditions, such as the Buddhist exercises, that create an experience which neither distinguishes between subject and object nor recognizes distinct objects. The second element in the transition to a partial instrumentalism is the choice by the physicist of one or another of these metaphysical traditions, and then the turning of the choice into a boundary condition for research. And this choice of metaphysical traditions, furthermore, is one between different sets of facts, because there are no tradition-independent facts.

Feyerabend then states that the choice of metaphysical traditions is a choice of "forms of life". Realism itself is thereby relativized to prior choices proceeding from cultural and social values. This is because a people decide to regard those things as real, which play an important role in the form of life they prefer. Thus the decision about what is real and what is not, begins with a choice of one or another form of life, and a people reject a universal criticism affirming a realistic interpretation of theories not in agreement with their chosen life form. Conversely realism merely reflects the preference for ideas accepted as foundational for their civilization and even for life itself. In this context instrumentalism is incidental to the choice of one or another theory for realistic interpretation. Instrumentalism is what is not culturally agreeable, and it no longer has the characteristics of a failure or defect. What has failed is not realism, but rationalism with its universalist criterion for realism. Feyerabend welcomes the failure of rationalists to explain science in terms of tradition-independent standards and methodologies, because it is a failure to put an end to attempts to adapt science to chosen forms of life. The failure of rationalism has freed science from irrelevant restrictions. He adds that it is furthermore in agreement with the Aristotelian philosophy, which also limits science by reference to common sense, except that in Feyerabend's philosophy the conceptions of the individual philosopher are replaced by the political decisions emerging from the institutions of a free society. This is Feyerabend's thesis of "democratic relativism". As of this writing Feyerabend's most mature and elaborate statement to date of his historicist and relativist philosophy is set forth in his *Farewell to Reason* (1987). In the "Introduction" to this book he says that science has undermined the universal principles of research, and he asks rhetorically: who would have thought that the boundary between subject and object would be questioned as part of a scientific argument, and that science would be advanced thereby? And yet, as he notes in his next sentence, this is precisely what happened in the quantum theory. Feyerabend explicitly states that he does not deny that there are successful theories using abstract concepts. What he denies is that knowledge should be based on universal principles or theories. Echoing Conant, perhaps without even recognizing so, Feyerabend says that science is a living enterprise as opposed to a body of knowledge, and that it is a historical process, although unlike Conant, Feyerabend's view is historicist and relativist, and also realist.

An important distinction that emerges from Feyerabend's historical relativist philosophy, is his distinction between "historical" or "empirical" traditions on the one hand and "theoretical" traditions on the other. This distinction is made in "Historical Background" in *Problems of Empiricism* and later in "Knowledge and the Role of Theories" and in "Trivializing Knowledge" in *Farewell to Reason.* His earlier philosophical views are clearly in the theoretical tradition, while his later views are clearly in the historical tradition. However, the distinction is not a fundamental one, because the thesis of his later view is that modern science with its theoretical tradition is a new historical tradition. All theoretical traditions are really historical traditions according to Feyerabend's later view. On the one hand the members of a theoretical tradition identify knowledge with universality, and they attempt to reason by means of a standardized logic. They distinguish the "real" world from the world of appearances, because they identify the reality with what their universal theories can describe as law-like and stable. And when their universal laws fail, the members of the theoretical tradition issue the "battle cry" stating: "we need a new theory!" In theoretical traditions true knowledge and logic are viewed as universal and independent of cultural traditions or regional circumstances. On the other hand the members of a historical tradition emphasize what is particular including particular regularities such as Kepler's laws. It produces knowledge that is restricted to certain regions, and which depends on conditions specifying the regions. And this knowledge is relative knowledge of what is true or false. Instead of using a standardized logic, they organize information by means of lists and stories, and they reason by example, by analogy, and by free association. They emphasize the plurality of knowledge, and consequently the history dependence and culture dependence of knowledge and of all logical standards. Feyerabend notes in this context that the complementarity thesis in modern quantum theory even contains the idea of relative knowledge, because of the relational character of quantum states. In a discussion on the semantical interpretation of theories in his "Knowledge and the Role of Theories" Feyerabend bases his historical relativism on an artifactual theory of the semantics of language. He rejects the idea that there is any truth that is capable of superseding or transcending all traditions and cultures, an idea that he traces to Parmenides. He argues that this belief confounds the properties of ideas with their subject matter. The subject matter remains unaffected by human opinions, and the erroneous implications is that scientific statements describing the subject matter are supposed to be expression of facts and laws, which exist and govern events no matter what anyone thinks of them. He maintains that the statements themselves are not independent of human thought and action; they are human products. They were formulated with great care to select only the "objective" ingredients of our environment, but they still reflect the peculiarities of the individuals, groups, and societies from which they arose. For example the validity of Maxwell's equations is independent of what people think about electrification. But it is not independent of the culture that contains them; it needs a very special mental attitude inserted into a very special structure combined with quite idiosyncratic sequences of historical developments.

Theoretical traditions are opposed to historical traditions in intention, but not in fact. Scientists trying to create a knowledge that differs from "mere" historical or empirical knowledge, succeeded only in finding formulations which seemed to be objective, universal and logically rigorous, but which in fact are used and interpreted in use in a manner that conflicts with the properties the formulations only seem to have. Modern science is a new historical tradition that has been carried along by a false consciousness. Feyerabend similarly criticizes the metaphysics of scientific realism of the theoretical traditions of science. Scientific realism accepts as real only what is lawful or may be connected by laws, and thereby regards the real to be what exists and develops independently of the thoughts and wishes of researchers. Feyerabend argues that connecting reality with lawfulness is to define reality in a rather arbitrary manner. Moody gods, shy birds, and people who are easily bored would be unreal, while mass hallucinations and systemic errors would be real. The success of science can not be a measure of the reality of its ingredients. Feyerabend notes that to support their view, the scientific realists say that while scientific statements are the result of historical processes, the features of the world are independent of those processes. But he argues that we either consider quarks and gods to be equally real, or we cease to talk about real things altogether. And he adds that to say that quarks and gods are equally real is not to deny the effectiveness of science as a provider of technologies and of basic myths; he intends only to deny that scientific objects and they alone are real. And he adds that the equal reality of quarks and gods does not mean that we can do without the sciences; he acknowledges we cannot.

Feyerabend's Criticism of Popper

Consider firstly Feyerabend's general view to Popper's philosophy. Initially sympathetic to Popper's philosophy, Feyerabend became one of its most relentless and truculent critics. In *Against Method* he rhetorically describes Popper's views as "ratiomania" and "law-and-order science". As his historical relativist philosophy became more mature, Feyerabend described the technical procedures of Popper's "critical rationalism" - the hypothesizing, testing, falsification, and new hypothesizing to produce new theories having greater empirical content - as merely rules of thumb, that cannot be taken as necessary conditions for science. Contrary to Popper, Feyerabend takes sides with Kuhn by maintaining that science is a historical tradition having practices that are not always recognized as explicit rules, and that may change from one historical period to the next. He compares understanding a period in the history of science to understanding a stylistic period in the history of the arts. In both science and the arts periods have an obvious unity, but it is one that cannot be summarized in a few simple rules, and the practices that guide it must be found by detailed historical studies. The general notion of such a unity, which Kuhn calls a "paradigm" and which Lakatos calls a "research programme", will therefore be poor in content. Feyerabend rejects the demands for precision made by some technical philosophers, saying that they are on the wrong track. Presumably this would include the criticisms by Shapere.

Consider secondly Feyerabend's specific criticisms of Popper's views on quantum theory. Feyerabend seems never to have been sympathetic to Popper's propensity interpretation, which represents the participation by the philosopher in the work of the physicist. Even while he was sympathetic to Popper's general philosophy, Feyerabend preferred to encourage physicists rather than to join them, as Popper did. Later when Feyerabend reconciled himself to the Copenhagen interpretation, he became explicitly critical of Popper's propensity interpretation. His criticisms of Popper are set forth in his "On A Critique of Complementarity" in *Philosophy of Science* (1968-1969), which he later had reprinted as "Niels Bohr's World View" in *Realism, Rationalism, and Scientific Method* (1981). Popper had offered two interpretations of the statistical quantum theory during his career. The earlier interpretation offered in *Logic of Scientific Discovery* involved a variation on the frequency interpretation of probability, and the later interpretation first advanced in his "Quantum Mechanics without the

Observer" (1967) was based on his propensity interpretation of probability. Feyerabend criticizes both these interpretations. Feyerabend rejects Popper's frequency interpretation of Born's statistical quantum theory. He admits that it is not unreasonable, if physicists already know what kinds of entities are to be counted as the elements of the collectives, and if they know that those elements are classical entities. And he agrees with Popper that one cannot draw inferences about the individual properties of the elements. But Feyerabend argues that Popper's view, that the elementary particle always posses a well defined value for all its magnitudes, i.e. position and momentum, is precisely what has been found to be inconsistent with the laws of interference and of the conservation laws. He therefore maintains that a new interpretation of the elements of quantum-mechanical collectives is needed, and that what is being counted as elements is not the number of systems possessing a certain well defined property; rather what is counted is the number of transitions from certain partly ill defined states into other partly ill defined states, as Bohr had maintained.

Feyerabend's criticism of Popper's propensity interpretation is similar. Popper viewed probability as a propensity, a physical property comparable to physical forces, and pertaining to a whole experimental arrangement for repeatable measurements. The wave function determines the propensity of the states of the particle, in the sense that it gives weights to its possible states. Thus in the two-slit experiment a change in the experimental arrangement such as shutting one of the slits, affects the distribution of the weights for the various possibilities, and thus produces a different wave function. Such a change in the experimental arrangement is analogous to tilting a pin board with the result that a new distribution curve of the rolling balls will differ from its shape, when the pin board is not tilted. Quantum mechanics is therefore viewed by Popper as a generalization of classical statistical mechanics of particles together with the propensity interpretation of probability. Feyerabend says that Popper's propensity interpretation is much more similar to Bohr's view, which Popper attacks, than to Einstein's view, which Popper attempts to defend. He says that Popper's thesis that the experimental conditions of the whole physical setup determine the probability distribution, is precisely Bohr's relational thesis, when Bohr proposed defining the term "phenomenon" to include the whole experimental arrangement.

But Feyerabend's thesis is furthermore that Bohr's idea of complementarity goes beyond the propensity interpretation by attributing to the experimental arrangement not only probability but also the dynamical variables of the physical system, notably position and momentum. Therefore Popper's thesis that a change in experimental conditions implies a change in probabilities alone, is not adequate to account for the kind of changes involved in the two-slit experiment. In other words complementarity asserts the relational character not only of probability, but of all dynamical magnitudes. Feyerabend agrees with Popper that a change of experimental conditions also changes probabilities, but he also says that what led to the Copenhagen interpretation is not merely the fact that there is some change in distribution with a change of experimental arrangement, but the kind of change encountered: trajectories which from a classical view are perfectly feasible, are forbidden to the particle. This is because the conservation laws apply not only on the average, so that one could postulate a redistribution without asking for some dynamical cause, but furthermore they apply in each single interaction. Thus a purely statistical redistribution is inadequate; each single change of path must be accounted for. Bohr's resolution consists of the renunciation of particle trajectories, the denial that particles posses well defined position with well defined *momenta* according to the indeterminacy relations. Feyerabend maintains that Popper confused classical waves with quantum waves, because he neglected the dynamics of the individual particle and construed quantum theory as pure statistics. Popper's claim that the reduction of the wave packet is not an effect characteristic of quantum theory, but rather is an effect of probability in general, is incorrect in Feyerabend's view. And Popper's claim that duality is the "great quantum muddle" is in Feyerabend's words nothing but a piece of fiction.

Feyerabend also has a number of other specific criticisms of Popper's philosophy of science, which are summarized in "Historical Background" in *Problems of Empiricism,* the second volume of Feyerabend's collected papers. There are eight such specific criticisms, which may be summarized as follows:

1. Feyerabend notes that theory exchange has not always proceeded by falsification. Noteworthy examples include the transition from the celestial theory of Ptolemy and Aristotle to that of Copernicus, and the transition from Lorentz's theory to Einstein's theory of special relativity. In these cases there were no refuting facts to explain rejection of the preceding theory.
2. The meaning of a hypothesis often becomes clear only after the process

that led to its elimination has been completed. The force of this objection seems to be that falsification brings about meaning change, that the decision to accept a test outcome as a falsification is also a decision that affects the semantics of the language involved in the test. Feyerabend elaborates on this thesis in his "Trivializing Knowledge", a paper criticizing Popper's philosophy in *Farewell to Reason*, where he says that the content of theories and experiment are constituted by the refutations performed and accepted by the scientific community, rather than being the basis on which falsifiability can be decided and refutation determined. He exemplifies this point with the stereotypic theory "all ravens are black", and he says that while a white raven falsifies this theory, the refutation depends on the reason for whiteness. A decision must be made as to whether a raven whose metabolic processes make it white, or whose genetic make up has been altered to make it white, or which has been dyed white, constitutes a falsifying instance. Feyerabend says that such decisions are not independent of falsification. He also uses this example to illustrate Lakatos' philosophy of science in "Popper's Objective Knowledge" a critical review of Popper's book in *Problems of Empiricism*. Here he states that what is needed is some insight into the causal mechanism that brought about whiteness, a theory of color production in animals. He also notes that this illustration shows the need for alternative theories in the process of testing.

3. The transition to a new theory may involve a change of universal principles, which breaks the logical links between the theory and the content of its predecessor. This break produces the semantic incommensurability, that Feyerabend has discussed at length in *Against Method* and earlier papers. Incommensurability is not only the principal basis for his historical relativism, which Popper opposes, but it is also inconsistent with Popper's thesis of scientific progress through increasing empirical content and verisimilitude.

4. Feyerabend rejects Popper's thesis of increasing content for reasons in addition to the occurrence of semantic incommensurability. This is a criticism that Feyerabend discusses at length in *Against Method*, where he states that a new period in the history of science commences with a "backwards movement" to a theory with less empirical content, that gives scientists the time and freedom needed for developing the main thesis of the new theory in greater detail, and for developing related auxiliary sciences. Scientists are persuaded to follow this backward movement by such "irrational" means as propaganda and *ad hoc* theories, that sustain a blind

faith in the new theory until it turns into what comes to be regarded as sound knowledge. This is what Feyerabend saw in Galileo's defense of the Copernican theory, where the relevant auxiliary science needing further development at the time was optics.

5. A closely related criticism of Popper's philosophy is Feyerabend's thesis that *ad hoc* adaptation of a theory may be the right step to take. The *ad hoc* adaptation may be made either to the theory or to the statements of observation. In Popper's philosophy these *ad hoc* adaptations are objectionable as content-decreasing stratagems. But Feyerabend maintains that they disguise the inadequacy of a new theory until the relevant auxiliary sciences can be developed, so that refutation ultimately does not occur.

6. The demand that the scientist look for refutations and take them seriously, will lead to an orderly development only in a world in which refuting instances are rare and turn up at large intervals. But this is impossible if theories are surrounded by an ocean of anomalies, unless we modify the stern rules of falsification using them only as rules of thumb, and not as necessary conditions for scientific procedure. Feyerabend frequently states elsewhere in his literary corpus that strict falsification would wipe out science as it presently exists, and would never permit it to have come into existence.

7. Popper's demand for increasing content makes sense only in a world that is infinite both quantitatively and qualitatively. On the other hand in a finite world containing a finite number of basic qualities or elements, the aim is firstly to find these elements, and then secondly to show how novel facts can be reduced to them with the help of *ad hoc* hypotheses. He adds that genuine novelty counts as an argument against the methods that produce it. Feyerabend gives no further explanation of what he means by this peculiar criticism, nor does he give any reference to any other part of his corpus for explanation.

8. Finally Feyerabend objects that content increase and the realistic interpretation of the idea that brings it about, restrain human freedom.

Feyerabend's Philosophy of Science

Of the four basic topics that may be considered in philosophy of science (aim of science, scientific explanation, scientific criticism, and scientific discovery) the place to begin an overview of Feyerabend's philosophy of science is with the topic of scientific criticism. Central to his philosophy is the thesis that the choice of scientific theories is not fully decidable empirically, and this thesis is the basis for his attack on Popper's falsificationism or "critical rationalism". Feyerabend also attacks Kuhn's sociological thesis of how the empirical undecidability is resolved. The arbitrariness in criticism permitted by this empirical indeterminacy has been described in various ways. Conant called it "prejudice", Kuhn called it "paradigm consensus", and Feyerabend called it "tenacity". Conant was simply dismayed by the phenomenon he observed in the history of science, but he took it more seriously than did his contemporaries, the Positivist philosophers, who preferred to dismiss it as simply unscientific. Conant found that prejudice is too frequently practiced by contributing scientists to be dismissed so easily. He also explicitly admitted the strategic role of his own prejudices in his preference for a historical examination of science.

Kuhn did not merely accept prejudice as a frequent fact in the history of science. He saw it as integral to science because of a sociological function that it performs within a scientific community, a function that is a condition for scientific progress. Prejudice, which Kuhn had earlier referred to as the "problem of scientific belief", is the sociologically enforced consensus about a paradigm, that is necessary for the scientific community to function effectively and efficiently for solving detailed technical problems referred to by Kuhn as "puzzles". Without the consensus the community could not marshal its limited resources for the exploration or "articulation" of the promises of the paradigm. In Kuhn's concept of science professional discipline becomes synonymous with conformity to prevailing view defined by the paradigm. The phase during which this conformity is a criterion for criticism and is effectively enforced by sociological controls, is "normal" science.

Feyerabend rejects Kuhn's thesis that prejudice functions by virtue of a sociologically enforced uniformity. In Feyerabend's view any such uniformity is indicative of stagnation rather than progress. Instead, prejudice understood as the principle of tenacity is strategically functional, because it has just the opposite effect that Kuhn thought: it promotes diversity and theoretical pluralism, which in Feyerabend's view are necessary conditions for scientific progress. It might be said that Feyerabend views Kuhn's sociological thesis of normal science as an instance of the fallacy of composition, the fallacy of incorrectly attributing to a whole the properties had by its component parts. Houses need not have the rectangular shape of their component bricks, and scientific professions need not have the monomaniacal prejudices of their members. The prejudice or tenacity practiced by the member scientist performs a function that does not obtain if his whole profession were unanimously to share in his prejudice or his tenaciously held view.

The process by which the individual scientist's tenacity is strategically functional is counterinduction. Its strategic functional contribution occurs because of Thesis I, which says that theory supplies the concepts for observation. Tenacious development of a chosen theory results in the articulation of new facts, which enhance empirical criticism. New facts produced by counterinduction can both falsify currently accepted theories and revitalize previously falsified theories. The revitalization occurs because the new facts occur in sciences that are auxiliary to the falsified theory. This possibility of revitalization justifies the scientist's prejudicial belief in a falsified theory, his irrational rejection of falsifying factual evidence.

The topic of scientific criticism leads to the topic of the aim of science. Popper has a well defined and explicit thesis of the aim of science. The aim of science in his view is the perpetual succession of conjectures and refutations in which each successive conjecture or theory can explain both what had been explained by its falsified predecessor and the anomalous cases that falsified the predecessor. The new theory is therefore more general than its predecessor, while it also replaces and corrects its falsified predecessor. Popper saw the process of refutation as involving a deductive procedure having the logical form of *modus tollens*. And because it is a procedure in deductive logic, it is not subject to cultural or historical change. Popper admits that application of the logic in the sense of experimental identification of the falsifying instances may be problematic and may take several years. But he maintains that the logic of falsification isolates the conditions for scientific progress, and that it represents adequately how science has proceeded historically, when it has proceeded successfully. He maintains that this procedure may be said to have become institutionalized, but its validity, which is guaranteed by deductive logic, does not depend on its institutional status. Its validity is ahistorical, and will never be invalidated by historical or institutional change; it is tradition independent.

Both Kuhn and Feyerabend deny that Popper's vision of the development of science is historically faithful. The principal deficiency in the Popperian vision is its optimistic assessment of the decidability of falsification. Not only do they view the range of nondecidability of scientific criticism to be greater than Popper thinks, but they also view it as having an integral role in the process of scientific development. This nondecidability gives the scientist a range of latitude which he is free to resolve by his strategic choices. Kuhn and Feyerabend disagree on which aims influence this choice, but they agree that they are historical or institutional in nature and may change. Furthermore, such changes involve semantical changes, that introduce an additional dimension to the scientist's freedom of choice, when they involve an incommensurable discontinuity. Kuhn views incommensurable change as characteristic only of occasional scientific revolutions, with sociologically enforced consensus resisting such change and defining the aim of science during the inter-revolutionary periods of normal science. Feyerabend also views incommensurable changes as infrequent, but he does not regard the interim periods as an enforced consensus contributing to scientific progress, but instead views normal science as Kuhn defined it as an impediment to progress. He therefore advocates a much more individualistic aim of science, which he refers to as scientific anarchy. Ironically both Popper and Feyerabend explicitly reference Marx's call for "revolution in permanence", but their meanings are diametrically opposed. Popper means perpetual conjectures and refutations occurring within an enduring institutionalized logical framework for conclusive refutation, while Feyerabend means perpetual institutional change with no controlling tradition-independent framework.

Feyerabend's discussion of scientific explanation contains much more criticism of other philosophers' views than elaboration of his own views. From the outset of his professional career he criticized the deductive-nomological concept of scientific explanation and of logical reductionism advocated by the Logical Positivists. Initially Feyerabend also considered Bohr's concept of explanation to be a higher kind of Positivism, but he later preferred to view Bohr as a kind of historicist philosopher, because of Bohr's distinctive relationalist interpretation of complementarity in quantum theory. As it happens, Bohr was sufficiently naive a philosopher, that Positivist, neo-Kantian, and historicist characterizations can all find support in his works.

For nearly the first two decades of his career Feyerabend subscribed to Popper's philosophy of science, which contains a concept of scientific explanation requiring universal statements. Popper's philosophy of explanation also contains the idea of deeper levels of explanation, where the depth is determined by the scope or extent of universality of the explanation. Initially Popper proposed his thesis of verisimilitude, according to which the deeper explanations are said to be closer to the truth. Later he reconsidered the idea of verisimilitude, but he continued to describe explanations as having greater or lesser depth according to the extent of their universality, and he also continued to describe the universal laws and theories occurring in explanations as having greater or lesser corroboration, because science cannot attain truth in any timeless sense of truth. After Hanson had persuaded Feyerabend to reconsider the merits of the Copenhagen interpretation of quantum theory, Feyerabend rejected Popper's concept of explanation by logical deduction from universal laws, and instead accepted historicism. He was led to this conclusion by his incommensurability thesis and by the nonuniversalist implications he found in Bohr's relationalist interpretation of quantum theory. Popper had stated that scientific theories are merely conjectures that may be highly corroborated, but may never be true in any timeless sense. Feyerabend furthermore says that theories have an even more historical character, since the complementarity thesis in quantum theory demonstrates their regional character. Complementarity makes quantum theory nonuniversal at all times, because it is conditional upon mutually exclusive experimental circumstances; it is not even temporarily universal. Feyerabend thus concluded that universal science, science containing universal laws and theories, is only apparently universal, and that it is actually a special and recent historical tradition.

Feyerabend's historicist philosophy of scientific explanation is in need of greater elaboration. For example he has yet to relate his views to the genetic type of explanation that is characteristic of historicism. Although this type of explanation had been dismissed by Positivists as merely an elliptical deductive-nomological explanation, it was discussed seriously by Hanson in "The Genetic Fallacy Revisited" in *American Philosophical Quarterly* (1967). Hanson distinguishes different levels of language, one for historical fact and one for conceptual analysis. He says that the distinction differentiates history of science from philosophy of science, and that the genetic fallacy consists of the attempt to argue from premises in the historical level to conclusions in the analytical level. Perhaps at a future time Feyerabend will relate his historicism to the problem of the genetic fallacy, and perhaps characterize fallacious from valid genetic arguments.

The topic of discovery may be taken to refer either to the development of new theories or to the development of new facts. Feyerabend's thesis of counterinduction is a thesis of the development of new facts. Thesis I enables the scientist to use the concepts supplied by new theory to make new observations. Counterinduction is a thesis of observation according to the artifactual philosophy of the semantics of language which Feyerabend set forth in his Thesis I. It is unfortunate that Feyerabend never examined Heisenberg's use of Einstein's admonition for reinterpreting the Wilson cloud chamber observations as an example of counterinduction. But Feyerabend never references anything written by Heisenberg, and it is unlikely that he has any appreciation for the differences between Heisenberg's and Bohr's philosophies of quantum theory. He addresses the problem of developing new theories in "Creativity" in his *Farewell to Reason.* In this brief article he takes issue with what other philosophers have often called the heroic theory of invention, the idea that creativity is a special and personal gift. He criticizes Einstein for maintaining a variation on the heroic thesis. Einstein wrote that theory development is a free creation, in the sense that it is a conscious production from sense impressions, and that theories are fictions, which are unconnected with these sense impressions, even though theories purport to describe a hidden and objective world. Feyerabend maintains that at no time does the human mind freely select special bundles of experience from the labyrinth of sense impressions, because sense impressions are late theoretical constructs and not the beginnings of knowledge. Feyerabend expresses much greater sympathy for Mach's treatment of scientific discovery. Mach advanced the idea of instinct, which Feyerabend contrasts with Einstein's idea of free creation. Mach offered an analysis of the process, according to which instinct enables a researcher to formulate general principles without a detailed examination of relevant empirical evidence. Instinct seems not as such to be inherent, but rather is the result of a long process of adaptation, to which everyone is subjected. Many expectations are disappointed during this process of adaptation, and the human mind retains the results of consequently altered behavior. These daily confirmations and disappointments greatly exceed the number of planned experiments. They are used to correct the results of experiments, which are in need of correction because they can be distorted by alien circumstances. Therefore, according to Mach empirical laws developed from principles proceeding from instinct are better than laws developed from experiment. In concluding his discussion of the topic of creativity Feyerabend advocates a return to wholeness, in which human beings are viewed as inseparable parts of nature and society, and not as independent architects. He rejects as conceited the view that some individuals have a divine gift of creativity. Feyerabend therefore apparently subscribes to the social theory of invention, as would be expected of a historicist.

Comments and Conclusion

Feyerabend's philosophy of science began with an agenda for modern microphysics: an attempt to show how a realistic microphysics is possible. Initially the conditions that he believed a realistic microphysics must satisfy were taken from Popper's philosophy of science, and these conditions are contained in Popper's idea of universalism. However, there is an ambiguity in Popper's use of "universalism", and that ambiguity was not only brought into Feyerabend's agenda while he had accepted Popper's philosophy; it was also operative in his philosophy after he rejected Popper's philosophy, because he rejected universalism in *both* senses. The first meaning of "universal" refers to the greater scope that a new theory should have relative to its predecessors, and the second meaning refers to the logical quantification of general statements. Bohr's insistence upon the use of classical concepts for observational description in quantum theory experiments makes quantum theory inconsistent with both these meanings of "universality". Thus Feyerabend's later acceptance of Bohr's interpretation of the quantum theory led him to reject universalism in both of Popper's senses, and consequently to advance his radical historicist philosophy of science. Quite apart from his acceptance of Bohr's use of classical concepts, Feyerabend had adequate reason to reject universalism in Popper's first sense: it is gratuitously reductionist and there are historic exceptions to it. Feyerabend notes explicitly in his *Against Method* for example that Galileo's theory of motion is less universal than that of the preceding theory, Aristotle's doctrine of the four causes, which explained qualitative change as well as mechanical motion.

But Feyerabend also believes that his Thesis I with its dependence on universal logical quantification cannot be applied to quantum theory because of Bohr's semantical thesis of complementarity, which is duality expressed with classical concepts, and he therefore rejects universalism in the sense of universal logical quantification. This rejection involves a semantical error that is made by many philosophers including the Logical Positivists and the Copenhagen physicists. That semantical error consists of implicitly regarding the meanings of descriptive terms or variables, or even larger units of language, as unanalyzable wholes. What is needed in order to see how universal logical quantification is consistent with duality without complementarity, is a semantical metatheory of meaning description which enables an analysis of the semantical composition of the meanings of the descriptive terms. Consider such an analysis:

Since Quine's rejection of the analytic-synthetic distinction, and notwithstanding the fact that he rejected analyticity altogether, the distinction may still be viewed as a pragmatic one instead of a semantic one, such that any descriptive universally quantified statement may be viewed as either analytic or synthetic. Consider a list of universally quantified affirmations having the same subject term and which are believed to be true. The concepts associated with the descriptive terms predicated of the common subject by the several statements in the list exhibit the composition in the complex meaning of the subject term. The meaning of the subject term may therefore be said to have parts consisting of the predicating concepts, and the meaning need not be viewed wholistically. Consider in turn the relations that may obtain among the concepts that are universally predicated in the affirmations having the common subject term. These predicate terms may or may not be related to each other by other universal statements. If any of the predicate concepts are related to one another by universally quantified negative statements, then the subject term common to the statements in the list is equivocal, and the predicate concepts related to one another by universal negations are parts of different meanings of the equivocal subject term. Otherwise the subject term common to the statements in the list is univocal, whether or not the predicate concepts may be related to one another by universally quantified affirmations, and the predicate concepts are different parts of the one meaning of the univocal subject term.

Terms are univocal; concepts are vague. All concepts are vague, but vagueness may be reduced in discrete increments by adding information. Adding universal statements to the list, reduces the vagueness in their common subject term by clarifying the meaning of the shared subject term with respect to the added predicate concepts. Adding universal negations relating concepts predicated of the common subject clarifies the meaning of the subject term by showing equivocation. Adding universal affirmations relating the concepts predicated of the common subject, clarifies the meaning of the subject term by revealing additional structure in the meaning of the common univocal subject term, and makes a deductive system. Now turn to microphysics: In all scientific experiments including microphysical experiments, the relevant descriptive language is dichotomously divided into those universal statements that are presumed for testing and those universal statements that are proposed for testing. The former are called test design statements and the latter are called theory statements. The distinction between them is pragmatic, because it depends upon the functions of the statements in testing. A given descriptive term occurring in a theory may be viewed as a subject term occurring in a list of universal affirmations with the list dichotomously divided into test design and theory statements. The subject term is thus common to the test design statements and to the theory statements. The dual analytic-synthetic nature of the statements makes the common subject term have part of its semantics supplied by the descriptive terms predicated of it by the test design statements, which are presumed true for the test. And this part of the term's semantics remains unchanged through the test, so long as the distinction between theory and test design statements remains unchanged.

However, the part of the term's semantics supplied by the theory statements changes; it is different before and after the test. Before the execution of a test of a theory, all interested scientists who agree to the test design, presumably also agree that the universal statements describing the test design are true independently of the theory, such that if the test outcome is an inconsistency between the test design statements and the theory statements, then it is the theory that is to be viewed as falsified. This independence of test design statements that is required for contingency in the test, precludes the test design statements from either implying or denying the theory to be tested or any alternative that addresses the same problem. Therefore for the cognizant scientific profession the semantical parts defined by the test design statements before test execution must be vague with respect to the theory. This amounts to saying that the theory does not define any part of the semantics of its constitutent terms. However it may happen that the originating proposer and the supporting advocates of the theory may have such high confidence in their theory, that for them the theory may have come to supply part of the semantics for its constituent terms even before the test. After the test is executed in accordance with its test design, the test design statements and the theory statements are either consistent or inconsistent with one another; and they therefore either characterize the same observed instances or they do not. If the test outcome is an inconsistency between the test design statements and the proposed theory, then the theory is falsified, and since it is therefore no longer believed to be true, it cannot contribute to the semantics of its constituent descriptive terms even for the proposer and advocates of the theory. But if the test outcome is not a falsifying inconsistency between theory and test design statements, then for each common term the semantics contributed by the two sets of statements are parts of one meaning complex of the univocal descriptive term, and they identify the same instances. Furthermore, the additional characterization supplied by the semantics of the theory statements resolves the vagueness that the meaning of the descriptive term had before the test for those who did not share the confidence had by the theory's proposer and its advocates. However, the original proposer and the supporting advocates of the theory have options if the test outcome was a falsification. They may choose to reverse the status of the test design statements and theory statements, such that the theory assumes the role of defining the subject of the test, and the test design is rejected as an adequate or appropriate description of the phenomenon under investigation. This prejudice or tenacity may occasion what Feyerabend calls counterinduction.

While the vagueness in the concept associated with the common subject term is reduced by a nonfalsifying test outcome, the vagueness in the concepts predicated of the subject term by the two sets of statements are not necessarily resolved in relation to one another merely by the nonfalsifying test outcome. Any resolution of the vagueness in these predicate concepts requires that these concepts furthermore be related to one another by additional universal statements, as is the case were the statements formerly used as independent test design statements are able to be incorporated into a deductive system with the theory after the test. The resulting deductive system may make test design statements logical consequences of the theory, but with the theory tested and not falsified, this loss of independence of the test design statements is no longer important. In a physical theory this would amount to deriving from the theory a new set of laws applicable to the functioning of the apparatus and physical procedures of an experiment, and described by the test design statements. This is possible in the case of relativity theory, but not in the case of quantum theory. In cases where description of the apparatus and physical procedures in terms of the laws constituting the theory is possible, then before the experiment the description in the test design statements of the apparatus' functioning with laws other than those of the tested theory must result in what retrospectively may be called "errors", and these errors must be small relative to the effect that the apparatus is used to produce or detect in the nonfalsifying test of the theory. And the concepts associated with predicates in the original test design statement must be viewed as vague relative to the terms in the theory with respect to the more precise meanings they may later receive from the definitive role of the nonfalsified theory, after the test design statements are made derivable from the theory. This vagueness means that before the test the concepts associated with the vocabulary used in the test design statements assumed the semantical status that Heisenberg called the physicist's "everyday" concepts.

Feyerabend's Thesis I requires that the test design statements, which describe the macrophysical experimental set up, must be incorporated into a deductive system consisting of the microphysical quantum theory. And since this derivative type of description has not yet been achieved, he accepts Bohr's complementary thesis, which is the description of the microphysical phenomena with classical concepts. As it happens, contrary to Bohr's instrumentalist thesis but consistent with Heisenberg's semantical views, the microphysical phenomena can be described with the variables in the mathematical expressions of the quantum theory and without classical concepts, but there is no quantum description of the functioning of the macrophysical apparatus by means of laws logically derived from the quantum theory. Thus at the conclusion of the first section of "Trivializing Knowledge" in *Farewell to Reason* Feyerabend says, that though Popper rejects reductionism, the variety of entities Popper admits to be real can be admitted as parts of the same world only if the theories that constitute them can be united in a way precluded by the incommensurability that Feyerabend finds in the relative knowledge in Bohr's complementarity thesis of quantum theory. He then concludes that science is not a theoretical tradition expressed as deductive systems, as he says Popper assumes, but rather is a historical tradition.

But contrary to Feyerabend, relativism is not the exclusive alternative to reductionism; the choice between classical and quantum macrophysical descriptions is a false dichotomy. The universal test design statements, such as those describing the experimental set up, need not say anything about the fundamental constitution of matter; that is what the microphysical theory describes. The semantics supplied by these test design statements may remain vague about this subject for an indefinite time after the test, just as they had to before the test was performed and while its outcome was not yet known. After the test the semantics supplied by the tested and nonfalsified quantum theory provides further resolution of the concepts associated with these terms common to both test design and theory statements, but the semantics supplied by the macrophysical predicates of the test design statements may retain their vagueness indefinitely until a reductionist macrophysical quantum theory is developed, since the concepts associated with these terms are not unanalyzable wholes, but rather are complexes of semantic values. If Feyerabend's Thesis I were modified such that after the nonfalsifying test outcome the theory as a set of universal analytic-synthetic statements defines only part of the semantics of its constituent descriptive terms, then Thesis I becomes applicable to quantum theory. The application of the modified semantical principle implies that the test design-defined part of the meaning complex associated with the theory's descriptive term is not properly called "classical", because it makes no microphysical claims. Before the test it is vague with respect to any microphysics, and Heisenberg's term "everyday" is appropriate to describe the concepts associated with these terms. But after the test outcome is known the whole meaning complex constituting each concept is more properly called a "quantum" concept, given that the quantum theory is not falsified, because the quantum theory resolves vagueness by the addition of the quantum theory-defined meaning parts to the whole meaning complex. And it is for this reason Heisenberg used quantum concepts when he described the *observed* free electron in the Wilson cloud chamber, and those quantum concepts were resolved by the context supplied by his matrix mechanics.

In summary, semantical analysis reveals that duality need not be expressed in classical terms by Bohr's complementarity principle, because the semantics of the descriptive terms used for observation are not simple, wholistic, or unanalyzable, and because prior to testing the semantics of these terms cannot imply an alternative description to that set forth by the quantum theory, in order for testing to have the contingency that gives it its function as an empirical decision procedure in the practice of science. Feyerabend was closer to the mark with the first of his two approaches to realism in microphysics set forth in his "Complementarity" (1958), and he might have retained universalism in quantum theory had he ignored the reducionist program of Ludwig, developed a metatheory of semantical description, and appropriately modified Thesis I. With appropriate modification as described above the application of Feyerabend's Thesis I to the quantum theory need not imply historical relativism, the rejection of the validity of universal quantification. The quantum theory with its quantum postulate, its duality thesis, and its indeterminacy relations has no need for Newtonian semantics. It is a universal theory with a univocal descriptive vocabulary, and it is not semantically unique in empirical science because of internal incommensurability. Had Feyerabend considered Heisenberg's realistic philosophy of the quantum theory, he would probably not have been driven to advocate his incommunsurability and historical relativist theses, in order to implement his realistic agenda for microphysics. Then instead of speaking of the Galileo-Einstein tradition, he could have referenced the Galileo-Einstein-Heisenberg tradition including Heisenberg's pluralistic thesis.

Consider further Feyerabend's incommensurability thesis, which is central to his historical relativism. Rejecting the naturalistic theory of the semantics of language including the language of observational description permits dispensing altogether with classical concepts in quantum theory, and thereby with incommensurability within the quantum theory. But Feyerabend sees incommensurability in Bohr's complementarity thesis only as a special case, a case that is intrinsic to a single theory due to the use of classical concepts. Most often Feyerabend treats incommensurability as a relation between different theories, and he maintained incommensurability even before he adopted Bohr's interpretation of quantum theory. In his earlier statements of the thesis he says that two theories are incommensurable, if they can have no common meaning, because there exists no general concept having an extension including instances described by both theories; the two theories therefore cannot describe the same subject matter, and therefore are incommensurable. In *Against Method* he also referenced Whorf's thesis of linguistic relativity to explain incommensurability in terms of covert resistances in the grammar of language. There he maintains that these covert resistances in the grammar of an accepted theory not only lead scientists to oppose the truth of a new theory, but also lead the scientists to oppose the presumption that the new theory is an alternative to the older one. He considers both the quantum theory and the relativity theory to be incommensurable in relation to their predecessor, Newtonian mechanics. However, he offers no evidence for his implausible historical thesis, that the advocates of Newtonian physics had failed to recognize that either quantum theory or relativity theory is an alternative to Newtonian physics at the time of the initial proposal of these new theories.

Feyerabend furthermore maintains that since incommensurability is due to covert classifications and involves major conceptual changes, it is hardly ever possible to give an explicit definition of it. He says that the phenomenon must be shown, and that one must be led up to it by being confronted with a variety of instances, so that one can judge for oneself. Feyerabend's concept of incommensurability suffers from the same kind of difficulty as Kuhn's concept of paradigm. Readers of Feyerabend must rely on his identification of which transitional episodes in the history of science are to be taken as involving incommensurability and which ones do not, just as Kuhn's readers must rely on the latter's identification of which transitional episodes are transitions to a new and incommensurable paradigm and which ones are merely further articulations of the same paradigm. Although the two philosophers do not hold exactly the same views on the nature of incommensurability, and while they disagree about Kuhn's thesis of normal science, they both refrain from developing a metatheory of semantical description that would enable their readers to individuate theories and thereby to characterize semantical continuity and discontinuity through scientific change. Feyerabend's recourse to the Wittgenstinian-like view that incommensurability cannot be defined but can only be shown, may reasonably be regarded as an obscurantist evasion in the absence of such a semantical metatheory.

The semantics of the Newtonian and relativity theories that Feyerabend says are incommensurable may be examined by considering their synthetic statements analytically. By way of example consider one of the more famous empirical tests of Einstein's general theory of relativity, the test that had a formative influence on Popper. This test known as the "eclipse experiment" was undertaken by two British astronomers, Sir Arthur Eddington of Cambridge University and Sir Frank Doyle of the Royal Greenwich Observatory. The test consisted of measuring the gravitationally produced bending of starlight visible during an eclipse of the sun that occurred on May 29, 1919, and then comparing measurements of the visible stars' positions with the different predictions made by Einstein's general theory of relativity and by Newton's celestial mechanics. The test design included the use of telescopes and photographic equipment for recording the telescopic images of the stars. Firstly reference photographs were made during ordinary night darkness of the stars that would be visible in the proximity of the eclipsed sun. These photographs were used for comparison with photographs of the same stars made during the eclipse. They were made with the telescope at Oxford University several months prior to the eclipse, when these stars would be visible at night in England.

Then the astronomers journeyed to the island of Principe off the coast of West Africa, in order to be in the path of the total solar eclipse. During the darkness produced by the eclipse they photographed the stars that were visible in the proximity of the sun's disk. They then had two sets of photographs, the earlier set displaying images of the stars unaffected by the gravitational effects of the sun, and the later set displaying images of the stars near the edge of the disk of the eclipsed sun and therefore produced by light rays affected by the sun's gravitational influence. The stars in both sets of photographs that are farthest from the sun in the eclipse photographs are deflected only negligibly in the eclipse photograph. And since different telescopes were used for making the two sets of photographs, reference to these effectively undeflected star images was used to determine an overall magnification correction. But correction furthermore had to be made for distorting refraction due to atmospheric turbulence and heat gradients. The distortions are large enough to be comparable to the effect being measured. But they are also random from photograph to photograph, and the correction can be made by averaging over the many photographs. Such are the essentials of the design of the Eddington eclipse experiment. The amount of deflection calculated with the general theory of relativity is 1.75 arc seconds. Eddington's findings showed a deflection of 1.60 $\pm$ 0.31 arc seconds. The error in these measurements is small enough to conclude that Einstein's general theory is valid, and that the Newtonian celestial mechanics can no longer be considered valid. Later experiments have reduced the error of measurement, thereby further validating the relativity hypothesis. In this experiment the test design statements include description of the optical and photographic equipment and of their functioning, of the conditions in which they were used, and of the photographs of the measured phenomenon made with these measurement instruments. These statements have universal import, since they describe the repeatable experiment, and are presumed to be true characterizations of the experimental set up. The theory statements are also universal, and each theory shares descriptive variables with the same set of test design statements. If the test design statements are viewed as analytic statements, then any descriptive variable occurring both in a test design statement and in either theory has a univocal semantics with part of its meaning contributed by one or several test design statements. This semantics is shared by both theories, and it makes the theories semantically commensurable.

Feyerabend has maintained that theories are incommensurable, because there is no concept that is general enough to include both the Euclidian concept of space occurring in Newton's theory and the Reimannian concept occurring in Einstein's theory. In fact the common part of the meanings in the semantics of the descriptive terms common to the two theories and to the test design statements, are not common meanings because of a more general geometrical concept. There is a common meaning because the test design statements are silent about the claims made by either theory, even as both the theories claim to reference the same instances that the test design statements definitively describe. Before the test this silence is due to the vagueness in the common part of the meaning of the terms shared by the theory statements and defined by the test design statements. In the case of the test design for Eddington's eclipse experiment, it may be said that before the test the meanings contributed by the test design statements are not properly called either Newtonian or Einsteinian. For purposes of describing the experimental set up, their semantics have the status of Heisenberg's "everyday concepts", that are silent about the relation between parallel lines at distances much greater than those in the apparatus.

After the test is executed and the nonfalsification of the relativistic theory and the falsification of the Newtonian theory are known outcomes of the test, the acceptance of the relativity theory is a pragmatic transformation giveing it the semantically defining status of an analytic statement, and the statements of the theory supply part of the semantics for each descriptive term common to the theory and the test design statements. This semantical contribution by the theory to each of these common descriptive variables may be said to resolve some of the vagueness in the whole meaning complex associated with each of these common terms, and thus the terms may be said to have Einsteinian semantics. But the semantics supplied to these terms by their test design statements is still vague, just as before the test. However, if the test design statements are consequently derived logically from the relativistic theory, then these common terms receive still more Einsteinian semantic values from the accepted relativity theory, which also contributes its additional structure to the semantics of the common terms. In this case everyday concepts may still describe the phenomenon, but the Einsteinian concepts are resolutions of the vagueness in the everyday concepts in the predicates in the test design statements. In either case, regardless of whether or not the test design statements describing the experimental set up can be logically derived from the relativity theory, no resolution of the everyday concepts to Newtonian concepts is involved either before or after the test, except for the convinced advocates of the Newtonian theory before the latter theory's falsification. But after the test outcome falsifying the Newtonian theory, even the most convinced advocates of the Newtonian theory must accept the semantically controlling role of the test design statements, or reject the test design itself.

Nonetheless some physicists inaccurately refer to the concepts in the test design statements of relativity theory as Newtonian concepts. This is because any relativistic effects in the test equipment are too small to be detected or measured, and therefore do not jeopardize the conclusiveness of the test. For example two different telescopes were used in the Eddington eclipse experiment to produce the photographs, one used before the eclipse and another used during the eclipse. Since the resulting two sets of photographs were compared, a correction had to be made for differences in magnification. But no correction was even considered for the different deflections of starlight inside the telescopes due to the different gravitational effects of their different masses even by those who believed in the relativity theory, because such differential relativistic effects were not empirically detectable. But the nonmeasurability or undetectability does not imply that the test design statements affirm the Newtonian theory. For the test to have any contingency the test design statements must be silent about the tested theory and any alternative to it. Consequently the concepts in the test design statements describing the phenomena were vague with respect to whether or not there is any relativistic effect introduced by the different masses of the two telescopes, and the concepts in the test design statements are too vague to be described as Newtonian or Einsteinian. This vagueness in the concepts in test design statements is indicated by a possible variation retrospectively called a measurement error, that is not due to failure to execute the test in conformity with the test design, and that is recognized only after the test outcome is accepted. There was such error in the Eddington experiment, but it was very small relative to the measured deflection of starlight by the sun's gravitational force through interplanetary distances. This inaccuracy due to vagueness is relative to the other concepts in the test design statements, and it must be distinguished from the vagueness relative to the concepts in the theory. The meaning parts or semantic values defined by the test design statements are vague with respect to those defined by the theory statements, but this vagueness does not affect the measurement accuracy, since the condition of independence precludes the theory statements being used for measurement.

In addition to Bohr's complementarity thesis and his own incommensurability thesis, Feyerabend is led to his radical historicism by the view that whether in philosophy of science or in any social science, cultural views and values including the criteria and research practices of empirical science are inseparable from historical conditions. In its radical variant it says that particular historical circumstances do not function to supply initial conditions for universal theories describing recurrent aspects of human social behavior, but rather preclude the validity of universals altogether. The persuasive objection to this historicism is that concepts are inherently universal (or as Popper says, all terms are disposition terms). The metatheory which proposes using synthetic universal statements analytically for semantical description, which also enables exhibiting semantical continuity through scientific change through history, is a variation on this old but valid objection to this old philosophy. However, Feyerabend's historicism enjoys a novel plausibility that could not be admitted by philosophies from Platonism to Positivism, which advance a naturalistic philosophy of the semantics of language. Platonic Ideas, Aristotelian forms and simple apprehensions, Romantic intuitions, and Positivistic phenomena, sensations, sense data, and operationalist definitions are all variations on the myth of "the given". The scientific revolutions of the twentieth century have forced philosophers, and specifically Pragmatists, to affirm that meaning and belief are mutually conditioning, and in this sense are relativized to one another. But this relativism is conditioned by universal statements used to describe the real world. The real world is what imposes constraints on this mutual conditioning in language, that makes falsification possible, and that reveals the real world to us. Given any selected set of concepts, only some statements can be maintained; and conversely given any selected set of stated beliefs, only some concepts may be maintained. The selection of truths is negotiable among interested scientists, but truth conditioning expressed in universal statements linking initial conditions and test outcomes is not negotiable once test design statements are chosen. New experiences anomalous to our universal beliefs force revisions of those universal beliefs and therefore of their semantics. In empirical science the locus of the semantical revision is a proposed universal hypothesis conditioned upon chosen universal test design statements.

The evolution of thinking from Conant's recognition of prejudice in science to Feyerabend's counterinduction thesis has brought to light an important limitation in Popper's falsificationist thesis of scientific criticism. In this respect Feyerabend's philosophy of science represents a development beyond Popper, even after discounting Feyerabend's radical relativism. Popper had rejected the Positivists' naturalistic philosophy of the semantics of language, and maintained that every statement in science can be revised. But the paradigmatic status he accorded to Eddington's 1919 eclipse experiment as a crucial experiment had dissuaded Popper from exploring the implications of the artifactual semantics thesis, implications which he identified with essentialism and therefore rejected. He saw that the decidability of a crucial experiment depends on the scientist "sticking to his problem", which is to say that the scientist should not redefine his problem by reconsidering any experiment's test design, especially after the test outcome has been a falsification of the proposed theory. Such animadversions in Popper's view have no contributing function in the development of science; they are objectionable because they are *ad hoc* content-decreasing stratagems, merely evasions. But the prejudiced or tenacious response of a scientist to an apparently falsifying test outcome does have a contributing function in the development of science, as Feyerabend illustrates in his examination of Galileo's arguments for the Copernican cosmology. Use of the apparently falsified theory as a "detecting device" by letting his prejudicial belief in the geocentric theory control the semantics of observational description enabled Galileo to reinterpret observations previously described with the equally prejudiced alternative semantics built into the Aristotelian cosmology. This was also the strategy used by Heisenberg, when he reinterpreted the observational description of the Wilson cloud chamber experiment with the semantics of his indeterminacy relations pursuant to Einstein's anticipation of Feyerabend's Thesis I, i.e. that theory decides what the scientist can observe. As it happens, the cloud chamber experiment was not designed to decide between Newtonian and quantum mechanics; the water droplets are very large in comparison to the electron, and the produced effect admits easily to either interpretation. But the counterinduction strategy could also have been used by tenacious Newtonians who chose to reject the findings from Eddington's eclipse experiment. The artifactual status of the semantics of language permits the dissenting scientists to view the falsifying test outcome as a refutation of one or several test design statements rather than as a refutation of the Newtonian theory. Or more precisely, what some scientists view as definitive test design statements, others may decide to view as falsified theory.

Feyerabend recognizes that there are semantical consequences to counterinduction. In "Trivializing Knowledge" he states that the "contents" of theories and experiments are constituted by the refutation performed and accepted by the scientific community, rather than being the basis on which falsifiability can be decided and refutation can be carried out as Popper maintains. He considers the stock theory "All ravens are black", and states that while a white raven falsifies the theory, the refutation depends on the reasons for the anomalous raven's whiteness. Earlier in his "Popper's Objective Knowledge" he gives the same example, and says that the decision about the significance of the anomalously white raven depends on having a theory of color production in animals. But his discussion by means of this stock theory pertains more to the factors that motivate a scientific community to decide between test design and theory statements, than to a description of the semantics resulting from that decision. Feyerabend has no metatheory of semantical description for characterizing the "contents" of theories and experiments. In this respect Feyerabend's philosophy suffers the same deficiency as Popper's.

RUSSELL HANSON AND OTHERS ON THE SEMANTICS OF DISCOVERY

Norwood Russell Hanson (1924-1967), born in New Jersey, was a U.S. Marine Corps fighter pilot during the Second World War, who earned the rank of major and was awarded the Distinguished Flying Cross and the Air Medal for flying combat missions over Japan. Afterward he studied at the University of Chicago, Columbia University, and Yale University in the United States, and then studied at both Oxford University and Cambridge University in England. He received a Ph.D. from Oxford in 1955 and a Ph.D. from Cambridge in 1956, and was afterward a fellow at the Institute for Advanced Study at Princeton. He accepted a faculty appointment at Indiana University in 1957, and was founder and chairman of Indiana University's Department of History and Logic of Science from 1960 to 1963. He then accepted a professorship on the philosophy department faculty of Yale University, which he had at the time of his premature death at the age of forty three in a crash of his private airplane in 1967. His principal works are *Patterns of Discovery* (1958) and *The Concept of the Positron* (1963). At the time of his death he left an uncompleted textbook in philosophy of science intended for first-year college students, which was edited by Willard C. Humphreys, a former student of Hanson, and then published as *Perception and Discovery* (1969). Hanson also wrote many articles, and a complete bibliography of his publications can be found in *Boston Studies in the Philosophy of Science*, Volume III (1968).

Bohm's Agenda for Future Microphysics

Hanson takes very seriously the question of the interpretation of the modern quantum theory, and he quite truculently defends the Copenhagen interpretation. In Appendix II of his *Patterns of Discovery* he notes that while for most practical microphysical problems Born and Schroedinger would have made the same theoretical calculations, nonetheless their alternative interpretations organized their thinking differently and therefore influenced their future research work in very different ways. After 1930 Born was led to work on collision behavior, on the statistical analysis of scattering matrices, while Schroedinger pursued investigation of the so-called ghost waves of the elementary particles. The interpretations, therefore, are important because they influence the agenda for future research. But Hanson does not view all interpretations to be equally worthy of consideration. One that he considers to be particularly unfortunate is the "hidden-variable" interpretation developed by David Bohm. Hanson explicitly criticizes Bohm's views, and understanding Hanson's philosophy of microphysics is aided by considering Bohm's agenda for future microphysics and Bohm's hidden-variable interpretation. David Bohm (1917-1992) was born in Wilkes-Barre, PA, and received his doctorate in physics from the University of California. He taught physics at Princeton, and then moved to England in 1957. He was professor of theoretical physics from 1961 at Birkbeck College, University of London, where he was professor emeritus from 1983 until his death in 1992. A brief biography may be found in the "General Introduction" in *Quantum Implications* (ed. B.J. Hiley and F. David Peat, 1987).

His principal statement of his "hidden-variable" interpretation of quantum theory is set forth in his book *Causality and Chance in Modern Physics* (1957), in which he reports that the interpretation was originally stimulated by a discussion with Einstein in 1951. The initial statement of the interpretation was published in 1952 in two articles in the *Physical Review*. He then found that his interpretation had been anticipated in important respects by de Broglie (1892-1987) in 1927. De Broglie's interpretation had been criticized severely and he had consequently abandoned it, but Bohm had further developed the thesis enough that the fundamental objections confronting de Broglie had been answered; Bohm's interpretation was shown to be consistent with all the essential characteristics of the quantum theory. Additional suggestions were made by Vigier, a colleague of de Broglie, and de Broglie then returned to his original proposals, since he believed that the decisive objections against them had been answered. Bohm and Vigier published a joint paper setting forth the interpretation in the *Physical Review* in 1954, and De Broglie wrote a "Foreword" to Bohm's 1957 book. Bohm also wrote another statement of the interpretation in 1957 titled "A Proposed Explanation of Quantum Theory in Terms of Hidden Variables at a Sub-Quantum Mechanical Level" in *Observation and Interpretation.* During the nearly forty years that Bohm Advocated his hidden-variable interpretation of quantum theory, the several published statements offer some variation, and even in the same paper he has offered more than one version.

Bohm's hidden-variable interpretation is an agenda for future microphysics, and his book *Causality and Chance* gives at least three related objectives in this agenda. His first objective is the apparently modest one of demonstrating that an alternative to the Copenhagen interpretation is possible, in the sense that it is not the only one that is consistent with the formalism and measurements of modern quantum theory. He states this objective not only because he has another interpretation in mind, but also because he maintains that the development of alternative views is important for the advancement of science, while advocates of the Copenhagen interpretation deny that any alternative view proposing a subquantum level of magnitude is conceivable. For example in his "Questions of Principle in Modern Physics" (1935) in *Philosophical Problems of Quantum Physics* Heisenberg states that the uncertainty principle must be taken as a question of principle making other formulations into false and meaningless questions, just as in relativity theory it is supposed that it is in principle impossible to transmit signals at speeds greater than the velocity of light. But Bohm maintains that without alternatives the physicist is constrained to work along accepted lines of thought in the hope that new experimental developments or lucky and brilliantly new theoretical insights eventually will lead to a new theory. In contrast Bohm maintains that one of the functions of criticism in physics is to suggest definite alternative lines of research that are likely to lead in a correct direction. This objective is particularly attractive to the philosopher of science, Paul Feyerabend, once an advocate of Bohm's interpretation, who still says that creating alternatives is necessary for the advancement of science.

Bohm's second objective is to propose an interpretation of the history of physics which shows successful precedents for the research strategy represented by his hidden-variable interpretation of quantum theory. The strategy consists of postulating that there exists a level of physical magnitude below the quantum level of magnitude represented by Planck's constant. This subquantum level contains new types of qualitative phenomena (entities or fields) governed by more deterministic laws than are known to exist at the quantum level of magnitude. The existence of this postulated subquantum level of microphysical phenomena is denied by the Copenhagen interpretation, and since the there is as yet no experimental evidence for such subquantum phenomena, the theory that postulates it is said to have "hidden" variables (where if it need be said it is the phenomena in the subquantum level of physical reality and not the symbolic variables in the equations of the theory that are understood to be hidden!).

Bohm opposes his historical interpretation to another that he calls "mechanism", a term that is unfortunately ambiguous in both philosophical and scientific usage, but which has a specific and somewhat elaborate meaning in his book. According to the objectionable mechanistic philosophy of physics the qualitative diversity of things in the world can be reduced completely, without approximation, and in every possible domain of science, to nothing more than the effects of some definite and limited general framework of quantitative laws, which are regarded as absolute and final. Prior to the development of quantum theory these quantitative laws were assumed to be deterministic; then with the development of the Copenhagen interpretation of quantum theory these laws were assumed to be indeterministic. Hence there are both deterministic and indeterministic varieties of mechanism; in the former causal laws are thought to be fundamental, while in the latter probability laws are thought to be fundamental. Indeterministic mechanism prevails today, because physicists have accepted Heisenberg's thesis that the indeterminacy principle represents an absolute and final limitation on our ability to define the state of things by any kinds of measurements now or ever possible.

In *Causality and Chance* Bohm maintains that both causality and chance are fundamental and objective, and that both determinism and indeterminism are merely idealizations. Thus he departs from Einstein's determinism. He rejects the subjective interpretation of probability, which says that chance is a result of human ignorance. And he rejects the idea common to both deterministic and indeterministic varieties of mechanism, that there is only one general framework of laws and a limited qualitative diversity. Bohm maintains that there are levels of depth or orders of magnitude with each level having its own laws and qualitative diversity. In the history of physics revolutionary developments have occurred when laws and qualities at a higher level are explained by those of a lower level. Experiments may disclose a breakdown of an entire scheme of laws by the appearance of chance fluctuations not originating in anything at the higher level, but instead originating in qualitatively different kinds of factors at a lower level. For example in classical physics a particle such as an electron follows the classical orbit only approximately, while in a more accurate treatment it is found to undergo random fluctuations in its motions arising outside the context of the classical level. Thus Bohm concludes by way of analogy and on the basis of his nonmechanistic interpretation of the history of science, that there is a still deeper level, a subquantum level, that in turn explains the randomness that is detected at the quantum level of magnitude. It is noteworthy that Bohm's hidden-variable interpretation is not just another interpretation of quantum theory motivated by prior ontological commitments, but rather is what may be called a heuristic, a discovery strategy, or a research programme which he says may be used as a guide for future research intended to develop a new microphysical theory.

Finally Bohm's third objective is that the hidden-variable interpretation used as a guide for future research for a new microphysical theory, is capable of resolving what he calls the "current crisis" in quantum physics. This crisis manifests itself in Dirac's relativistic quantum theory, when the wave equation is applied to the description of particle scattering with very high energies and at short distances. For the Schroedinger wave equation to be used in such applications, it must be corrected by a type of *ad hoc* mathematical adjustment called "renormalization". Furthermore the behavior of very high energy particles in experiments reveals that there exist many new kinds of particles not previously known, and that they are unstable, decaying into one another and creating other particles. Nothing like this is accounted for by current quantum theory. To Bohm these problems for the current quantum theory suggest that "elementary particles" are not really elementary, and that the concept of a subquantum level enables the physicist to consider a whole range of qualitatively new kinds of theories approaching the currently accepted theory only as approximations that hold in limiting cases. He believes that the current crisis in quantum theory portends a revolution in microphysics, and that the hidden-variable interpretation offers a superior guide for research that promises to resolve the crisis.

These three objectives of Bohm's agenda represent successively more ambitious claims. The first claim is merely that an alternative to the Copenhagen semantical interpretation describing a subquantum level of magnitude is conceivable in the sense that it is consistent with the data and formalism of quantum theory. The second claim states more ambitiously that the history of physics reveals that a postulating of lower levels of magnitude supplies an analogy that is a productive strategy to guide new research. The third claim is still more ambitious; it states that a new scientific revolution in microphysics is at hand, and that the hidden-variable semantical interpretation will produce a new microphysical theory that will resolve the current crisis in quantum theory. As de Broglie said in the closing sentence of his "Foreword" to *Causality and Chance* (1957), Bohm's book comes at exactly the right time.

Bohm's Hidden-Variable Interpretation of Quantum Theory

Consider next a brief overview of the hidden-variable interpretation, Bohm's means for implementing his three-point agenda for future microphysics. On the hypothesis that there exists a subquantum-mechanical level containing hidden phenomena, the statistical character of the current quantum theory is viewed as originating in random fluctuations of new kinds of entities existing at this lower subquantum-mechanical level. Heisenberg's indeterminacy principle and his particular statistical treatment of it, is something that exists only in the context of the higher, quantum-mechanical level. Bohm believes that by broadening the context to include a subquantum-mechanical level, it becomes possible to diminish the indeterminacy below the limits set by Heisenberg's principle. Bohm states that subquantum processes may be found in the domain of very high energies and of very short distances, even though at lesser energies and at greater distances the high degree of approximation permitted by the laws of the quantum-mechanical level means that the entities at the subquantum level cannot be playing a very significant role in quantum-level events. Bohm illustrates what the indeterminacy principle would mean in terms of a subquantum-mechanical level by means of an analogy with Brownian movement. A smoke particle is subject to random fluctuations originating in collisions with the atoms that exist at a lower level of magnitude. As a result of these random collisions the motion of the particle cannot be completely determined by the position and velocity of the particle at the level of the Brownian motion itself. Bohm cites a 1933 paper by the German physicist, R. Furth, who shows that the lack of determination is not only qualitatively analogous to that obtained in the quantum theory, but is also quantitatively analogous to the form of the indeterminacy relations.

Thus, for a short time interval with random fluctuations of a given magnitude in the mean position and a given magnitude in the mean momentum, the magnitudes satisfy a relationship involving a constant that depends on the state of the gas, and is formally analogous to Heisenberg's indeterminacy relation involving Planck's constant. In his hidden-variable interpretation Bohm postulates that associated with each electron there is a small body that is so small, that at the quantum-mechanical level it can be approximated as a mathematical point, just as in the earliest forms of atomic theory the atom was so described. Then he assumes that associated with this particle there is a quantum-level wave that oscillates in a subquantum-level field, which satisfies the Schroedinger wave function, and that exercises a quantum-mechanical force on the particle. The precise nature of the quantum-mechanical force exerted by the field on the body is not important, because a great variety of kinds of forces could lead to essentially the same results at the quantum level. The quantum-level force produces a tendency to pull the particle into regions where the subquantum-level field has its strongest intensity, as described by Born's probability distribution. But this tendency is also resisted by random motions analogous to Brownian motions, that originate at the subquantum level. The origin of these motions is not important; it is sufficient they have the property such that the average of their motions satisfies the Schroedinger wave equation, and that they are communicated to the particle. The net effect of the quantum-level force and the subquantum-level random motions in the subquantum field, is to produce a mean distribution in a statistical ensemble of particles, which favors the regions where the quantum-level force field is most intense, but which still leaves some chance for a typical particle to spend some time in the regions where the field is relatively weak. This result is analogous to the classical Brownian motion of a particle in a gravitational field, where the random motion which tends to carry the particle into all parts of the container is opposed by the gravitational field, which tends to pull it towards the bottom of the container.

Bohm then proposes an alternative explanation of the two-slit experiment. When the particle passes through a slit, it follows an irregular path, because it is affected by subquantum random motions. After a large number of particles have passed through the slit system, a pattern forms with particles accumulating where the field intensity is greatest due to the effects of the quantum force, as described by Born's probability distribution. The pattern is different if two slits are open than if only one is open. The closing of one of the two slits influences even those particles that pass through the open slit, because it influences the quantum-level force felt by the particle as it moves between the slit system and the screen. Thus the hidden-variable interpretation can explain how the appearance of the wave-particle duality originates, while the Copenhagen interpretation requires acceptance without further discussion of the fact that electrons enter the slit system and appears at the screen with an interference pattern.

Bohm also comments on Heisenberg's gamma-ray microscope thought experiment. He maintains that Heisenberg's indeterminacy principle should not be regarded as primarily an external relation between the measuring apparatus and the particle, and therefore as expressing the impossibility of making measurements of unlimited precision in the quantum domain. Rather it should be regarded as primarily an internal characteristic of the particle itself, and therefore as expressing the incomplete degree of self-determination characteristic of all entities that can be defined in the quantum-mechanical level. When the physicist measures processes at the quantum-mechanical level, the process of measurement will have the same limits on its degree of self-determination as every other process at this level. On the other hand subquantum-mechanical processes involving very small intervals of time and space will not be subject to the same limitations as those of the quantum-mechanical processes. Bohm criticizes analogies between quantum phenomena and Brownian motion which describe the wave and particle as basically different entities that interact in a way that is not essential to their modes of being. He says that this does not seem to be very plausible, because the fact that wave and particle are never found separately suggests that they are both different aspects of some fundamentally new kind of entity, which is likely to be quite different from a simple wave or particle. But he adds that the analogy with the Brownian movements admits to more than one variation. If in the classical case, the Brownian motion were viewed not as a motion of particles but as a motion of very fine droplets of mist, then the indeterminacy of the droplets in a vapor at its critical temperature where the distinction between liquid and gaseous states disappears, is a fluctuation in which the droplets are always forming and disappearing. This is an indeterminacy of the very existence of the droplets.

Similarly at the quantum level it may be found that the very mode of existence of the electron will be found to be indeterminate. The fact that the electron shows its characteristic wave-particle duality in its behavior suggests that the particle showing this critical opalescence is the relevant concept of particle. It may explain high energy transformations, the creation and destruction of elementary particles. In the Copenhagen interpretation it is very difficult to consider this problem; because the subquantum level is inadmissible, one is restricted to making blind mathematical manipulations with the hope that somehow one of these manipulations will lead to a new and correct theory. But if the subquantum level is admitted, then processes with very high energy and very high frequency may be faster than the processes taking place at the lower level. In such cases the details of the lower level would become significant, and the current formulation of the quantum theory would break down. The creation of a particle such as a meson may thus be conceived as a well defined subquantum-level process, in which the field energy is concentrated in a certain region of space in discrete amounts, while the destruction of the particle is just the opposite process. At the quantum-mechanical level the precise details of this process are not significant, and therefore can be ignored; this in fact is done in the current quantum theory, which discusses the creation and destruction of particles as merely a kind of "popping" in and out of existence with special creation and destruction operators in the mathematics. However, with very fast high energy processes the results may well depend on these subquantum-mechanical details, and if this should be the case, then the current quantum theory would not be adequate for the treatment of such processes.

Bell on the Hidden-Variable Theory

The hidden-variables thesis of quantum theory has its contemporary advocates, one of whom is J.S. Bell, a theoretical physicist associated with CERN in Geneva, Switzerland. In 1987 Bell published his collected papers under the title *Speakable and Unspeakable in Quantum Mechanics*. In a paper titled "Introduction to the Hidden-Variable Question" (1971) Bell discusses his motivations for defending and developing the hidden-variable thesis. His first reason, and the one that he finds most compelling, is the possibility of a homogeneous account of the physical world, which is to say, a single uniform ontology for microphysical and macrophysical domains based on classical concepts, instead of the schism described by Heisenberg. Bell denies that there is a boundary between classical and quantum worlds. He agrees with Einstein that the wave function is an incomplete and provisional microphysical theory. His second motivation concerns the statistical character of quantum mechanical predictions. Once the incompleteness of the wave function is suspected, then the seemingly random statistical fluctuations may be viewed as determined by the extra hidden variables, which are hidden because at the present time physicists can only conjecture their existence. His third motivation is the peculiar character of some quantum-mechanical predictions considered in the famous *gedanken* experiment formulated by Einstein, Podolsky and Rosen (EPR), and a refinement proposed by Bohm in 1957, in which a Stern-Gerlach device is used to produce deflections in opposite directions by particles moving simultaneously from a source in opposite directions. The experiment permits the observer to know in advance the result of measuring one particle's deflection by observing the other's deflection even at great distance. The implication is that the outcomes of such measurements are actually determined in advance by variables over which the physicist has no control, but which are sufficiently revealed by the first measurement, that he can anticipate the result of the second. Therefore, contrary to the Copenhagen view there is no need to regard the performance of one measurement as a causal influence on the result of the second distant measurement, and the description of the situation can be "local".

Heisenberg's indeterminacy principle says no quantum-mechanical state can be dispersion free for every variable. On the other hand the hidden-variable theory says that all observations are fully determined, such that each quantum-mechanical state must correspond to an ensemble of states each with different values of the hidden variables, and only the component states are dispersion free. Therefore, one way to formulate the hidden-variable problem is to search for a formalism permitting such dispersion-free states. Bell proposes such a formalism, which he says provides an explicit causal mechanism by which operations on one of the two measuring devices in the apparatus can influence the response of the other, distant device. This is the opposite of the resolution hoped for by EPR, who had envisaged that the first device could serve only to reveal the character of the information already stored in space and propagating in an undisturbed way toward the other detecting equipment.

In "Speakable and Unspeakable in Quantum Mechanics" (1984) Bell says that the problem of quantum theory is not how the world can be divided into the speakable macrophysical apparatus, which we can talk about, and the unspeakable quantum system, which we cannot speak about.

The problem is how to explain the consequences of events at one place propagate to other places faster than light, which is in gross violation of relativistic causality. Most notably he says that specific quantum phenomena which require such superluminal explanation have been realized in the laboratory by Aspect, Dalibard and Roger, who published their findings in 1982. Bell says that there exists an apparent incompatibility at the deepest level between the two fundamental pillars of contemporary physical theory, and that a real synthesis of quantum and classical theories requires not just technical developments but radical conceptual renewal.

Hanson on Perception, Observation and Theory

Hanson defends the Copenhagen interpretation, and he criticizes the hidden-variable interpretation and Bohm's agenda. Before turning to Hanson's defense and criticisms, consider firstly his meaning of "the Copenhagen interpretation" of modern quantum theory. In *Concept of the Positron* and elsewhere Hanson distinguishes the "Copenhagen interpretation" from what he calls the "Bohr interpretation", and his view may be compared with that of Feyerabend's. In "The Quantum Theory of Measurement" in *Observation and Interpretation* (1957) and in "Niels Bohr's Interpretation of the Quantum Theory" in *Current Issues in the Philosophy of Science* (1961) Feyerabend maintained that there is no single Copenhagen interpretation, but that there are multiple interpretations due to different philosophical ideas that distinguish them. He says that there are dogmatic elements in Bohr's approach, i.e. classical concepts. He also says that Bohr's interpretation contains the philosophical thesis, that in future physics it will be impossible to invent a new conceptual scheme that will replace the classical concepts in the quantum domain. Although Feyerabend rejects this philosophical thesis, he maintains that classical concepts cannot just be excluded, because introducing a new set of concepts is not something that occurs independently of and prior to the construction of theories. However, Hanson rejects efforts by philosophers such as Feyerabend to include what Feyerabend admits are the dogmatic elements of the Bohr interpretation in the Copenhagen interpretation. The dogmatic elements consist particularly in what Hanson calls Bohr's "naive epistemology". Perhaps it could be said with caution that with the rejection Bohr's "naive epistemology" Hanson's philosophy of quantum theory is one that Heisenberg might have formulated, had Heisenberg rejected Bohr's epistemological ideas which he included in his doctrine of closed-off theories, and instead followed through on Einstein's admonition, that theory decides what the physicist can observe. In setting forth his own philosophy of physics Hanson makes no explicit mention of Heisenberg's doctrine of closed-off theories, nor does he maintain that the validity of Newtonian mechanics or of quantum mechanics is permanent; he does not accept a naturalistic theory of the semantics of language, and does not speak of the "forms of perception" as does Bohr.

Hanson maintains that in microphysics all the limitations placed on our conceptions of what the microphysical world is like, are really limitations arising out of the linguistic features of the formal languages available. Such is particularly the case with Heisenberg's uncertainty relations. The uncertainty relations and Heisenberg's thought experiment involving a gamma-ray microscope are often said to state limits to the possibility of observation within microphysics. Hanson says that this is true in an unsuspecting way: there never have been nor could there ever be experiments or observations pertinent to the establishment of the uncertainty relations, because these relations are the conceptual or logical consequence of the language of quantum theory. In the formalisms for modern quantum physics there is a logicolinguistic obstacle to any attempt to describe with precision the total state of an elementary particle, and if there is a logical limit to such a description, then there is *ipso facto* a limit to such observation. The conceptually impossible is observationally impossible. Hanson's thesis is that theory is integral to observation, or as he also says, observation is "theory-laden".

This is an ironic implication from Einstein's admonition to Heisenberg that it is the theory that decides what the physicist can observe. Hanson's is the same philosophy of observation that Einstein told Heisenberg in 1925, and that Heisenberg used to develop the uncertainty relations. But Hanson was not led to develop his philosophy of observation by reflection on Heisenberg's autobiographical chronicles, in which Heisenberg relates his discussion with Einstein and the use that he made of it. Hanson identified Heisenberg's views on observation with those of Bohr, which Heisenberg included in his explicit and systematic philosophy. Nor was Hanson led to develop his philosophy in response to Feyerabend's criticisms of Bohr's "dogmatic" interpretation of quantum theory; Hanson's philosophy of observation was developed many years before. His philosophy of observation was drawn from Wittgenstein's *Investigations* and from the *gestalt* psychology. It is necessary, therefore, to consider briefly Wittgenstein's ordinary-language philosophy and Hanson's use of it in his philosophy of science.

Ludwig Wittgenstein (1889-1951) was a somewhat reclusive individual who wrote a somewhat unsystematic philosophy of language in a somewhat obscure style, and who is thought to have anticipated certain ascendant trends in philosophical thinking. In fact Wittgenstein seems twice in his lifetime to have anticipated successfully the ascendant trend of philosophical thought with his two principal works: firstly his *Tractatus Logicus-Philosophicus* (1922) and then later his *Philosophical Investigations* (1953). The latter explicitly includes a repudiation of the thesis of the former, yet each work gathered its own retinue of sympathetic interpreters and devout disciples. Both the *Tractatus* and its author attracted the attention of Schlick and his Vienna Circle (except for Carnap, who after one meeting with Wittgenstein was unimpressed). But in spite of Schlick's invitations to join the Vienna Circle, Wittgenstein remained aloof from them, just as he remained aloof from all other sublunar affairs. About thirty years later the *Investigations* inspired philosophers who were becoming disillusioned with the technical pedantics of Logical Positivism, and its thesis occasioned the formation of a new philosophy of language. Conventionally historians of philosophy now refer to the two opposing dogmas in these two books as the "ideal-language tradition" and the "ordinary-language tradition" respectively. The ideal-language view set forth in the *Tractatus* has a reformist flavor, which accords special status to symbolic logic, such as may be found in Russell's *Principia Mathematica*. The *Tractatus* advanced an ideal (not metaphysical Idealist) interpretation for symbolic logic, consisting of what is called a picture-theory semantics. This is one of many variations on the naturalistic theory of the semantics of language, and it is also the most naive. This first book also advanced a constructionalist view of language. It described all sentences in the ideal language as consisting of elementary sentences, which in turn consist of semantically independent names of simple objects. All nonelementary sentences are constructible from the elementary ones. The former are said to be "truth functional", which means that the truth of the constructed compound sentences depends completely on that of their component elementary sentences.

As a result of this semantical atomism and logical constructionalism, the understanding of any sentence ultimately reduces to knowing its logical structure and what its constituent names refer to. This is a variation on the mechanistic philosophy of the semantics of language, which is called "logical atomism". The principal argument in defense of the ideal-language tradition is that ordinary language is unsuitably vague and misleading for philosophy, just as it is unsuitable for empirical sciences like modern physics, which rely on mathematics. The initial attractiveness of symbolic logic was the expectation that it could serve philosophy as mathematics serves physics. In philosophy of science this programme evolved into the Logical Positivist reductionist programme of Carnap and followers such as Feigl and Hempel, in which the controlling agenda was the logical reduction of theories to a semantically significant observation language, in order to demonstrate the meaningfulness and semantics of the scientific theories.

But experience with the reformist efforts of the ideal-language philosophers, notably the Logical Positivists, led some younger philosophers to charge that ideal languages are even more unsuitable than ordinary language for philosophy, and that philosophical analysis should be directed toward the examination of colloquial language. The outcome was a new folk philosophy that was self-consciously naive. Wittgenstein anticipated this reaction, perhaps because it was also his own reaction to his own *Tractatus*, and was led to develop his ordinary-language philosophy. Early statements of his new philosophy were set down in a set of notebooks published as *The Blue and Brown Books* (1958), and the more mature statement is the *Investigations*. The latter work describes philosophy as a kind of empirical linguistics, and its main themes are (1) the variety of uses of language, (2) the need for the philosopher to consider statements not in isolation but in the context that occasions their utterances, and (3) the definition of meaning in terms of usage. Wittgenstein maintained that the problems of philosophy originate in philosophers' misunderstanding of certain crucial terms such as "know", "see", "free", "true", "reason"; and that the resolution of these problems requires examination of the uses of these words as they occur in ordinary language discourse. The later Wittgenstein seems clearly to have rejected the naturalistic theory of the semantics of language. He asks rhetorically in the *Investigations*, that if the formation of concepts can be explained by facts of nature, then should the philosopher not be interested not in grammar, but rather in that in nature which is the basis of grammar. He answers that the philosopher is not interested in natural science or in natural history, and he affirms an artifactual theory of the semantics of language stating that a concept is comparable to a style of painting. But the artifactual theory that he accepts seems to be a wholistic one, since he states in the opening pages of *The Blue and Brown Books* that understanding a sentence means understanding a language.

Hanson was of the generation of philosophers who took their professional educations after the Second World War, and he was also one of those who looked to Wittgenstein's new philosophy to rise above the inadequacies of the Logical Positivist philosophy of science. Hanson was not an ordinary ordinary-language philosopher; he was firstly a philosopher of science, and if there was an "ordinary language" of interest to him, it was the language ordinary to contemporary physics including most notably microphysics. He was specifically drawn to Wittgenstein's comments in the *Investigations* about "seeing", in order to re-approach the subject of observation in physics, which modern quantum theory had made so problematic.

Hanson's discussions about observation and theory are set forth in *Patterns of Discovery*, in "Observation and Interpretation" in *Philosophy of Science Today* (1967), and in *Perception and Discovery*. Hanson rejects the Positivist view that seeing is merely a matter of predetermined sensations, sense data, phenomena, or retinal reactions in the eye, and that interpretation is something added to the predetermined perception as a second and discrete step in the perceptual process. He maintains instead that "there is more to seeing than meets the eye", and he follows Wittgenstein's view that interpretation is an integral component of seeing instead of something forced on it. The significance of this point is that perception is not predetermined and fixed by nature but is variable, and he illustrates this variability in perception by using both Wittgenstein's and others' ambiguous drawings that admit to reversible interpretations. He explicitly invokes *gestalt* psychology (something that Wittgenstein did not do), to explain the reversiblity of interpretations of ambiguous drawings as changes in the conceptual organization of what is observed.

In this context Hanson references Duhem's example in *The Aim and Structure of Physical Theory* of the visiting layman to a physicist's laboratory. The layman would have to learn physical theory before he could observe what the trained physicist observes. Duhem had described this commonplace state of affairs in terms of his Positivist semantics of observation and theory. But Hanson is a critic of Positivism, and he does not maintain any such two-tiered semantical thesis, as Duhem had. He maintains that the laboratory situation reveals that the elements of the laboratory in the visitor's field of perception are not organized as they are for the trained physicist. Physical theory provides the physicist with patterns within which data appear intelligible; it is what makes possible observation of phenomena as being of a certain kind and as related to other phenomena. The differences between *gestalten* are due to differences in previously acquired background knowledge, knowledge which involves language.

Hanson is therefore led to follow Wittgenstein's ordinary-language analysis, because examination of commonly used locutions in colloquial discourse reveals the relation between language and the variability of interpretation in observation. The locution "seeing as" reveals that seeing is to see an object as a certain kind of thing, which is brought out by the verbal context in which the locution occurs. The text supplies the interpretation. But his thesis is still stronger than merely stating that language reveals an interpretation; he invokes the locution "seeing that" to exhibit a necessary role for language in interpretation. The idea of "seeing that" explains the relation of "seeing as" and the observer's background knowledge: to see something as a certain kind of thing is to see that it behaves in a certain known and expected manner. The "seeing that" locution supplies a statement of the background knowledge, which can be true or false. Seeing is therefore a "theory-laden" activity in the sense that the seeing is interpreted by reference to our background knowledge. Without a linguistic component to seeing, nothing we saw could be relevant to our knowledge. Before the wheels of knowledge can turn relative to a given visual experience, some assertive or propositional aspect of the experience must have been advanced. Only statements can be true or false; visual experiences must be cast into the form of a language to be considered in terms of what we know to be true, i.e. in terms of our theories.

To illustrate his thesis that perception is theory-laden, Hanson uses the example of the second-century and the seventeenth-century astronomers who both look at the dawn. They both have the visual experience of the sun, but they do not see the same thing, because each believes different astronomical theories, the former, Ptolemy, believes in the geocentric theory, the latter, Galileo, in the heliocentric theory. Nevertheless, it can still be said that they see the same thing, since the sun could be described by both as a brilliant yellow disk. Hanson calls this latter kind of description "phenomenal seeing", but he maintains contrary to the

Positivists that such phenomenal seeing is not the ordinary way of seeing. It is something that requires special effort, because seeing is normally interpretative, or it is something that occurs when the observer is confronted with a new seeing experience, in which case what is seen cannot be characterized by reference to his background knowledge. Observation in science aims to pass beyond the phenomenal seeing occurring in the case of the new experience, and to get the visual experience to cohere against a background of established knowledge.

Furthermore, Hanson's thesis is not only that language is necessary for the interpretation that is integral to perception, but also that the logicogrammatical form of the language used for description exercises a formative control over the interpretative thinking that occurs in perceiving. Just as seeing may be stated locutions which are "that..." clauses, so too can facts and theories. For this reason Hanson says that Ptolemy could not express in the second century what were facts for Galileo fifteen centuries later. Physical concepts are intimately connected with the formalisms and notations in which scientists express them, including the formalisms used today in contemporary microphysics. The dependence of physical concepts on the mathematical formalisms is a very strategic consideration in Hanson's rejection of attempts by Bohm and Feyerabend to propose interpretations of the uncertainty relations and the Schroedinger wave function, that are alternatives of the Copenhagen interpretation of modern quantum theory. For Hanson the Copenhagen interpretation is precisely that interpretation which is supplied by the formalism of the modern quantum theory, because contrary to both the Positivists and to Bohr, it is the formalism that supplies the intelligible patterns and conceptual organization in perception for the observations relevant to microphysics.

Hanson further follows Wittgenstein when he maintains that the meaning of a sentence is its use, and that there are multiple uses for a sentence. Thus he states that the laws and theories of physics have many uses, and not just one, as most philosophers have maintained. The contingently empirical status of a statement is one of the uses of the theory in science. Another is to make the phenomena cohere in an intelligible way, such that empirical disconfirmation does not result in the negation of the concept described by the theory, but rather results in no coherent concept at all. The dynamical laws of classical physics, for example, are a system of propositions that are empirically true, and the fundamental propositions on which the system rests are empirically true. But these fundamental propositions are also treated as axioms, such that the system delimits and defines its subject matter. Then nothing describable within the system could refute its law statements; disconfirmatory evidence counts against the system as a whole, and it only shows that the system does not hold, where formerly it was thought to hold. Hanson calls this use of laws and theories "functionally *a priori*". These ideas are reminiscent of Heisenberg's comments in "Questions of Principle in Modern Physics", in which he says that it is not the validity but only the applicability of classical laws, which is restricted by modern relativity and quantum physics. Hanson does not reference Heisenberg, but his thesis of the "functionally *a priori*" use of laws and theories is in this respect similar to Heisenberg's doctrine of a closed-off theories, with the noteworthy exception that Hanson does not reserve certain axiomatic systems such as classical mechanics for observation in physics, as does Heisenberg in his explicit philosophy of physics. Heisenberg's philosophy of observation in his doctrine of closed-off theories does not admit the variability in perception that Hanson's philosophy asserts. Instead in his explicit philosophy Heisenberg followed Bohr's thesis that there are "forms of perception", that are found only in colloquial language and in its refinements in classical physics.

Hanson's semantical investigations sometimes took a turn away from the wholistic approach of *gestalt* psychology. In the chapter on classical particle physics in *Patterns of Discovery* he considers the idea that the meanings of some names have their properties built into them, such that falsification of statements predicating those properties of the named substances is effectively impossible. And in "Newton's First Law: A Philosopher's Door into Natural Philosophy" in *Beyond the Edge of Certainty* (1965), he states that "rectilinearity", "motion *ad infinitum*", and "free force", are interdependent conceptions within classical mechanics, in such a way that it is possible to treat the idea of uniform, rectilinear motion *ad infinitum* as itself built into the notion of free force, as part of the latter's semantical content. The terms in Newton's first law are semantically linked: the meaning of some of its component terms "unpacks" sometimes from one or two of the others, but then sometimes the meaning of these unpacks from that of the first. Which are the "contained" and which are the semantical "containers" can affect the logical exposition of any mechanical theory built thereon. These are semantical decisions which guarantee that in different formalizations of Newton's theory, different meaning relations will hold between the law's constituent terms. The term "unpack" in connection with semantical analysis is a phrase used by the Pragmatist philosopher William James, although Hanson does not reference James. It is unclear if Hanson ever thought of this type of semantical analysis as an alternative to his frequent recourse to *gestalt* psychology. Nevertheless it is an alternative approach in semantical analysis, because it is not wholistic: on the *gestalt* thesis it is not possible to "unpack" a *gestalt* into its component parts, because the *gestalt* is more than a mechanical assembly of its parts. In his discussions of quantum theory Hanson never exploited this mechanistic or logical analysis of meanings into component parts.

Hanson on the Copenhagen Interpretation and Discovery

Hanson rejects all three of the objectives in Bohm's agenda for future physics. His argument against Bohm's third objective, that a future hidden-variable theory will resolve the difficulties in current quantum theory, is that Bohm and other advocates of alternatives to the Copenhagen interpretation offer nothing but promises. In *Quanta and Reality* Hanson calls Bohm's proposal a "congeries of excitingly vague, bold-but-largely-formless, promising-but-as-yet-unarticulated, speculations". The Copenhagen interpretation on the other hand is a "working theory" however imperfect it may be, and a speculation is never an alternative to a working theory.

Hanson's argument against Bohm's first objective, that an alternative to the Copenhagen interpretation is possible, is similar to his criticism of the third objective. Hanson denies that an alternative to the Copenhagen interpretation is possible until a new mathematical quantum theory formalism is developed, because on Hanson's thesis the Copenhagen interpretation is not a semantics supplied by related philosophical or metaphysical ideas about the subject, but rather is the semantical interpretation resulting from the logico-grammatical form of the formalism. Therefore contrary to physicists such as Bohm and Lande, and contrary to philosophers such as Feyerabend and Popper, the Copenhagen interpretation even after disengagement from Bohr's naive epistemology, is not just one of several alternative semantical interpretations; it is that one which is defined by the relationships in the mathematical formalism. With his distinction between the "Bohr interpretation" and the "Copenhagen interpretation" Hanson places himself in agreement with Bohm and Feyerabend, when the latter maintain that the quantum theory is not permanently valid. Hanson agrees that the current quantum theory may be superseded. But he still considers the wave-particle duality to be the defining characteristic of the Copenhagen interpretation and as integral to the formalism. Because he maintains that the Copenhagen interpretation is defined by the logico-grammatical form of the mathematical formalism itself, he defends it as the only interpretation that "works". He therefore says that in the absence of any algebraically detailed and experimentally adaptable alternative, the Copenhagen interpretation represents the conceptual possibilities currently open to practicing physicists, and that it will not be abandoned until it is completely replaced by an alternative, completely detailed, algebraically articulated theory.

Bohm's second objective in his agenda for future physics is that the history of physics suggests (contrary to a "mechanistic" thesis, as he uses that term) that the future microphysical theory will describe phenomena at the lower level of magnitude than does the current quantum theory, and that his proposal of a hidden-variable theory of the subquantum level may serve as a heuristic for future microphysics. In fact the idea of developing a heuristic for future scientific discovery or theory development is closely related to Hanson's interest, and Hanson does not attack Bohm's second objective in terms of Bohm's antimechanistic historical thesis. But he has his own historical thesis influencing his views on scientific discovery. His analyses are greatly influenced by the Cambrian physicist Paul A. Dirac. Dirac (1902-1984) was a theoretical physicist at Cambridge University, who shared the Nobel Memorial Prize for physics in 1933 with Schroedinger. Dirac had published a methodological statement on the future of physics in his "The Evolution of the Physicist's Picture of Nature: An account of how physical theory has developed in the past and how, in the light of this development, it can perhaps be expected to develop in the future" (*Scientific American*, May, 1963). In this brief paper Dirac contrasted the theory development approaches of Schroedinger and Heisenberg. Dirac was much more sympathetic to the former's approach, according to which the development physical theory should be guided by the aesthetics of the mathematics of the theory, in contrast to the latter's approcah in which a mathematical formalism is developed to fit the available data. However, this is not the issue in Dirac's views that influenced Hanson, who was actually much more sympathetic the Heisenberg's approach in which theory originates with the experimental data. Hanson was influenced by Dirac's historic accomplishment, the "transformation theory" developed by Dirac in 1928, which not only combines relativity and quantum mechanical descriptions of electron properties, but also enables physicists to exhibit the wave-particle duality by transforming mathematically the wave description into the quantum description and vice versa. Both in his "Copenhagen Interpretation of Quantum Theory" in the *American Journal of Physics* (1959) and in his chapter "Interpreting" in *Concept of the Positron*, Hanson states that objections to the Copenhagen interpretation arise from a failure to appreciate the historical and conceptual role it had played in Dirac's 1928 paper, and he reports that in conversation with Dirac, Dirac told him that the Copenhagen interpretation figured essentially in his development of his relativistic quantum field theory, and not as merely a philosophical afterthought appended to the algebra. This personal conversation with Dirac more than anything explains why Hanson maintains that the Copenhagen interpretation is integral to the formalism of the quantum theory. He argues against Feyerabend that even if it were possible to have a minimum statement of quantum theory with no more interpretation than is required barely to describe the facts, this is what Dirac felt he had, and Dirac's paper would not have been the paper that it actually was, had its assumptions been purified of the Copenhagen interpretation, as Feyerabend advocates. But for his thesis of scientific discovery Hanson turned not to Dirac, but to the American Pragmatist philosopher, Charles S. Peirce.

Retroductive Logic and Semantical Constraints in Discovery

Hanson was influenced not only by Wittgenstein but also by the American philosopher, Charles S. Peirce, the founder of Pragmatism, although he did not accept Peirce's views on observation. In his "How to Make Our Ideas Clear" (1878) Peirce set forth his pragmatic maxim, which anticipated Bridgman's operationalist theory of the semantics of observation by nearly fifty years. Peirce's maxim says that our conception of the practical effects that we conceive an object might have, is the whole of our conception of that object. And in his discussions of abduction Peirce distinguishes observed facts from judgments of fact, and says that observations have to be accepted as they occur, while judgments of fact are controllable. According to C. S. Peirce's theory of scientific discovery, hypotheses are judgments of fact expressed in propositions, and all such propositions are additions to observed facts, which are sense impressions of singular events associated with particular circumstances. The addition to observed facts by propositions Peirce calls practical knowledge, and it is something that is controllable and subject to error. Hypotheses are the result of inference, and Peirce distinguishes inductive and abductive types of inference. Abduction, or retroduction, involves both the simple interrogation or framing of the hypothesis, and also the entertaining or selecting of one hypothesis by testing its ability to account for surprising facts. The difference between abduction and induction is that the former involves guesswork and originality, while the latter only tests a suggestion previously made. Once the hypothesis is formulated, abduction is an inference that satisfies the following form: (1) the surprising fact, *C*, is observed; (2) but if *A* were true, then *C* would be a matter of course; (3) hence, there is reason to suspect that *A* is true. Peirce says that Kepler's development of his three laws is the greatest piece of retroductive reasoning ever performed. He rejects J.S. Mill's view that Kepler merely generalized on Tycho's data, and that there was no reasoning in Kepler's procedure. Peirce maintains that at each step of Kepler's investigation, Kepler had a theory which approximated the data, that Kepler modified his theory to make it closer to the observed facts, and that the modifications were never capricious. He adds that given a choice between two hypotheses, the simpler is preferable, where simplicity is to be understood not as a logical simplicity but rather as an instinctive simplicity, because unless man has a natural bent in accordance with nature's, he has no chance of understanding nature at all.

In his chapter on theories in *Patterns of Discovery* Hanson rejects both the hypothetico-deductive and the Positivists' inductive accounts of scientific discovery. He rejects the inductivist thesis that scientific theories are developed by an enumerating and summarizing of observable data, as the Positivists maintained for the development of empirical generalizations; he states that empirical laws explain, they do not simply summarize. He also rejects the hypothetico-deductive thesis that scientists start from hypotheses, as Popper maintained, for the development of theories; he says that scientists do not start from hypothesis, but rather they start from data. The initial inference is not from higher level hypotheses to observations, but the other way around. Hanson's views on the logic of scientific discovery were significantly influenced by Peirce, even if Hanson's views on the semantics of observation were not Positivist like Peirce's. Hanson discusses Peirce's abductive or retroductive inference in his *Patterns of Discovery* and in several articles. The article setting forth his most mature views on retroduction is "Notes Toward A Logic of Discovery" in *Perspectives on*

Peirce (ed. Bernstein, 1965), which includes summaries of Hanson's earlier papers. The logic of retroduction pertains to the scientist's actual reasoning, which proceeds from an anomalous situation to the delineation of a kind of explanatory hypothesis that fits into an organized pattern of concepts. In *Patterns of Discovery* Hanson refers to the pattern of concepts as a conceptual *gestalt*, which functions to make the anomalous situation appear intelligible. The conceptual *gestalt* supplies the semantics for the theory or hypothesis. In Hanson's philosophy the semantics of observation is variable, while in Peirce's it is fixed and uncontrollable.

The formal criteria for the retroductive logic of discovery are the same as those for the hypothetico-deductive logic of explanation. They both contain the same elements: a hypothesis, statements of initial conditions, and the conclusion deductively derived from the hypothesis and statements of initial conditions. One difference between them is the direction of the inference. In the hypothetico-deductive logic the inference is from the hypothesis and statements of initial conditions of an experiment, to the statements describing the observed outcome of the experiment as a conclusion. This process is used for experimental testing, and if the results are not anomalous, it also serves as the logic of the explanation of the resultant phenomenon. But in the retroductive logic the direction of inference is in the opposite direction. The statement reporting an observed experimental outcome describes an anomaly relative to what is expected, and the problem is one of finding the hypothesis that is capable of functioning in a hypothetico-deductive account that will explain the anomalous situation as occurring as a matter of course. But the difference between the hypothetico-deductive and the retroductive types of inference is not just a matter of the directionality of the inference. They are also different because the former is determinate, while the latter is not. In hypothetico-deductive inference consistent premises must produce consistent and unique conclusions, while in the retroductive inference there may be many alternative and mutually inconsistent hypotheses that are able to explain deductively the formerly anomalous test outcome from the same set of statements of initial conditions. From this nondeterminate character of retroductive inference Hanson concludes that retroduction cannot yield a uniquely specific and detailed hypothesis. But he maintains that it can yield an indication of the type of hypothesis that is most plausibly to be considered as worthy of serious attention. And the decision about what type of hypothesis is the most plausible depends in turn on the structure of presently accepted theories and on the "shape" of the most reliable conceptual frameworks that highlight hypothesis types for the problem solver. Therefore, much as it is only against the background of the intelligible and the conceptually comprehensible offered by existing theories that the anomalies stand out at all, so it is also in these same terms that the scientist comes to know which types of hypotheses will do the job and which do not. Reflection on this analysis reveals why Hanson defends the Copenhagen interpretation, understood as the semantics that is defined by the formalism of the quantum theory: The Copenhagen interpretation is the type of hypothesis that (in Hanson's view) will most plausibly resolve the current anomalies to Dirac's relativistic quantum theory, just as it had enabled Dirac to develop his quantum theory in 1928.

Hanson also maintains that the conceptual *gestalten* constituting the semantics for currently accepted theories not only supply some guidance for the creation of new theories, but also offer what he calls "conceptual resistance" which must be overcome for scientific discoveries. The development of a new theory requires a new *gestalt* just as in the reinterpretation of the ambiguous drawing, and similarly there is a resistance to such a change. In *Patterns of Discovery* Hanson illustrates this in the historical episode in which Kepler developed the theory that the orbit of Mars is elliptical. In formulating this theory Kepler had to reject the traditional belief held since Aristotle, that the orbits of the planets are circular because unlike sublunar motions their motions are perfect. This is also the thesis in Hanson's most significant historical analysis set forth in his *Concept of the Positron*. This work is original historical research in which Hanson interviewed several physicists including the three principals in the episode: Carl Anderson, P.A.M. Dirac, and P.M.S. Blackett. All three physicists discovered the positron, but only Blackett recognized that the particle discovered experimentally by Anderson was the same one that was postulated theoretically by Dirac. Anderson made photographs of electron tracks in the cloud chamber, and he concluded that one of them showed a positive electron, because the change of the particle was positive while its mass was too small to be a that of a proton. Dirac had published his theoretical paper on the positron in 1931, a year before Anderson's photograph. Hanson states that one reason that Anderson did not recognize any connection between his cloud chamber experiments and Dirac's quantum theory, is that such experiments rely on concepts that are largely classical in nature such as track-leaving particles. But the greatest conceptual constraint, the one that led many physicists to reject the idea of the positive electron, was in the semantics of the concept of the electron. That semantics was such that an intimate association between the electron and the proton, and between the two basic units of electricity, negative and positive, made the very idea of a particle other than a proton or an electron very difficult to conceive. Just as positive/negative exhaust the totality of electrical charge, so too the proton/electron was thought exhaust the totality of charged particles, since the proton and the electron came to be thought of not only as carrying the charge but as being the charge. Hence there was a conceptual resistance to the idea of a third charged particle built into the structure of classical electrodynamics and the elementary particle theory. Dirac's 1928 paper offered a relativistic quantum theory that was Lorentz-invariant, but it also contained negative energy solutions that could not be eliminated. Originally he had hoped that these strange solutions could be construed as protons, and then he thought of them as "vacancies" which are positive charge solutions with the mass of the electron. This constituted the gradual development of his prediction of the existence of positive electrons before they were observed. In 1933 Blackett and Occhialini reported that the Anderson particle and the Dirac particle are the same thing, by using the new photographic technique in which the particles took their own photographs.

Hanson's Philosophy of Science

It is impossible to separate Hanson's ideas about the aim of science from his ideas about scientific discovery. His concept of science is what he calls "research science", as opposed to what he calls "almanac science". In his "Introduction" in *Patterns of Discovery* he states that in a growing research discipline, inquiry is directed not to rearranging old facts and explanations into more elegant formal patterns, but rather to the discovery of new patterns of explanation. The idea that observation is theory-laden is strategic to this purpose. Then in the chapter titled "Observation" he states that the scientist aims to get his observations to cohere against a background of established knowledge. This kind of seeing is the goal of observation. And similarly in the last chapter titled "Elementary Particle Physics", the area of contemporary physics that is presently a research science, he states that intelligibility is the goal of physics, by which he means the conceptual struggle to fit each new observation of phenomena into a pattern of explanation. Often the pattern precedes recognition of the phenomena, as Dirac's theory of 1928 preceded discovery of the positron, the antiproton, and the antineutron. But then Dirac's pattern was itself the outcome of an effort to find a suitable explanation for prior phenomena, namely a unified, relativistically invariant theory of electron spin, which would give the correct fine structure formula, explain the Zeeman effect of the doublet atoms, describe the Compton scattering, and supply a model of the hydrogen atom.

Hanson offers an evolutionary perspective on scientific explanation. In the third chapter of *Concept of the Positron* he states that the concept of scientific explanation has experienced a historical evolution, that follows upon the historical development of physics. Leibniz denied that Newton's theory offers explanation, even though he admitted that it offers acceptable predictions. Today the concept of explanation advanced by the Positivists, such as Hempel, are based on the concepts of Newton's physics including notably the deterministic thesis that explanation implies deterministic prediction. The concept of explanation implied in the nondeterministic quantum theory is not yet accepted. Hanson states that if just after Leverrier had predicted the existence of the planet Neptune in 1847, a time when Newtonian physics had reached its apex, some physicist who had proposed a new theory that explained all that Newton's theory explained and furthermore explained several minor flaws in Newton's theory, the new and better theory would have been viewed as merely a predictive device, not an explanation. But if Newton's theory then began to show major weaknesses, while the new theory succeeded where Newton's had failed, still these accomplishments would decide nothing. The scientists would begin to show increasing reliance on the new theory, yet it would not be accepted as an explanation. All the same, younger physicists would develop the new theory further. Finally if Newton's physics had begun to fall apart while the new theory opened up new branches of science, focused on problems never before perceived, fused disciplines previously thought to be distinct, and sharpened experimental techniques to an unprecedented degree, then the very pattern of thinking in an inquiry properly called "scientific" would reflect the new physics with its new concept of scientific explanation; to be able to cope with a scientific problem at all, would be to have become able to build it into the conceptual framework of the new physics.

Hanson distinguishes three stages in this process of the evolution of a new concept of explanation; they are the "black box", the "grey-box", and the "glass box". In the first stage, the stage of the black box, there is an algorithmic novelty, a new formalism, which is able to account for all the phenomena that an existing formalism can account for. Scientists use this technique, but they then attempt to translate its results into the more familiar terms of the orthodoxy, in order to provide understanding. In the second stage, the stage of the grey box, the new formalism makes superior predictions in comparison to the older alternative, but it is still viewed as offering no understanding. Nonetheless it is suspected as having some structure that is in common with the reality it predicts. In the third stage, the stage of the glass box, the success of the new theory will have so permeated the operation and techniques of the body of the science, that its structure will also appear as the proper pattern of scientific inquiry. Hanson says that quantum theory is still in the second stage, because scientists have not yet ceased to distinguish between the theory's structure and that of the phenomena themselves. Explanation is customarily thought to be explanatory, because it describes the causes of the phenomena explained. Therefore, the concept of causality also undergoes the kind of evolution that occurs with the concept of explanation. In the chapter titled "Causality" in *Patterns of Discovery* Hanson says that cause words are theory-laden; they are the details in an intricate pattern of concepts. Causes are connected with effects, but only because theories connect them, not because the universe is held together with a cosmic glue. Questions about the nature of causation are to a large degree questions about how certain descriptive terms in definite contexts coupled together compliment and interlock with a pattern of other terms. The elements of explanation, causation, and theorizing become worked into a comprehensive language pattern.

Hanson's discussion of scientific criticism is principally concerned with the topic of crucial experiments. He takes up the topic in a chapter in *Concept of the Positron* in which he discusses the different concepts of light in the history of physics, and he discusses it again later in a special chapter in *Perception and Discovery*. Hanson's rejection of the idea of crucial experiments has its basis in his thesis that observation is theory-laden. A commonly referenced example of a crucial experiment is Foucault's 1850 crucial test between the wave and particle concepts of light. In that experiment Foucault demonstrated that light travels more rapidly in air than in water. According to the doctrine of the crucial experiment the corpuscular hypothesis should have been banished forever. But this has not happened. The photoelectric effect and the Compton effect can only be explained on a corpuscular theory of the nature of light. The experiments are not crucial, because the observations are important only against the assumptions, theories, and hypotheses that are in the balance before the experiment is performed. One of the assumptions is that light cannot be both wave and particle. The crucial test is a test of the alternative hypotheses together with all of their assumptions, just as in ordinary scientific observation there is a pure registration or sensation plus all of the assumptions necessary to give those sensations meaning. If we were forced to revise our assumptions, then the crucial experiment must be interpreted, so that it need not decide against one of the hypotheses. Some of the most profound revolutions in modern science have consisted not in the criticisms of old hypotheses, but in the criticism of the assumptions underlying the hypotheses. Crucial experiments are crucial against some hypothesis only in relation to a stable set of assumptions, which we do not wish to abandon. But no set of assumptions is permanently valid. Hanson says that crucial experiments are out of the same bag as pure observations and uninterpreted facts; they are philosophers' myths.

Wittgenstein said language has many uses. Hanson's discussions of crucial experiments pertains only to theories that may intelligently be disconfirmed. Although in principle all statements of science are testable and can be falsified, in practice theories often have another use or function. Following Wittgenstein's thesis that language may have many uses, Hanson maintains that theories functioning as "pattern statements" supplying a conceptual *gestalt* will not yield an intelligible statement negating the theory, if the theory is viewed as disconfirmed. This is because the theory gives the phenomena their intelligibility; and this explains why scientist will not reject a theory even while they recognize the existence of anomalies that are not intelligible in the theory. What scientists do in practice is to attempt to save the theory with small modifications or wait until a new and more adequate theory is proposed that explains all that the old theory explains as well as the anomalies to the old theory.

Anomalies do not make scientists give up intelligibility. It is for this reason that physicists have not given up the Copenhagen interpretation in spite of the anomalies confronting Dirac's theory. Thus Hanson opposing Bohm in the "Postscript" chapter in *Quanta and Reality*, states that dropping orthodox quantum theory right now would be to stop doing microphysics altogether. Then Hanson immediately adds that should the heretics (Bohm *et al*) succeed in accounting for everything that orthodox theory now describes, and do so without the divergence difficulties and the renormalization nuisance even without the uncertainty relations and the irreducibly statistical laws, should they do all this, then physicists of the world will be at their feet, and science will have ascended to a new plane of power and fertility.

Hesse on Models and Analogy

Quanta and Reality (1962) is a collection of discourses initially cast as a radio series by the B.B.C. in 1961. It includes a dialogue involving Bohm, a "Postscript" commentary by Hanson, and an article, "Models and Matter", by the Cambridge University philosopher of science, Mary B. Hesse. Hanson's comments are generally critical of Bohm; Hesse's are more sympathetic. This alignment among the participants is not limited to the specifics about the contemporary quantum theory; it divides along issues about the semantics of scientific theories in general and also about the role of semantics in scientific discovery.

On Hanson's view the semantics of a theory involves only the mathematical formalism and the measurements that the equations of the formalism relate. The relations expressed by the theory form the conceptual *gestalt*, which is the semantics of the theory. And in the case of quantum theory the Copenhagen interpretation with its wave-particle duality thesis is integral to the formalism of the quantum theory. Furthermore the semantics of the quantum theory so understood is strategic to the further development of microphysics, as indicated by the fact that Dirac relied on it for his development of his field quantum theory. Hanson does not deny that there may also be other language about the microphysical domain explained by the equations of the quantum theory, language that does not contradict the quantum theory, but he views such supplementary language as "merely philosophical" and not as part of the theory itself. He places Bohr's "naive" epistemology in this category of supplementary philosophical language. Opponents to the Copenhagen interpretation agree with Hanson that semantics has a strategic role in scientific discovery. But they do not agree that the Copenhagen interpretation is integral to the formalism of the theory. They disagree not only because some of them propose alternatives to the wave-particle duality, but also because in general they maintain that there is more to the semantics of theories than just the formalism and measurement data. The source of this additional semantics that they say is found in all theories, is (vaguely speaking) the imaginative and often nonliteral use of language that is characteristic of theories in physics. This imaginative language may involve analogies and metaphors, and the distinctively additional semantics is called a "model". This is one of several common meanings for the term "model". Sometimes it has a more general meaning, which is whatever may consist of the semantics of the syntactical formalism of a theory. But in this context the term "model" functions to articulate the different views of the issue. Unlike Hanson, Hesse views the ideas of waves and particles as "models" for quantum theory, and her view proceeds from a sophisticated examination of these questions.

Hesse's views about semantics are influenced by her former mentor at Cambridge, R. B. Braithwaite, a Logical Positivist philosopher of science. But their views are not the same. In his *Scientific Explanation* (1953) Braithwaite distinguishes two ways in which a calculus may receive a semantical interpretation. Firstly a calculus may receive an interpretation that makes it a scientific theory containing theoretical terms, when the logically posterior consequences determine the meaning of the theoretical terms in the representation in the calculus of the premises. Theoretical terms are thus said to have "indirect" meaning, since their meanings are determined by their contexts in relation to one another and to the sentences expressing directly testable propositions which the experimentalist can derive from them. In other words, the meanings of the theoretical terms are indirect, because they receive their semantics contextually and not ostensively. Yet at the same time Braithwaite also maintains that a good theory is capable of growth, such that it must be an alternative way of describing the empirical generalizations upon which it is based. Therefore the meanings of the theoretical terms are not limited by being explicitly defined. Braithwaite does not explain how theoretical terms satisfy this last criterion simply by contextual use.

The theory may furthermore be given an interpretation in the second way that Braithwaite calls a "model", and which he does not view as a literal interpretation for the domain explained by the theory. In the model the logically prior premises determine the meaning of the terms occurring in the representation in the calculus of the conclusions. Thus Braithwaite says that theories and models have different epistemological structures, even when they have the same calculus. For example according to Braithwaite the solar system may serve as a model for the hydrogen atom, even though it is understood that the atom is not literally to be taken as a solar system.

Braithwaite says that thinking of theories by means of models is always "as-if" thinking, e.g. thinking of the atom as if it were a solar system. He also says that for the physicist, Schroedinger's wave function is exhaustively interpreted in terms of its use in the calculus of the quantum theory, and he says in a footnote that no one supposes that the wave function denotes a wave in any ordinary sense of wave. Modern quantum theory therefore does not have any model, in Braithwaite's view.

Hesse's views are not quite the same as Braithwaite's. Unlike Braithwaite, Hesse does not distinguish the semantics of theoretical terms from models. In fact for Hesse the models supply some of the indirect meaning had by the theoretical terms, and furthermore extrapolation on the basis of the models explains how the theories grow, so that Hesse's interest in the semantics of theoretical terms gets her into the topic of scientific discovery. And Hesse believes that the concepts of wave and particle supply modern quantum theory with two contrary models. In her *Models and Analogies* (1953) Hesse compares two earlier conflicting protagonists in the semantical issue about models. One is Duhem and the other is Campbell. In his *Aim and Structure of Physical Theory* Duhem had argued a view similar to Hanson's, that the physical theory consists only of the equations and measurements, and that even if models based on analogy with more familiar phenomena may have some heuristic value for developing a new theory, still these models are not part of the theory itself and may be discarded after the theory is constructed. On the other hand in his *Physics, The Elements* (1920) N. R. Campbell argued that analogically based models are not dispensable aids, but are indispensable to a theory because they assist in the continuous extension of the theory. The motivating intent behind this view is that scientific theories are not static museum items, but rather are always growing as part of the growth of science; and this latter view is the one that Hesse accepts. In her own examination of analogies and models Hesse distinguishes three parts to an analogy, which she calls "the positive analogy", "the negative analogy", and "the neutral analogy". The positive analogy consists of those aspects of the familiar phenomena that are known to apply to the phenomenon explained by the theory. These include the similarities that may have occasioned recognition of the analogy in the first place. The negative analogy consists of those aspects of the familiar phenomena that are known not to apply or known to be irrelevant to the phenomenon explained by the theory. The neutral analogy is the most important from the viewpoint of scientific discovery; it consists of those aspects of the familiar phenomena whose relevance to the problematic phenomena in the domain of the theory is presently unknown, and therefore whose explanatory potential for further development of the theory is not yet known. The further theoretical exploration of the problematic phenomena will be guided by the neutral analogy; it is a key, as it were, to scientific discovery. In "Models and Matter" Hesse says that in quantum theory the wave and particle models are such that what is positive analogy in the one model is negative analogy in the other. She also says without elaboration that in the two models there are still features that physicists cannot classify as either positive or negative, and that it is because of these features that the particle and wave models are yet essential. If physicists were forbidden to talk in terms of models at all, then they would have no expectations at all, and would be imprisoned forever inside the range of existing experiments. Hesse does not explicitly endorse Bohm's analogically based Brownian model for future quantum physics, but unlike Hanson she views the semantics of theories based on analogy to be strategic to scientific discovery.

Hesse on Metaphor

In addition to analogy Hesse considers another figure of speech known as metaphor in her "Explanatory Function of Metaphor" in *Logic, Methodology and Philosophy of Science* (ed. Bar-Hillel, 1965), where she states that her views are significantly influenced by the "interactionist" concept of metaphor proposed by Max Black in his *Models and Metaphors* (1962). Black's interactionist view is opposed to the "comparison" view. On the comparison view the metaphorical statement is nonliteral, because if it is taken literally it is a false statement, and it can be restated as an exhaustive list of similes, which are literal statements expressing the similarities implied in the metaphor. In other words in rejecting the comparison view Black rejects the thesis that metaphors are elliptical similes. In her paper Hesse distinguishes the primary system and the secondary system, where both systems may be taken as real or physical systems that are described literally. The metaphoric use of language to describe the primary system consists of transferring to it a word or words, that normally are used in connection with the literal description of the secondary system. In a scientific theory the primary system is the domain of the *explanandum* described in an observation language, while the secondary system is the domain of the *explanans* described either by observation language or by a familiar theory from which the explanatory model is taken. The explanation of the *explanandum* for the primary system consists of statements that metaphorically employ vocabulary describing the secondary system and that are applied to the primary system on the basis of some similarity or analogy.

In his statement of his interactionist view Black lays down a criterion for the literal equivalence of a metaphor: the metaphor can be re-expressed as an exhaustive list of statements expressing all the similarities in the metaphor as literal similes. Then he rejects the possibility of reducing the metaphor to such a list of similes, because such a list can never be exhaustive. This inexhaustibility is especially important to Hesse, because the possibility of indefinitely extending and explaining the metaphor constitutes the fruitfulness of the explanatory model containing the metaphorical language. But the thesis that metaphor cannot be reduced to literal language is not all there is to Black's interactive view of metaphor. The interactive thesis is called "interactive", because the metaphorical use of language is seen as changing the literal meanings of the words that are used metaphorically; there is an interaction of the meanings of the words in their descriptions of the primary and secondary systems. For example the metaphorical statement "Man is a wolf" makes wolves seem more human and men seem more vulpine. This is contrasted with the comparison thesis, which purportedly assumes that the literal description of both primary and secondary systems is unaffected by the metaphor, such that the meanings of the terms remain semantically invariant. In Hesse's view the semantical variance postulated by the interaction view of metaphor is relevant to scientific explanation, because metaphor changes the semantics of the observation language. Hesse sees this meaning variance in the observation language as contrary to the assumptions of the hypothetico-deductive account of explanation, in which it is assumed that descriptive laws pertaining to the domain of the explanandum remain empirically acceptable and semantically invariant in all changes of explanatory theory. She therefore advances the view that the deductive model of explanation should be modified and supplemented by a view of theoretical explanation as metaphoric redescription of the domain of the explanation.

The interactive view of metaphor advanced by Black and used by Hesse, is not the prevailing view. Conventionally metaphor is construed as an elliptical simile containing implicitly the idea of an underlying similarity that can be explicitly and literally expressed by a simile with the words "like" or "as". For example in his *Philosophy of Language* (1964) William P. Alston sets forth what might be taken as the comparison thesis of metaphor. Like Black and Hesse, Alston maintains that metaphor has an indeterminacy in it that is inexhaustible. But he also maintains that it is a mistake to believe that metaphorical and literal language are different kinds of meaning. On Alston's view the difference between metaphorical and literal language is one of degree, where literal language may be identified with established usage and metaphor is a new usage that is derived from established usage. All meanings are literal meanings, and the derived usage in a metaphor may be expressed literally with greater or lesser extent of explanation. When the new usage has become part of the established usage, then the metaphor has become a "dead" metaphor, because explanation of its derivation from established usage becomes an exercise in etymology. Furthermore unlike Black or Hesse, Alston does not say that metaphor must be capable of being reduced to an exhaustive list of similes, in order to be reduced to literal use, because there is indeterminacy in literal language as well as in metaphor. Alston references Friedrich Waismann's "Verifiability" in *Logic and Language* (1952) stating that literal words denoting physical objects have an inexhaustible vagueness which remains even after all attempts at clarification. This vagueness remains because in addition to actual cases of indeterminacy of application, one can think of an indefinite number of possible cases in which one would not know what to say. Waismann calls this inexhaustible vagueness the "open texture" of descriptive language. Alston denies that metaphor is simply vagueness, but he says that in both metaphorical and established language there is an inexhaustible indeterminacy due to the fact that it is impossible to decide in advance on every possible usage of a word. The conclusion to be drawn from this is that Black's criticism for the reduction of metaphor to literal language by means of an exhaustive list if similes is not a feasible criterion, because it would demand more determinateness of nonliteral language than of literal language. A weaker criterion therefore is in order. It would seem sufficient to require only that a metaphor be reexpressible with at least one simile that makes explicit an implicit underlying similarity, presumably but not necessarily the similarity that is intended by the speaker or writer initiating the metaphor. Furthermore semantical variability or meaning variance must be a property of both metaphorical and literal language, or it must be a property of neither, since the former is merely the elliptical expression of the latter.

These considerations are relevant to Hesse's thesis about metaphor in theoretical explanation in science. Her tacit premise is that meaning variance does not occur in literal language. On this premise the nonreducibility of metaphor to literal language is strategic to her rejection of the adequacy of the hypothetico-deductive thesis of theoretical explanation, and it is strategic to her reliance on metaphor to account for semantical change or meaning variance in the language for description of observed phenomena. On the other hand if as Alston says metaphor is reducible to literal language, then semantical variability must be a property of both metaphorical and literal language, or it must be a property of neither. And it is clearly a property of metaphor; otherwise there would be no "dead metaphors", thus indicating that the metaphorical use has become the established or literal use. Thus the reducibility of metaphor to literal language implies that metaphor cannot satisfactorily be used as an explanation of semantical change in science, even if it can serve to indicate that semantical change has occurred relative to currently established meaning. The theory-laden character of observation discourse resulting from theory revision is a much more general aspect of the semantics of language than just its metaphorical usage. The explanation of semantical change or meaning variance demands a general theory of semantical description for literal language. At the same time metaphor seems clearly to have a role in occasioning semantical change, and it may have a strategic value for the development of new theories in science. While Hesse considers metaphor principally in connection with the problem of scientific explanation, and offers interesting analyses of it, Bohm considers metaphor principally in connection with the problem of scientific discovery. But he has also revised his view of physics and his agenda for future physics since his earlier works of the 1950's.

Bohm's Revised Agenda for Future Physics

For more than thirty-five years following his initial 1952 statement of his hidden-variable interpretation Bohm continued to expound his views in philosophy of science and also in metaphysics and epistemology. In 1967 he wrote *Wholeness and the Implicate Order*, and in 1987 he wrote *Science, Order and Creativity*, in which he refers to a planned book, *Causal Interpretation of Quantum Mechanical Field Theory.* Bohm's statements that are most relevant to the subject of scientific discovery are to be found in *Science, Order and Creativity*, particularly in the introductory chapter and in the two succeeding chapters, which altogether take up about half of the book.

The introductory chapter is very important, because it reveals that Bohm has changed his agenda for future microphysics. Firstly he maintains that there is no difference between science and philosophy. While Hanson also states that physics is natural philosophy, Bohm's statement means something very different. Bohm explicitly rejects the prevailing view of the aim of physics, which he says is to produce mathematical formalisms that can correctly predict the results of experiments. He maintains that, since quantum theory and relativity theory were never understood adequately in terms of what he calls "physical concepts", physics gradually slipped into the practice of talking about equations. And he states that Heisenberg gave this practice an enormous boost with the idea that science can no longer visualize atomic reality in terms of physical concepts, and with the idea that mathematics is the basic expression of our knowledge of reality. Bohm maintains that the current emphasis on mathematics has gone too far. In stating that science is the same as philosophy, Bohm means that as philosophy had traditionally done, now science must unify knowledge instead of offering physicists a fragmentation like it has today. In times past there was a general vision of the universe, of humanity, and of man's place in the whole. But specialization in modern science became narrower and led eventually to the present approach, which is fragmentary. Bohm also opposes what he sees as another aim of modern physics, which is to analyze everything into independent elements that can be dealt with separately. This further contributes to fragmentation. Bohm believes that the time has come to change what is meant by "science". This change is to be implemented by a "creative surge", that will eliminate fragmentation.

These comments suggest a basic change in Bohm's agenda for future physics. The creative surge is not aimed at resolving what he formerly called the "current crisis" of contemporary mathematical field quantum theory with better equations, as he had earlier proposed could be achieved by his hidden-variable interpretation. Nor is it aimed at extrapolating into the future a directionality revealed in his interpretation of the history of physics; in fact it is quite the opposite. It is an effort to revise the aim of physics to recapture the metaphysical system building of the past. The only one of the three original objectives in his agenda for future microphysics that Bohm retains, is the creation of an alternative interpretation of microphysics, and that alternative is hidden-variable or "causal" interpretation.

Bohm on Perception and Metaphor

In the two chapters following the introduction to his *Science, Order and Creativity* Bohm sets forth his philosophy of science. His philosophy of science is not necessarily specific to his revised agenda for physics, and it is worthy of serious consideration independently on its own merits as an alternative to Hanson's views on creative scientific discovery. Bohm advances a theory of perception that is explicitly opposed to that of the Logical Positivists, and that is also characteristic of contemporary post-Positivist philosophy of science. It also reveals some influence from Einstein. On Bohm's view perception takes place in the mind and in terms of theories. The observational data obtained by Archimedes in his bath, had little value in themselves. What was significant was their meaning as perceived through the mind in an act of creative imagination. The principal change occurring in modern science is that this mental perception is more mediated through elaborate instruments that have been constructed on the basis of theories. Furthermore the questions that scientists ask arise from already existing knowledge.

Bohm's philosophy of perception is central to his views on scientific discovery. The development of science is now obstructed by fragmentation, which in turn is caused by subliminal rigidities in thought, that Bohm calls the "tacit infrastructure of scientific ideas". One example of the tacit infrastructure of scientific ideas is the Newtonian notions of space and time, that led Lorentz to preserve both the idea of the constancy of the velocity of light and the ideas of absolute space and time by explaining the anomalous results of the measurements of light by postulating changes in the measuring apparatus as the apparatus moves through the ether. Bohm notes that the tendency of the scientist's mind to hold to what is familiar is enhanced by the fact that the overall tacit infrastructure is interwoven in the institutions on which depends the professional security of the scientist. The means for breaking out of the tacit infrastructure of scientific ideas and to create new theories is metaphor. Unlike the Cambrian philosophers of science, such as Braithwaite and Hesse, Bohm does not consider the role of metaphor in scientific discovery to be relevant merely because of a residual

Logical Positivist concern with the semantical status of theoretical terms. Because Bohm maintains a philosophy of perception that is as contemporary as Hanson's, the semantics of theoretical terms is not problematic for him. For Bohm, therefore, the role of metaphor in scientific discovery is a central philosophical topic in its own right. Metaphor occasions what Bohm calls "creative perception", and he also refers to "metaphoric perception". Bohm defines metaphor as the simultaneous equating and negating of two concepts. Metaphoric perception brings together previously incompatible ideas in radically new ways. The unfolding of a metaphor, that equates different and even semantically incommensurable concepts can be very fruitful. In using the term "incommensurable" Bohm references Kuhn's sense of the term, which Kuhn in turn had gotten from Feyerabend. This role of metaphor in discovery is possible because perception in science takes place primarily through the mind and much less through the senses; the realm of physics is now that of perception through the mind, and theory dominates experiment in the development of the scientific perception of nature.

He also maintains that communication is essential to perception in science. He understands "communication" in a very broad sense to include the individual's own articulate mental dialogue with himself. The scientist engages in an inner dialogue with himself as well as with his colleagues, and in this dialogue he is disposed in his thinking by the social background. Insights enfolded in this inner dialogue must be unfolded by discourse with colleagues and eventually by publishing. Fragmentation may proceed to the point that communication becomes blocked, because the tacit infrastructure of ideas not only limits the individual, but also the whole scientific community in their creative acts of perception. Both paradigms and specialization may cause fragmentation in this way. Bohm considers communication to be so central to perception that he speaks of "perception-communication"

Bohm equates his thesis of the tacit infrastructure of scientific ideas with Kuhn's thesis of scientific paradigm. A paradigm is not just the articulate theory, but also the scientist's whole way of working, thinking, communicating, and perceiving with the mind. However, Bohm rejects Kuhn's thesis that "normal science" is without any creativity, and that "revolution" is completely discontinuous. Bohm maintains that semantic incommensurability can be overcome with metaphor. He furthermore says that revolution occurs when a new metaphor is developed, and normal science is the creative unfolding of that new metaphor. In Bohm's view there is much more creativity in normal science, than Kuhn admits. Bohm also criticizes Popper's thesis of falsifiability. He maintains that today an excessive emphasis is being placed on falsifiability in the sense that unless a theory can immediately or very shortly be falsified, then that theory cannot be regarded as properly scientific. A new idea with broad implications may require a long period of gestation before falsifiable consequences can be drawn from it. Fragmentation has occurred in modern microphysics between mathematical formalism and informal discourse in microphysics. Differences in the informal discourse gave rise to an issue between Bohr and Einstein, as well as among later physicists. This change in the language of physics occasioned by the development of quantum theory has led to a communication breakdown. Both Bohr and Einstein agreed on the mathematical formalism, but there is still no common informal language. Bohm believes that if Bohr and Einstein had been willing to entertain a free dialogue to eliminate the rigidities that block communication, then perhaps a new creative metaphor might have emerged for microphysics. In such a dialogue each person must be able to hold several points of view in a sort of active suspension, while treating others' views with the consideration he gives his own. This would lead to the intellectual free play needed for a new creative metaphor.

Bohm proposes his own causal or hidden-variable interpretation for consideration in this spirit. The interpretation of a formalism is something that is in the informal discourse, not in the measurements or the equations themselves. This view is fundamentally contrary to Hanson's. In Bohm's view all the available interpretations of the quantum theory, as with any other physical theory, depend fundamentally on implicit or explicit philosophical assumptions, as well as on assumptions that arise in countless other ways. The image of the "hard-nosed" scientist, who does not admit to the existence of the philosophical assumptions in the informal language, is just another example of the subliminal influence, that is exerted on scientists by the tacit infrastructure of ideas of the scientific community at large. It might be said that the elaborate metaphysics set forth by Bohm is his attempt to make explicit the philosophical ideas "enfolded" in his own hidden-variable interpretation of quantum theory.

Comment and Conclusion

Contrary to often expressed opinion the topic of scientific discovery has not been a neglected one in philosophy of science. The above survey reveals that many philosophers and scientists have addressed it with a semantical approach. But no attempt to develop a metatheory of scientific theory development using the semantical approach of examining figures of speech has yet succeeded, even though historic scientific discoveries seem to have resulted from the use of such figures of speech as analogies and metaphors. To date the only metatheories that are sufficiently universal to function as applicable procedures for scientific discovery are those based on the discovery-systems approach. But in his historical research Hanson had recognized and documented that semantics has operated as a constraint upon discovery, and he understood that this phenomenon implies the need for a reconsideration of the nature of scientific language, even though he himself had failed to set forth an adequate metatheory of semantical description. The following somewhat lengthy commentary is divided into four parts. Firstly Hanson's attempt at a logic of discovery is critiqued. Secondly Hanson's defense of the Copenhagen interpretation is considered in the context of semantical change in science. Thirdly Hanson's principal criticism of Bohm's views is viewed in historical retrospect. And finally some comments are given on Bohm's strategy of metaphor.

Turning to Hanson's attempt at a logic of scientific discovery, some comparisons with Popper suggest themselves as a outset. Both Popper and Hanson are characteristically post-Positivist philosophers of science. Just as Einstein states that theory decides what the physicist can observe, so too Popper states that observation is "theory-impregnated" and Hanson states that observation is "theory-laden". Both philosophers of science reject the view that there is an observation language having a semantics that is predetermined by the natural processes of perception, even though they do not agree on the nature of perception: Popper distinguishes perception from objective knowledge, while Hanson reinterprets perception ("seeing") so that it includes variable interpretation. But both agree that observation does not supply empirical science with permanent foundations. However, Popper and Hanson pick up post-Positivism from opposite ends of the stick with little likelihood that they could ever meet in the middle. Popper's starting point is the logic of scientific criticism, and he rejects all claims that there can be a logic of scientific discovery, when he rejects inductivism as the consequence of the naturalistic theory of meaning. And for the same reason he rejects all meaning analysis. Hanson's starting point on the other hand is the logic of scientific discovery, and he does not hand off all such matters to psychology as Popper does. But for Hanson the logic of scientific discovery is not just a matter of formal logic, even though formally the retroductive inference has the same criteria as the hypothetico-deductive inference. For Hanson "logical" also contains the idea of the conceptual, which is to say, the semantical. Contrary to Popper, Hanson views semantical analysis to be integral to philosophy of science. In fact it could be said that in Hanson's philosophy of science semantics is integral to all its topics: to discovery, to explanation, to criticism, and to the aim of science. But in spite of the importance that Hanson places on semantics, he never used or developed a systematic philosophy of language. His principal inspiration was Wittgenstein's *Investigations*, a very unsystematic approach to philosophy of language. In his discussion of "seeing" Wittgenstein employed ambiguous drawings such as are commonly used in texts on *gestalt* psychology, and Hanson therefore developed a semantics of language based on the idea of the "conceptual *gestalt*". Unfortunately *gestalt* psychology is a very blunt instrument for semantical analysis; it is a wholistic approach to semantical description.

Hanson's philosophy of scientific discovery was also greatly influenced by Paul Dirac. When Dirac told Hanson that the Copenhagen interpretation figured essentially in his development of the formalism of his relativistic quantum theory, Hanson took the position that the Copenhagen interpretation (without Bohr's "naive epistemology") is that one unique semantical interpretation supplied by the formalism itself, and not some philosophical idea appended to the formalism. However, the *gestalt* semantics is not adequate to the defense of Hanson's view of that Copenhagen interpretation is integral to the formalism of the modern quantum theory. Had Dirac said just the opposite of what Hanson reports he said about the Copenhagen interpretation's relation to the formalism of quantum theory, the *gestalt* semantics would have been neither more nor less serviceable for a semantical analysis of quantum theory, because the conceptual *gestalt* does not enable the philosopher of science to separate or to distinguish the semantics that may in some way or another be integral to the quantum theory's formalism, from that which may not be integral to the formalism but instead may be viewed as in some way or another appended to the formalism as what Hanson calls "mere philosophy". In fact Hanson's *gestalt* semantics does not even offer him a basis for his distinction between the "Copenhagen interpretation" and the "Bohr interpretation". The wholistic character of the "conceptual *gestalt*" makes it impossible to partition the semantics of the quantum theory into parts, such as those parts that are integral to the formalism and those parts that are not, or those parts that are properly called the "Copenhagen interpretation" and those parts that are called the "Bohr interpretation".

The wholistic character of the conceptual *gestalt* also affects Hanson's attempt to explain scientific discovery. On the one hand the conceptual *gestalt* offers "conceptual resistance" to any change to a new *gestalt* and therefore to any new theory. In other words it offers a semantical impediment to scientific discovery. On the other hand it also offers a guide to scientific discovery, because it informs the scientist of the kind of hypothesis that may emerge from the retroductive logic of scientific discovery. The semantics of the currently accepted theory may function in both of these contrary ways, but the *gestalt* psychology does not explain how. More specifically in connection with the modern quantum theory, the *gestalt* psychology does not explain why Hanson should be defending the Copenhagen interpretation as a guide instead of attacking it as an impediment to the discovery of a new and more empirically adequate quantum theory.

The reason for this problem is the basic fact that the wholistic *gestalt* cannot function in a logic of scientific discovery or in any other application of logic. Its wholistic character deprives the retroductive "logic" of any procedural character. Retroduction can only describe the conditions that the new *gestalt* must satisfy after it has been hit upon, which is to say that it is a statement of a scientific problem that the discovery must solve rather than a procedure for obtaining a solution. On the *gestalt* view the discovery itself is a transition that does not admit to a procedure, just as the transition from one interpretation of an ambiguous drawing to another does not admit to a procedure. Just as there could never be a mathematical formalism to describe the transition occurring in a substantial change as described in Aristotle's physics, so too there could never be a logical formalism to describe the transition occurring in a *gestalt* change in Hanson's philosophy. In both cases the transition from one form to the other is a substitution, which is instantaneous, whole and complete, and with no intelligible continuity to warrant calling it a process instead of a replacement.

Turn next to the second topic, Hanson's defense of the Copenhagen interpretation and his view that it is necessarily implied by the formalism of the theory. Since the topic of scientific discovery is not just formal logic, semantical issues arise. The central question is whether the semantics of physical theory is adequately specified by the equations of the theory together with the discourse describing the measurement apparatus and procedures used to obtain the data related by the equations, or whether there is furthermore additional discourse characterizing the domain of the equations and measurements. Hanson is one philosopher who rejects any semantical role in scientific explanation for any discourse other than the equations of the theory and the statements of experimental description and measurement procedures. He relegates any additional discourse to the status of "mere philosophy", and maintains that physicists who indulge in such philosophizing without also improving on the mathematical physics to produce an empirically superior formalism, are rightly not held in high esteem by their professional colleagues. Accordingly he maintains that the wave-particle duality, which he believes is the distinctive characteristic of the Copenhagen interpretation, is not some semantics added to the formalism of the quantum theory by those statements that he calls "mere philosophy", but rather is integral to the formalism of the theory. His basis for stating this position is Dirac's statement made personally to Hanson, that the wave-particle duality is integral to the formalism, and that it was strategic in his own development of his relativistic quantum theory.

However, there are a number of physicists and philosophers who disagree with Hanson's position, and they point to several reasons why one may believe that there is additional semantics than that supplied by the equations and statements of measurement procedures. The first is that both Bohr and Einstein accepted the same formalism and measurements, yet disagreed about the semantics of the theory. Bohm cites this fact in support of his thesis that physics and philosophy are the same. A second reason for maintaining that the semantics of the theories includes more than just the equations and statements of measurement procedures, is the thesis that the so-called philosophical discourse supplying the additional semantics, is also strategic for scientific discovery and thus for the future of the science. Hanson recognized the implications of different interpretations for Born and Schroedinger, but the philosophical differences between Einstein and Bohr are more fundamental; it is noteworthy that the issue between Einstein and Bohr about whether quantum theory is "complete" was about the future agenda for microphysics.

The issue's resolution seems to require recognition of the occurrence of a semantical change, a change that is accepted by some but not by others. Those who accept the Copenhagen interpretation are those who have accepted the semantical change, while those who propose alternatives are those who have retained the previously established semantics. For some physicists the associated "philosophical" discourse assumes a definitive role, that makes impossible the semantical changes required for acceptance of the new theory. In the first chapter of his *Concept of the Positron* Hanson gives a brief historical overview of the wave and particle theories of light. He notes that Newton did not have a semantics for the terms "wave" and "particle", that made the concepts dichotomous or mutually exclusive, when he proposed his theory of "fits". Only later did these concepts assume their dichotomous implications, when the experiments of Foucault, Frenzel, and Young were believed to have the force of crucial experiments, that enable the physicist to decide between the two characterizations. The concepts of wave and particle had undergone semantical change with the advance of physical experiment and theory.

By the twentieth century the wave-particle dichotomy had become very well established even though the discoveries of Planck and Compton enabled physicists to express the wave-particle duality mathematically prior to development of the modern quantum theory by Heisenberg and Schroedinger. Planck had discovered that the quantum of action has an energy that is equal to the radiation frequency multiplied by Planck's constant. Later the photon concept became established with Einstein's theory of the photoelectric effect. In 1923 Compton stated that the photons have a momentum that is equal to Planck's constant multiplied by the wave number of the radiation. The Compton equation is also expressed in terms of the wave length, which is the reciprocal of the wave number. And when a substitution is made by eliminating Planck's constant, the result is an equation expressing of the wave-particle duality, in which the particle aspect is expressed by the energy quantity and the momentum, and the wave aspect is expressed by the frequency and the wave number (or wave length) magnitudes. Interestingly in his *Conceptual Development of Quantum Mechanics* (1966) Max Jammer states that Bohr had come to his complementarity principle by consideration of these two equations, and he references a four-page postscript to a paper written by Bohr in 1925. This is a year before Heisenberg reports that Bohr developed his complementarity principle.

Yet in spite of this mathematical assertion that wave and particle are aspects of the same physical reality, the dichotomous concepts were retained by Bohr, because he retained the classical concepts in his complementarity principle, and relegated mathematical formalism to an instrumentalist status, even as he affirmed the wave-particle duality. The complementarity principle is a contradiction resulting from his belief in the naturalistic philosophy of perception, which in turn implies that the concepts of wave and particle like all classical concepts cannot be changed. And it is an example of the philosophical discourse defining the semantics in a way that is inconsistent with the semantics defined by acceptance of the theory. After some weeks of argument with Bohr, Heisenberg accommodated Bohr's complementarity thesis by accepting the idea that the wave particle duality is expressed by the uncertainty principle, save that the mathematical formalism of the uncertainty principle is consistent while the complementarity principle is inconsistent. Heisenberg made this accommodation, because he accepted Bohr's naturalistic philosophy of perception. Yet in so doing, he was himself philosophically inconsistent, since unlike Bohr he did not construe the formalism instrumentally. Instead, following Einstein's precedent with the Lorentz contraction, Heisenberg construed his formalism realistically when he affirmed the uncertainty principle.

The only way that the wave-particle duality can be affirmed consistently is to let the equations control the semantics of the terms "wave" and "particle", when and however these terms are associated with the formalism. This produces a semantical change in the meanings of the terms, with the result that the terms no longer stand for classical concepts and are therefore no longer dichotomous. Thus the equations function as definitions in which the concepts are correlatively but no longer dichotomously defined. This amounts to using the equations of the theory in a "functionally *a priori*" manner and as "pattern statements", as Hanson said, and letting the theory decide what is observed, as Einstein said.

Although Heisenberg maintained that the terms "wave" and "particle" have classical semantics as they occur in Bohr's complementarity principle, he may have been approaching this recognition of a semantical change, when in his "Questions of Principle" (1935) he said the restrictions on classical concepts as enunciated in the uncertainty relations acquire their "creative value" only by making them questions of principle, such that they can afford the freedom necessary for a noncontradictory ordering of experience. Given his doctrine of closed-off theories, his use of the term "restrictions" in this statement may be understood to mean redefinition. And in the light of his autobiographical description of his development of the uncertainty relations, his phrase "creative value" seems clearly to refer to the role of the mathematical equations in defining the semantics, when the formalism is used for observation, as in the case of his reconsideration of the vapor trails in the Wilson cloud chamber. In other words the formation of the new semantics is integral to the new scientific discovery. In this paper Heisenberg also states that the system of mathematical axioms of wave and quantum mechanics entitles the physicist to regard the question the simultaneous determination of position and impulse values as a "false problem", just as Einstein's relativity theory makes the question of absolute time a false question in the sense that they are devoid of meaning. Clearly the reason why Heisenberg perceives that such questions become devoid of meaning, is that the meanings have been changed by the in-principle maneuver of giving semantical control to the new equations.

Hanson reiterates Heisenberg's in-principle approach. In the chapter "Elementary Particle Physics" in his *Patterns of Discovery* Hanson states that one cannot maintain a quantum-theoretic position and still aspire to the day when the difficulties of the uncertainty relations have been overcome, because this would be like playing chess and yet hoping for the day when the difficulties of having but one king will have been overcome. But Hanson proceeds beyond Heisenberg. Heisenberg's theory of semantical change, his doctrine of closed-off theories, was not only intended to explain semantical change, but was also intended to explain semantical permanence for classical concepts used for observation. In contrast Hanson maintains in the same chapter cited above, that the uncertainty principle is built into every observation of every fruitful experiment since 1925. In Hanson's explicit and systematic philosophy of science, unlike Heisenberg's, the theory controls even the semantics of the language used for description of observed phenomena. He states how a theory has its "creative value" in ways that Heisenberg actually used and chronicled in his development of the uncertainty principle, but which Heisenberg did not incorporate into his explicit and systematic philosophy, the doctrine of closed-off theories.

The question of whether or not the duality thesis is asserted by the formalism of quantum theory depends on the views of the particular physicist. The physicist who buys the Copenhagen interpretation, also accepts the semantical control that the equations exercise over the concepts in the formalism. Another physicist who proposes an alternative to the

Copenhagen interpretation, one that implies the wave-particle dichotomy instead of duality, has elected to let other discourse exercise definitional control over the semantics of the terms, and has then decided to relate the concepts to the formalism in ways that may or may not affirm the wave-particle duality. More generally even philosophers and physicists who buy the Copenhagen wave particle duality, need not agree with Hanson that the duality is defined by the formalism. Bohm is the noteworthy example of one who accepts the duality thesis while also maintaining that it may but need not be the interpretation for the formalism; in his view all interpretations are controlled by what he calls the "informal" language. One outcome of the semantical change resulting from giving semantical control to the formalism of the theory, as Hanson advocates, is a complication in the problem of how empirical control is also exercised over the theory, such that independent evidence is possible and tautology is prevented. A possible escape clause in Hanson's philosophy of perception is his admission of theory-independent "phenomenalist seeing". But he never actually invokes this idea to defend the empiricism of science, even while he never doubts the empirical character of science. Instead he invokes Wittgenstein's idea of the multiple uses of language.

Thirdly consider Hanson's principal criticism of Bohm's hidden-variable interpretation of quantum theory. Hanson's criticism is that Bohm has not developed any new equations. Bohm's theory of "creative perception" is no less contemporary than Hanson's theory of perception that says that observation is theory-laden. But Bohm's agenda for future physics is very different from Hanson's. In his "Hidden Variables and the Implicate Order" in *Quantum Implications* Bohm admits that his proposed hidden-variable interpretation did not "catch on" among physicists, since it gives exactly the same predictions for all experimental results as does the Copenhagen interpretation, which he calls the "usual theory". Thus Hanson's critique of Bohm in his "Postscript" in *Quanta and Reality* twenty five years earlier has been vindicated at least to date by the behavior of the physics profession in the years that have elapsed. Physicists have no interest in alternative interpretations for their own sake, i.e. when they are not associated with new and empirically superior equations that solve problems which the current mathematical physics cannot solve. Initially Bohm proposed his interpretation as a heuristic for developing superior equations that could count as solutions to the renormalization problems in relativistic quantum physics, but later he revised his agenda and proposed a new aim of physics, that does not aim at improved mathematical equations. It is tempting to compare his proposal to revise the aim of physics with the strategy of the archer who, after missing the target, paints a new target around the place where the arrow landed. Where Bohm's hidden-variable thesis landed is in a metaphysics of nature, and that certainly is not what contemporary physics aims at. There is enough continuity in Bohm's philosophical thinking, that his wish to revise the aim of physics and redefine the meaning of "science" making it the same as philosophy, is not disreputably *ad hoc*. Nonetheless he did not initially set out to revise the aim of physics. And not surprisingly physicists have not followed him in his new agenda, because if they did, physics would make a three-hundred year regression to the metaphysical system building that prevailed previous to the historic Scientific Revolution.

Finally consider Bohm's strategy of metaphor. Their differences not withstanding, Bohm and Hanson have a common belief underlying their interests in scientific discovery. Traditionally it was thought that language has merely a passive role, such that firstly a discovery is made by observation of nature, and then language is employed to report the discovery. This view underlies the Logical Positivist's thesis that induction produces empirical generalizations. Until the post-Positivist period the only active role for language was in logical deduction, and deduction is not thought to involve semantical change. But Bohm and Hanson have both rejected the naturalistic philosophy of the semantics of language, that assigns to language such a passive role in scientific discovery. Instead they both embrace an artifactual philosophy of the semantics of language, and so give to language an active role that enables language manipulation to function as an instrument or heuristic, and thus make it a discovery strategy. Retroduction, analogy, and metaphor merely represent alternative semantical strategies. Unfortunately none of these strategies have yet been developed enough to be made procedural, and the muses of ancient Greek mythology are still as important to scientific discovery as they are to poetry and music. Proceduralization is the problem that to date has been addressed more effectively by the artificial intelligence systems of the information scientists than by the semantical analyses of the philosophers of science.

While metaphor is not serviceable for a discovery procedure, at least yet, it may be recognized as an outcome of mechanized discovery procedures because of the very unconventional statements generated as system outputs. As a linguistic phenomenon metaphor may be explained with the semantical thesis that the meanings of terms have a complex composition. For purposes of analysis metaphor may be viewed in the context of predication to form a sentence. Other modes of expression such as phrases or texts larger than sentences may reveal metaphorical use, but these may be transformed grammatically into the subject-predicate sentence form. One of the identifying features of a metaphorical description is, that if the term that is metaphorically predicated of a subject is taken in its conventional, i.e. literal sense, then the statement is false. Another is that it is not an unrecognized mistake; it is intentionally issued with no intention to deceive and with the intention of revealing something which is believed to be true. The central problem, therefore, is how the metaphorical description can be both true and false. One possible answer is that metaphor is a kind of equivocation, and this proposal seems inevitable so long as meanings are viewed as simple wholes, such that the metaphorical description is completely true on its one meaning and completely false on its other. A more suggestive way to formulation the question is to ask how the metaphorical predication can be partially true and partially false. This suggests an alternative to equivocation, because it suggests that meanings have parts. A metaphorical predication invokes only part of the meaning of the descriptive predicate, and excludes the remainder of the meaning complex. A speaker's conventional linguistic usage associates the entire meaning complex with the predicate term, and the metaphor is false if the term is predicated with its full and conventional semantics. But the initiator of the metaphor is able to detect that part of the meaning which is truly predicated of the subject, and he implicitly expects the hearer or reader to suspend part of the predicate's semantics, while the speaker or writer uses the portion that he wishes to use to describe the subject. A listener may or may not succeed in understanding the metaphorical use of the predicated term depending on his ability to detect the applicable part of the predicate's semantics. Some authors discussing metaphor render it as some kind of esoteric mode of speech, which cannot be "reduced" to literal language. But metaphors are explained in literal terms to the uncomprehending listener or reader. To explain the metaphorical predication of a descriptive term to a subject, is to list those sentences or clauses believed to be true of the subject, which may substitute for the predicated metaphor, and which set forth precisely those parts of the predicate's meaning that the issuer intends to be applicable. And the explanation may also be elaborated by listing those sentences or clauses that are not believed to be true of the subject, but which are conventionally associated with the predicated term when it is predicated literally. These negative sentences state what is intended to be excluded from the predicate's meaning complex in the metaphorical usage.

For example to explain the metaphor "Man is a wolf", the speaker may say "Man is a wolf, because man is ..., and man is ..., and...", where in the clauses he substitutes predicates that identify those characteristics of wolf, that he intends to be applicable to man. And if in this substitute predication he finds himself further using metaphorical descriptions, then the substitution process is repeated with other clauses, until the entire explanation is literal. The explanation may be elaborated for clarity by the sentence "Man is not a wolf, because man is not ..., and man is not ..., and...". Substitutions in this negative sentence use subordinate clauses that have predicates describing characteristics conventionally associated with wolves, but which the issuer of the metaphor does not believe can truly be predicated of men. The affirmative explanatory sentence sets forth those parts of the meaning associated with "wolf" that are intended to describe man in the metaphorical use of "wolf", and the negative explanatory sentence sets forth whatever parts of the conventional or literal meaning associated with "wolf" that the issuer intends to suspend for metaphorical purposes. Simile may be viewed in the same way. In simile, the terms "like" and "as" make more explicit the speaker's intent to use only part of the meaning of the descriptive term. The simile "Man is like the wolf" may be explained with the sentence "Man is like the wolf, because man is..., and man is..., and...". Semantical change for the term "wolf" occurs when the metaphorical predication becomes conventional, and this produces an equivocation. The equivocation consists of two literal meanings, and the second meaning is now a "dead metaphor". A dead metaphor is a meaning from which the suspended parts have become conventionally excluded to produce a second literal meaning. The dead metaphor may also be a change of meaning in which the first meaning becoming archaic. This may occur in some cases of scientific discovery or theory development. The new theory supersedes an old one, such that the old meaning becomes as archaic as the older theory, and the new meaning becomes the only conventional meaning applicable to the subject of the theory. However, the change cannot be a complete semantical change, or that fact that the new and old theory address the same subject could not even be detected. The semantical change applies only to some parts of the term's meaning with other parts providing the needed semantical continuity, usually supplied by statements of test design.

HERBERT SIMON AND OTHERS ON DISCOVERY SYSTEMS

Herbert Simon (1916-), born in Milwaukee, Wisconsin, entered the University of Chicago in 1933 where he received a BA degree in 1936 and a Ph.D. in political science in 1942. He was awarded the Nobel Memorial prize for economics in 1978. He spent his career as a faculty member at Carnegie-Mellon University in Pittsburgh, most of it in the Graduate School of Industrial Administration. As of this writing he is a faculty member in both the Psychology and Computer Science Departments, and is also a member of the University's board of trustees. His autobiography, *Models of My Life*, was published in 1991.

In his autobiography he reports that the most important years of his life were 1955 and 1956, when his interest turned from administration and economics to the psychology of human problem solving, and specifically to considering the symbolic processes that people use in thinking. He and his long-time collaborator, Alan Newell, had concluded that computers can be applied to imitating intelligence symbolically, instead of just numerically, an insight that Simon says was a crucial step required for genuine artificial intelligence to emerge. In 1956 his first artificial-intelligence system named **LOGIC THEORIST** used heuristic search methods to find proofs in Whitehead and Russell's *Principia Mathematica.* However, the fact that this system found proofs in formal logic is purely incidental; Simon rejects the view held by some that formal logic is the appropriate language for artificial-intelligence systems and that problem solving is a process of proving theorems. The significance of **LOGIC THEORIST** is its use of heuristic search methods and of symbol manipulation. Simon defines artificial intelligence as symbolic processing, and he defines cognitive psychology as understanding human thinking by modeling ordinary problem solving with artificial-intelligence systems. Since those days Newell and Simon have developed many artificial-intelligence systems, several of which are described in their *Human Problem Solving* (1972).

Simon views scientific discovery as a special case of human problem solving, and therefore maintains that it can be examined with the artificial-intelligence approach. However, his systems development work was not directed to scientific discovery until the late 1970's. His principal publications pertaining to scientific discovery are *Models of Discovery* (1977) which contains reprints of some of his earlier articles relating information processing concepts to scientific discovery, and most notably his *Scientific Discovery; Computational Explorations of the Creative Process* (1987) which describes several discovery systems that rediscovered various historic scientific laws and theories. To appreciate Simon's views on scientific discovery, it is necessary firstly to consider his views on human problem solving by artificial-intelligence systems, and to appreciate his views on human problem solving, it is informative to consider what he calls his most important contribution to economics, his postulate of bounded rationality. And to appreciate Simon's postulate of bounded rationality, it is helpful firstly to review its principal and still prevailing alternative, the neoclassical postulate of rationality.

Neoclassical Maximizing Rationality and Weber's Ideal Types

Simon proposes his thesis of bounded rationality as an alternative to two other concepts of rationality that have currency among economists. The principal alternative to Simon's view is the prevailing neoclassical postulate, which says that consumers are rational because they maximize their utility, and that producers are rational because they maximize their profits. The other alternative to Simon's is the rational expectations postulate, which is a distinctive extension of the neoclassical postulate of utility and profit maximization. The rational expectations view will be considered below in the discussion of the **BVAR** type of discovery system. The rational expectations postulate is an extended version of the neoclassical view, and Simon's critique of neoclassicism also applies to the rational expectations view. Before turning to Simon's bounded rationality postulate consider firstly the neoclassical postulate.

The neoclassical postulate of rationality has its origins in Adam Smith's doctrine of self interest set forth in his *Wealth of Nations* (1776), the seminal document for modern economics. Smith was greatly impressed by Isaac Newton's celestial mechanics. In his *Essay on the History of*

Astronomy Smith described Newton's celestial mechanics as the greatest discovery ever made by man, and Smith aspired to describe economic life as a harmonious mechanism, as Newton had done for the heavens. In Smith's system entrepreneurs' rational behavior in pursuit of their economic self interest unintentionally produces a beneficial and harmonious outcome for the national economy; this is his doctrine of the "invisible hand". However Smith's perspective was not closed or self contained. It was part of a larger moral universe of natural laws, which Smith had earlier described in his *Theory of Moral Sentiments* (1759). In Smith's natural-law philosophy the pursuit of economic self interest is morally constrained by men's natural sympathy for others and also by their natural desire for the approval of others. Such moral constraints on the pursuit of self interest were soon excluded from theoretical economics, and much later they came to be viewed as sociological or institutional structures instead of natural moral laws.

Almost one-hundred years after the *Wealth of Nations* a new development occurred in economic theory, which is now called the "marginalist revolution", and which might also be described the completion of Smith's agenda for a Newtonian economics. The term "marginal" means incremental or differential, and the incremental economic analysis lent itself to mathematical expression with the differential calculus developed by Newton. The result was an elegant mathematical rendering of economic theory, in which the rationality postulate became a matter of calculating the global maximization of consumer utility and producer profits by suitable values for the first and second derivatives of the relevant mathematically expressed demand functions. This theory of price determination describes the allocation of goods and services in an optimally efficient manner later called "Pareto optimality" after the Italian economist, Vilfredo Pareto. A half century later there was another revolution called the "Keynesian revolution" after the economist, John Maynard Keynes. Pre-Keynesian economic theory had assumed that the Pareto optimum resulting from rational maximizing behavior by each individual would also maximize income and output for the whole economy. In his *General Theory* (1936) Keynes set forth a new theory saying that individual maximizing behavior could result in less-than-full-employment equilibrium, which he said occurred during the Great Depression of the 1930's. This resulted in the distinguishing of microeconomic theory of price determination from the macroeconomic theory of income determination. Keynes thus revolutionized economic theory, but he did not explicitly attack the rationality postulate of classical economics. At the same time economists began applying statistical inference techniques to estimate equations with the macroeconomic data, that were being collected to describe national economic conditions. Both the availability of these data and the development of the computer occasioned the evolution of a specialty area in economics called "econometrics", although there were earlier Institutionalist economists whose statistical analyses of economic data have been called econometrics. Since Haavelmo, however, nearly all the econometricians have been neoclassical economists, who require that the selection of variables for the equations be justified by neoclassical theory. Until very recent years econometrics was exclusively the application of statistical inference techniques to economic data structured by neoclassical microeconomic and macroeconomic theory. Even today any econometric model that does not result from the *a priori* imposition of the neoclassical theory upon the data is derisively referred to as "atheoretical".

The above overview of the neoclassical rationality postulate reveals that it is not viewed by economists as just one of many alternatives; it has served as the foundation for modern economics since Adam Smith. Anyone attempting to overthrow this maximizing rationality postulate is attempting a new scientific revolution in economics that would be much more radical than any of the revolutionary developments in the history of neoclassical theory. Nevertheless, there have been dissenters such as the German Historicists and the American Institutionalists, and the reason for their dissent has always been the same: the empirical inadequacy and heroic unrealism of the neoclassical theory with its basis in its rationality postulates. Neoclassical theorists have not been unaware of these problems in their maximizing rationality postulates. Before turning to Simon's alternative, consider briefly Max Weber's thesis of the "ideal type", a semantical contrivance proposed to defend the neoclassical rationality concept against its critics. Simon does not refer to Weber, but Weber is a worthy contrast to Simon as a tautological treatment of rationality.

Weber's discussion of his doctrine of the ideal type or "*idealtypus*" can be found in English translation in *The Methodology of the Social Sciences* (Tr. by Shils and Finch, 1949), and principally in the chapters titled "'Objectivity' in Social Science and Social Policy" and "The Meaning of 'Ethical Neutrality' in Sociology and Economics". His doctrine of the ideal type contains ambiguities that have been noted by recognized

Weberian scholars such as Theodore Abel, Alexander von Shelting, and Talcott Parsons; and a translation of their relevant texts can be found in *Max Weber's Ideal Type Theory* (1969) by Rolf E. Rogers. However, these ambiguities in Weber's doctrine do not detract from consideration of the ideal type in the present context. Weber defined the ideal type as a mental construct that has two basic features: Firstly it involves one or several points of view. According to Weber's theory of knowledge this perspectivism is characteristic of all concepts including both natural science and social science concepts, because no concept can capture reality in all its potentially infinite variety of aspects. Weber explicitly rejects the copy theory of knowledge that he finds in the Historicist philosophy of social science, and he refers to the Historicists' claim of pure objectivity in science as the "naturalistic prejudice". In the present context it is noteworthy that the rational aspect of human behavior is a central aspect of reality that Weber includes in the ideal type concepts in pure economic theory. The second feature is that it involves a one-sided accentuation or intensification of the point of view in the ideal type. Nonrational considerations are not denied, but the maximizing postulate is made unrealistically extreme as a limiting case. Weber explicitly rejects the charge that the ideal type is a complete fiction, but he calls it "utopian" since historical concrete individuals do not completely conform in their behavior to the accentuated, maximizing rationality described by the ideal type. In other words individual instances not conforming to pure economic theory do not falsify the theory containing the ideal-type concepts, and as Weber states, the ideal type is not a hypothesis and is not tested by its application to reality. Weber says that the ideal type is used to compare theory to reality, in order to reveal by contrast the irrational aspects of human behavior. Pure economic theory utilizes ideal-type concepts exclusively, and it makes certain assumptions, notably the maximizing assumptions, which hardly ever correspond completely with reality, but which approximate it in various degrees. "Pure" neoclassical economic theory assumes the dominance of economic interests and precludes consideration of political or other noneconomic aspects in reality.

It is evident that the ideal type is a semantical contrivance, one that is comparable to Heisenberg's "questions in principle", Hanson's "functional *a priori*" and Feyerabend's "counterinduction", but with a problematic difference: it is pervasive and definitive of the whole science of economics. Every theory or law in any empirical science requires two sets of statements. The first set is the theory statements, that are contingent and are proposed for testing. The second set is the set of statements for characterizing the phenomena independently, and which must be presumed true for testing. The latter set of statements, called test design statements in experimental sciences, may also be the statements used to describe the measurement procedures in either experimental or nonexperimental sciences. Given these two sets of statements, the scientist confronted with the observation of an instance not conforming to what is predicted by a proposed theory has two options: He can reject the theory as falsified by the deviant instance, or he can reject the test design statements and revise the semantics of the characterization of the phenomena under consideration so as to save the theory. The first option is clearly empirical; the second may also be empirical, if the scientist recognizes that his test design statements have thereby assumed contingent status and become a theory. The ideal type is a semantical contrivance to achieve the second option but in a very problematic way: it does so without recognizing any presumed statements as new theory language. As Weber said, the ideal type is not a hypothesis. This pervasiveness relegates the rationality postulate and its neoclassical economic theory to the status of a tautology. All of the science's semantics is determined by the postulate, and there is no independent characterization of the subject matter of economic theory. Pure economic theory with its ideal-type concepts is applicable to the extent that it is applicable, and it is nonapplicable to the extent that it is not applicable. In other words all observed human behavior is rational wherever economic theory applies to it, and is irrational wherever the theory does not apply, and all such observed "irrational" behavior is excluded from economics. When the Weberian economist compares the ideal type with observed behavior to detect irrational behavior, he is not using it as a counterinductive "detecting device" as Feyerabend advocates. The Weberian will view the observed "irrational" behavior as a phenomenon to be excluded from neoclassical theory rather than as one to be investigated for a more empirically adequate economic theory.

Many contemporary economic theorists are only less explicit in their willfully dogmatic acceptance of neoclassical economic theory with its maximizing rationality postulate. They are recalcitrantly reluctant to dispense with the elegantly mathematical and uniquely determinate solutions to complex theoretical problems enabled by merely setting the first derivative of an economic equation to zero and checking second derivatives for suitable algebraic sign. The mathematical indeterminacy resulting from the rejection of the maximizing assumption of the neoclassical rationality postulate, implies that many problems formerly thought to have been solved after two hundred years of theoretical development in the neoclassical tradition, are not really solved at all. Professional economists who have dutifully labored and groveled for years to earn their doctorate credentials and have their papers published, do not welcome being advised that their purportedly empirical theory is only slightly less semantically vacuous than the mathematical differential calculus they use to express it. But not all economists are doctrinaire neoclassicals. Some are empiricists, who are willing to accept the first of the two options mentioned above, the option of rejecting a theory or law that is falsified by nonconforming instances. And some such as Simon are furthermore willing to reconsider even the assumed maximizing rationality postulate of the neoclassical economic theory, that these falsifications imply.

Simon's Postulate of Bounded Rationality and "Satisficing"

In his autobiography Simon relates that in what he calls his first piece of scientific work, a study in 1935 of public recreation in the city of Milwaukee, he saw a pattern that was the insight for what was to become his thesis of bounded rationality. For this study he was examining the budgeting process for the division of funds between playground maintenance, which was administered by one organization, and playground activity leadership, which was administered by another organization in the Milwaukee municipal government. He found that the actual budget allocation decision was not made as economic theory would suggest; what actually occurred was that the two organizations wanted more funds for their proper functions, and he generalized that people bring decisions within reasonable bounds by identifying with partial goals for which their own organizational units are responsible. This insight was taken up in Simon's Ph.D. dissertation (1942), which he later published as *Administrative Behavior* (1947), the book referred to by the Royal Swedish Academy of Sciences as an "epoch-making book", when they awarded him the Nobel Memorial Prize for Economics in 1978. In his autobiography Simon says that his entire scientific output may be described as a gloss on two basic ideas in *Administrative Behavior*: (1) human beings are able to achieve only a very bounded rationality, and (2) as a consequence of this limitation, they are prone to identify with subgoals. The first of these ideas is fundamental to Simon's critique of neoclassical rationality, and the second is fundamental to his theory of human problem solving.

In his autobiography Simon says that his "A Behavioral Model of Rational Choice" (1955) reprinted as chapter fourteen in his *Models of Man* (1987) is his first major step toward his psychological theory of bounded rationality. In that 1955 paper he says that the neoclassical concept of rationality is in need of fairly drastic revision, because actual human behavior in making choices does not satisfy three basic assumptions underlying neoclassical maximizing rationality. Those three assumptions are: (1) a decision maker has a knowledge of the relevant aspects of his environment, which if not absolutely complete, is at least impressively clear and voluminous; (2) a decision maker has a well organized, consistent, and stable system of preferences; and (3) a decision maker has a skill in computing, that enables him to calculate for the alternative courses of action available to him the one that will enable him to reach the highest achievable point in his preference scale. In his "Rational Choice and the Structure of the Environment" (1956) reprinted as chapter fifteen of *Models of Man* Simon proposes replacing the neoclassical postulate of maximizing with his more modest postulate that he calls "satisficing". This means that instead of optimizing, the decision maker's limited information and computational ability require that he adapt "well enough" to achieve his goals instead of optimizing.

The first chapter of his *Sciences of the Artificial* (1969) reveals that Simon recognizes exactly the same things about neoclassical rationality that Weber identified as the two basic features of the ideal-type concept. Firstly like Weber's thesis of viewpoint in the ideal type, Simon calls neoclassical rationality an "abstract idealization", because it selectively directs attention to the circumstances of the decision maker's outer environment for his adaptive behavior. Also in the chapter "Task Environments" in his *Human Problem Solving* (1972) he says that it is the task that defines the "point of view" about the environment, an idea that is comparable to Weber's thesis that it is "interests" that determine one's point of view. Secondly, just as Weber said that the accentuated rationality in the ideal type is "utopian", Simon calls neoclassical rationality "heroic" to describe its unrealistic character, and later in 1983 in his *Reason in Human Affairs* again without referencing Weber, he describes optimization as "utopian". But unlike

Weber, he does not relegate to the status of the "irrational" all the decision making that does not conform to the neoclassical ideal type of rational behavior. Instead Simon sees the empirical inadequacy of neoclassical rationality as good reason for replacing it with his behaviorally more realistic concept of bounded rationality.

In the second chapter of his *Sciences of the Artificial* and then in his "From Substantive to Procedural Rationality" in *Models of Bounded Rationality* Simon uses the phrase "substantive rationality" for the neoclassical maximizing rationality, which considers only the decision maker's goals and outer environment. And he uses the phrase "procedural rationality" for the satisficing psychological cognitive procedures including the decision maker's limited information and computational abilities consisting of what Simon calls the decision maker's inner environment. The study of cognitive processes or procedural rationality is interesting only when the substantively rational response is not trivial or instantly obvious; it is usually studied in situations in which the decision maker must gather information of various kinds, and must process it in different ways to arrive at a reasonable course of action for achieving his goals.

Simon refers to the Pareto optimality described in the economists' theory of general equilibrium, which combines the individual maximizing choices of a host of substantively rational economic actors into a global optimum for the whole economic system, as the "ideal" market mechanism. Then he says that there is also a "pragmatic" market mechanism described by the Nobel laureate economist Frederich von Hayek, that is more modest and believable, because it strives for a measure of procedural rationality by tailoring decision-making tasks to the realistic computational capabilities and localized information with no promise of optimization. He quotes at length a passage from Hayek's "The Uses of Knowledge in Society", *American Economic Review* (1945), in which Hayek asks: what is the problem we wish to solve when we try to construct a rational economic order? And Hayek answers that the economic calculus does not describe the optimization problem, since it is a problem of the utilization of knowledge that is not given to anyone in its totality. The price system is a mechanism for communicating information, and the most significant fact about it is the economy of knowledge with which it operates, that is, how little the individual participants need to know in order to be able to take the right action. Simon maintains that it is Hayek's "pragmatic" version that describes the markets of the real world, and that the substantive rationality of neoclassical theory is worthless, since it is not backed up by executable maximizing algorithms. He says that consumers and business firms are not maximizers, but rather are satisficers. They accept what is "good enough" because they have no choice. The rationality that they actually use is a satisficing procedural rationality. Examination of the limits of rationality leads to consideration of the price system mainly as an institution that reduces the amount of nonlocal information which economic actors must have to make "reasonable", i.e. satisficing, decisions.

Bounded Rationality, Institutionalism, and Functionalism

Simon's description of the real-world market-determined price system as pragmatic and as an institution, places him in the worthy intellectual company of the American Institutionalist school of economic thought, even though he does not identify himself as such, and so a few background comments about this school of economics are in order. In his "Introduction" to his *Types of Economic Theory* the Institutionalist economist, Wesley C. Mitchell, says that there have been different types of economic theory not only because there have been different types of problems, but also because there have been different conceptions of human nature. The American Institutionalist school was a revolt within the American economic profession, which rejected the formal and abstract deductivism in neoclassical economics and instead appealed to experience. It had its roots in the Pragmatist philosophy, the only philosophy indigenous to the United States, which itself was a revolt in the American philosophy profession, and which rejected the natural-law and utilitarian traditions in European academic philosophy.

The founding father of American Institutionalism is the iconoclastic if not also the somewhat eccentric economist, Thorstein Veblen (1857-1929). In his "Why is Economics not an Evolutionary Science?" in his *The Place of Science in Modern Civilization* (1919) Veblen characterized the neoclassical economists' hedonistic psychology as describing man as a "lightening calculator" of pleasures and pains, who passively responds to his environment and is unchanged by the environment. Veblen rejected this conception of man and proposed instead an anthropological conception, in which the individual's psychology is formed by institutions prevailing in the community in which he lives, and which evolves. He also therefore proposed that economics itself is an evolutionary science that employs a "genetic" type of theory, which describes the cumulative cultural growth of economic institutions, instead of the "taxonomic" type of theory used by neoclassical economists like the Austrian school. He rejects the Austrian's *ad hoc* attempts to save their natural-law explanations from deviant facts by invoking "disturbing factors". He also explicitly references Charles Darwin, and rejects the German Historicist school as pre-Darwinist for offering only enumeration of data and narrative accounts instead of a genetic theory.

Another noteworthy example of an Institutionalist economist is John R. Commons (1862-1945). In his *Institutional Economics* (1934) Commons states explicitly that he is following the Pragmatist philosophy of Charles S. Peirce. In the second volume of this book Commons discusses Weber's ideal type concepts, and he criticizes their fixed and unchanging character. Commons states that the utopian character of the ideal type only becomes more utopian as scientific investigation advances. Instead of the ideal type, Commons proposes the "changeable hypothesis", that takes into account new factors revealed to be relevant in the investigation, that and retires from consideration old factors revealed to be irrelevant. This amounts to demanding that economics be more empirical. Weber had explicitly denied that the ideal type is a hypothesis. Commons says that use of hypotheses makes less utopian the utopias that our minds create. Commons does not explicitly propose revising the maximizing assumption in the neoclassical rationality postulate. A typical institutionalist, he maintains that in addition to economic interactions described by neoclassical economics there are other, namely institutional, factors that are also operative in determining the outcomes of economic transactions. In both his earlier works and again in his final work, The *Economics of Collective Action* (1950, 1970), he proposes a more adequate psychology, which he calls a "negotiational psychology" as opposed to the hedonist psychology of the utilitarians. He also calls it an objective and behavioristic psychology instead of the subjective psychology of pain and pleasure, because it is the psychology of language, duress, coercion, persuasion, command, obedience, propaganda, of physical, economic and moral powers. He distinguishes three types of transactions: (1) bargaining transactions, which occur in the market, and which is the type treated in neoclassical economic theory, (2) managerial transactions, which occur between levels in organizational hierarchies, and (3) rationing transactions, which are agreements about apportioning, such as occur in budgeting decisions. He says that all three types have "futurity", that is, they require some security that future outcomes occur as expected so that expectations can operate as working rules, and that the three types are functionally interdependent. The institutionalist perspective focuses on the second and third types of transactions, because these represent collective action in control of individual action. Commons is particularly interested in the social control exercised by courts over the working rules in bargaining transactions. Perhaps it might be added that there are at least as many theories of bounded rationality, as there are bounding institutional factors seen by institutionalist economists.

Simon elaborates on the relation of institutions to his thesis of satisficing bounded rationality in his " Rationality as Process and as Product of Thought" (1978) reprinted in his *Models of Bounded Rationality*. He does not explicitly refer to the academic literatures of either Pragmatist philosophy or Institutionalist economics, but instead draws upon the "functionalist" type of explanation often found in the sociological literature. He references the *Encyclopedia of the Social Sciences* (1968) in which functionalism is defined as an explanation of how major social patterns operate to maintain the integration or adaptation of larger social systems. More formally stated, functional explanations are about movements of a system toward stable self-maintaining equilibria. Most notably Simon states that there is no reason to suppose that the attained equilibria are global maxima. In other words, functional explanation describes satisficing behavior. In this paper he furthermore maintains that functional analyses are not focused on quantitative magnitudes as are found in price theory, but are focused on qualitative and structural questions, and typically on the choice among a small number of discrete institutional alternatives. Particular institutional structures or practices are seen to entail certain desirable or undesirable consequences. A shift in the balance of consequences, or in the awareness of them, may motivate a change in institutional arrangements. Like economic sociologists, who recognize the underlying role of economic institutions, Simon argues that economists have not actually limited themselves to maximization analyses, but have utilized such qualitative functional analyses when they seek to explain institutions and behavior that lie outside the domain of price theory, distribution, and production. In his autobiography he says that most of the conclusions that neoclassical economists draw do not depend on the assumption of perfect rationality, but derive from auxiliary institutional assumptions that are required, in order to reach any conclusions at all. And in his *Reason In Human Affairs* (1983) Simon says that markets do not operate in a vacuum, but are part of a larger framework of social institutions, which provide the stable environment that makes rationality possible, because the institutions supply reliable patterns of events.

In "Rationality as Process..." Simon states that the characterization of an institution is almost never arrived at deductively from consideration of the function that it must perform for system survival. Functional analysis is not deductive like theoretical neoclassical economics. Rather an institution is a behavior pattern that is empirically observed, and observation of the pattern occasions the question of why it persists, that is, what function it performs. Institutions can be observed in every society, and their existence is then rationalized by the argument that its function is requisite. But Simon comments that this kind of reasoning may demonstrate that a particular behavioral pattern is a sufficient condition for performing an essential social function, but it cannot demonstrate that the particular pattern is a necessary one. Alternative patterns may be functionally equivalent, since they serve the same need. In other words there may be many satisficing institutional patterns for accomplishing the same social goal.

Human Problem Solving, Cognitive Psychology, and Heuristics

Simon's theory of procedural rationality is his theory of human problem solving, and it is elaborately set forth in his *Human Problem Solving* (1972) co-authored with Allen Newell. This nine-hundred page *magnum opus* took fourteen years to write. During this period Simon also wrote a briefer statement, *Sciences of the Artificial* (1969), and several articles since reprinted in his *Models of Discovery* (1977), an anthology of many of his previously published papers. Much of *Human Problem Solving* consists of detailed descriptions of problem-solving computer programs, none of which pertain to scientific discovery. Nonetheless his views on human problem solving are relevant to methodology of science, because he considers scientific discovery to be a special case of human problem solving. At the outset of *Human Problem Solving* the two collaborating authors state that their aim is to advance understanding of how humans think by setting forth a theory of human problem solving. The concluding section of the book sets forth a general statement of their theory, which is based on the computer programs described in the body of the book and presented as empirical evidence for the theory. They state that the specific opportunity that has set the course for their book is the development of a science of information processing, more recently called computer science. Their central thesis is that explanation of thinking can be accomplished by means of an information theory, and that their theory views a human as a processor of information, an information processing system. They say that such a description of the human is not just metaphorical, because an abstract concept has been developed of an information processor, which abstracts from the distinctively mechanical aspects of the computer. The authors compare the explanations in information science to the use of differential equations in other sciences such as classical physics. An information theory consisting of computer programs is dynamic like differential equations, because it describes change in a system through time. Such a theory describes the time course of behavior, characterizing each new act as a function of the immediately preceding state of the system and its environment. Given at any time the memory contents characterizing the system's state at that moment, the program determines how the memory contents will change during the next computing cycle and what the contents will be at the end of the cycle. The fundamental methodological problems of theory construction and theory testing are the same in the two types of theory. The theory is tested by providing a specific set of initial and boundary conditions for the system, by using the equations or program to predict the resulting time path, and by comparing this predicted path with the actual path of the system. The advantage of an information-processing language over the mathematical languages for formulating a theory of thinking, is that an information processing language takes symbolic structures rather than numbers as values of its variables.

The information theory about human thinking and problem solving is a theory in cognitive psychology. Newell and Simon note that their theory is concerned with performance, specifically with the performance of intelligent adults in our own culture, while psychologists have traditionally been more concerned with learning. In his autobiography as well as elsewhere Simon distinguishes cognitive psychology from both the *gestalt* and the behavioristic approaches to psychology. He rejects the black-box approach of the behaviorists and especially of B.F. Skinner, who maintains that the black box is empty. Simon also rejects the reductionist version of behaviorism, according to which complex behavior must be explained in terms of neurological processes, and he also rejects the neurological modeling approach of the psychologists who use parallel connectionist networks or neural nets type of computerized explanations. Newell and

Simon propose a theory of symbols located midway, as it were, between complex thought processes and neurological processes. Simon acknowledges a debt to the Gestaltists and their allies, who also recognize a layer of constructs between behavior and neurology, but Simon rejects the Gestaltists' wholistic approach to these constructs. Simon proposes an explicitly mechanistic type of explanation of human thinking and problem solving in terms of information processing.

Simon defines human thinking as a system of elementary information processes, organized hierarchically and executed serially. Simon relies on the concept of hierarchy as a strategy for managing complexity. He defines a hierarchical system as one that is composed of interrelated subsystems, each of which in turn is hierarchical in structure down to a lowest level consisting of an elementary subsystem. In human problem solving hierarchy is determined by the organization of subgoals, which is the second idea that Simon said is basic to his entire scientific output. Hierarchical organization is common in computer systems; applications programs are written in compiler and interpreter languages such as **FORTRAN** and **BASIC**, and these languages in turn contain reserved words that are names for macros, which are subsystems in the compiler library, which in turn contain lower level subsystems, and so on down to a basic level consisting of elementary systems in binary code. For the specifically problem-solving type of human thinking Simon has analyzed information processing into a few basic concepts. The first of these is the "task environment", by which he means the problem-solving processor's outer environment as viewed by the problem solver to produce a "problem space", together with the goal that orients the problem solver to his task environment. The problem space is the inner environment consisting of the processor's internal representation of the outer task environment, and in which the problem solving activities take place. Simon maintains that there is no objective representation of the task environment independently of some processor's problem space. Furthermore it is the task or goal that defines the "point of view" about the problem-solving processor's outer environment, and that therefore defines the problem space. Simon calls this defining process an "input translation process". Thirdly in addition to task environment and problem space, Simon introduces the concept of "method". A method is a process, that bears some "rational" relation to attaining a problem solution, as formulated and seen in terms of the internal representation, which is the problem space. Here the term "rational" is understood as satisficing in the sense that a satisfactory as opposed to an optimal solution is achieved. In the mechanical processor, the method is the computer program, and most of Simon's theory of problem solving pertains to the method.

Simon distinguishes three types of method. The first is the recognition method, which can be used when the solution is already in the processor's memory, and artificial-intelligence systems using this method rely on large stores of specific information. Computer programs using this type of method contain a conditional form of statement, which Simon calls a "production". In a production, whenever the initial conditions are satisfied, the consequent action is taken. And when the conditions of several alternative productions are satisfied, the conflicts between them are resolved by priority rules. In his autobiography Simon notes that productions have become widely accepted to explain how human experts make their decisions by recognizing familiar cues directly, and that productions have been used for the "expert systems" in artificial intelligence. Experts, both human and mechanical, do much of their problem solving not by searching selectively, but simply by recognizing the relevant cues in situations similar to those they have experienced before. It is their wealth of experience that makes them experts. The second type of method is what Simon calls the generate-and-test method. In this method the computer system generates a problem space, and has as its goal to find or to produce a member in a subspace identified as a solution by a test. The generality and weakness of this method lies in the fact that the generation and test procedures are independent, so that the amount of search is very large. Simon typically portrays this method as requiring a search that is so large, that it cannot be carried out completely, and so must proceed in a random manner. To address this problem of innumerable possibilities Peirce had advanced his logic of abduction, which postulates a natural light or instinctive genius for producing correct theories. Simon advances instead his theory of heuristics, the third type of problem-solving method, which exploits the information in the task environment as that task environment is represented internally in the processor by the problem space. In the heuristic search, unlike the generate-and-test method, there is a dependence of the search process upon the nature of the object being sought in the problem space and the progress being made toward it. This dependence functions as a feed back that guides the search process with controlling information acquired in the process of the search itself, as the search explores the internalized task environment. This method is much more efficient than the generate-and-test method, and it explains how complex problems are solved with bounded rationality, both human and mechanical.

These alternative methods represent different artificial-intelligence research programmes, software development vs hardware development, i.e. knowledge vs speed. The generate-and-test method is dependent on fast hardware; the heuristic search method is dependent on efficient software design. Researchers preferences for one or another of the methods are affected by developments in hardware technology. The hardware preference is known as the "brute force" approach, and as the technology has advanced, it has enabled the implementation of artificial-intelligence systems that offer little new software but greatly improved performance for the extensive searching of very large problem spaces. For example the *Wall Street Journal* (30 April 1990) reported that a group of five Carnegie-Mellon University graduate students with IBM Corporation funding have developed a multiprocessor chess-playing system named "Deep Thought", that exhibits grand-master performance with superhuman speed. The system does not represent any noteworthy software development either in chess-playing search heuristics or in expert chess-playing strategies. Instead it explores the huge chess-playing problem space more quickly and extensively than can the human grand-master limited by human bounds of rationality.

On Scientific Discovery and Philosophy of Science

In the 1960's and early 1970's, many years before Simon and his colleagues at Carnegie-Mellon University had actually written functioning computerized discovery systems simulating historic discoveries, Simon had written articles claiming that scientific discovery is a special case of human problem solving. In these articles he related his human problem-solving approach for discovery, to views published by various philosophers of science. The articles are reprinted in his *Models of Discovery*, where he comments in his "Introduction" that the subject of scientific discovery and of creativity generally has always been surrounded by dense mists of romanticism and downright knownothingness. In his "Scientific Discovery and the Psychology of Problem Solving" (1966) Simon states his thesis that scientific discovery is a form of problem solving, i.e. that the processes whereby science is carried on can be explained in terms that have been used to explain the processes of problem solving. Problem-solving thinking is an organization of elementary information processes organized hierarchically and executed serially, and consisting of processes that exhibit large amounts of highly selective trial-and-error search based on rules of thumb or heuristics. The theory of scientific discovery is a system with these characteristics, and which behaves like a scientist. Superior problem-solving scientists have more powerful heuristics, and therefore produce adequate solutions with less search or better solutions with equivalent search, as compared with less competent scientists. Science is satisficing, and to explain scientific discovery is to describe a set of processes that is sufficient, and just sufficient, to account for the amounts and directions of scientific progress that have actually occurred. Furthermore, for every great success in scientific discovery there are many failures, and a theory to explain scientific discovery must predict innumerable failures for every success.

In this paper Simon also takes occasion to criticize the philosophy of science literature. He maintains that this literature tends to address the normative rather than the descriptive aspects of scientific methodology, and that philosophers are more concerned with how scientists ought to proceed, in order to conform with certain conceptions of logic, than with how they do in fact proceed. And, he adds, their notions of how scientists ought to proceed focuses primarily on the problem of induction. He concludes that the philosophy of science literature has little relevance to the actual behavior of scientists, and has less normative value than has been supposed. But he finds two exceptions in the philosophy of science literature: N.R. Hanson and T.S. Kuhn. He says that both of these authors have made significant contributions to the psychology and sociology of scientific discovery, and that they have been quite explicit in distinguishing the process of discovery from the traditional canons of "sound" scientific method. He also says that he has made much use of the views of both of these philosophers. Simon's principal commentary on the philosophy of Hanson is his defense of Hanson against the view of Popper in "Does Scientific Discovery Have a Logic?" (1973). He notes that Popper rejects the existence of a logic of scientific discovery in Popper's *Logic of Scientific Discovery* (1934), and he says that Popper's view is opposed by Hanson in the latter's *Patterns of Discovery* (1958) and by Peirce. Peirce used the term "retroduction", which Simon says is the main subject of the theory of problem solving in both its normative and positive forms. Simon observes that Hanson made his case by historical examples of scientific discovery, and that he placed great emphasis on discovery of perceptual patterns. In this 1973 article as well as in his earlier "The Logic of Rational Decision" (1965) Simon distinguishes heuristic search from induction, and defends the idea of a logic of scientific discovery in the sense that norms can be derived from the goals of scientific activity. He defines the logic of scientific discovery as a set of normative standards for judging the processes used to discover or test scientific theories, where the goal from which the norms are derived is that of discovering valid scientific laws. And Simon emphasizes that the heuristic strategy does not guarantee results or success, and he therefore argues that he has not smuggled any philosophical induction axiom into his formulation of a logic of discovery, and that such a logic does not depend on the solution of the problem of induction. Simon states that discovering a pattern does not involve induction or extrapolation. Induction arises only if one wishes to predict and to test whether or not the same pattern will continue to obtain when it is extrapolated. Law discovery only means finding patterns in the data that have been observed; whether or not the pattern will continue to hold for new data that are observed subsequently, will be decided in the course of predictive testing of the law, and not in discovering it. It may be noted that after Simon's colleagues had created functioning discovery systems based on heuristic search, Simon often described some of those systems as using inductive search. However, in his *Scientific Discovery* Simon explicitly rejects the search for certainty associated with attempts to justify inductivism. Simon subscribes to Popper's falsificationist thesis of scientific criticism, and in his "Ramsey Eliminibility and the Testability of Scientific Theories" (1973) reprinted in his *Models of Discovery* Simon proposes a formalization of Popper's requirement that admissible theories be falsifiable. This formalization is his "FITness" criterion, which is a neologism containing an acronym for "Fit and Irrevocable Testability". According to this requirement a theory should be admissible for consideration if and only if (1) in principle it is falsifiable by a finite set of observations, and (2) once falsified, additional observations cannot rehabilitate it. In a footnote at the end of this paper Simon adds that the FITness criterion is only a requirement for falsifiability, and that it says nothing about the disposition of a theory that has been falsified, such that the FITness criterion is compatible with what Lakatos calls "methodological falsificationism". Simon's stated reason is that methodological falsificationism permits a falsified theory to be saved by modifying the auxiliary hypotheses that connect it with observables. In his "Methodology of Scientific Research Programmes" in *Criticism and the Growth of Knowledge* (1970) Imre Lakatos distinguished "dogmatic falsificationism" from "methodological falsificationism", and within the latter type he further distinguished "naive" and "sophisticated" subtypes. Simon's reference to auxiliary hypotheses in his footnote suggests he believed that his FITness criterion is of the "sophisticated" subtype. But he later apparently reconsidered; in *Scientific Discovery* he called his FITness criterion "naive falsification", and gave two reasons: Firstly the criterion postulates that there are wholly theory-free observation sentences, whose truth status can be determined by observations that have no theoretical presuppositions. Secondly his criterion is too strict to be applied to any theory. Simon maintains that there are no wholly theory-free observation sentences.

Simon's comments on Kuhn's philosophy are principally concerned with Kuhn's distinction between normal and revolutionary science. Kuhn maintained that the revolutionary transition is a *gestalt* switch, while Simon defends his own view that heuristic search procedures apply to revolutionary changes as well as to normal science. In his "Scientific Discovery and the Psychology of Problem Solving" Simon says that his theory of scientific discovery rests on the hypothesis that there are no qualitative differences between the processes of revolutionary science and those of normal science, between work of high creativity and journeyman work respectively. Simon points to the fact that trial and error occur in both types of work. He argues that trial and error are most prominent in those areas of problem solving where the heuristics are least powerful, that is, are least adequate to narrow down the problem space, such that the paths of thought leading to discoveries often regarded as creative might be expected to provide even more visible evidence of trial and error than those leading to relatively routine discoveries. Later in his *Scientific Discovery* Simon develops the idea of the amount of trial-and-error search into the distinction between "strong" methods, which he says resemble normal science, and "weak" methods, which resemble revolutionary science. He identifies expert systems based principally on productions, where there may be almost no search needed for problem solving, as paradigmatic cases of strong methods exemplifying normal science.

Simon's argument that trial and error is used in all types of discovery, is his defense of the heuristic method. But method is only one aspect of his theory of problem solving; there is also the definition of the problem space. He acknowledges that scientific work involves not only solving problems but also posing them, that correct question asking is as important as correct question answering. And he notes that Kuhn's distinction between normal and revolutionary science is relevant to the relation of question asking and question answering. In the 1966 article Simon identifies the problem space, which is the problem solver's point of view of the outer environment, with Kuhn's idea of paradigm, and he identifies the definition of the problem space with the process of problem formation. Firstly Simon notes that normal science need not pose its own questions, because its questions have already been formulated for it by the current paradigm produced by the most recent scientific revolution. The problem space is thus given by the current state of the science; the problematic case is the scientific revolution, which establishes the new paradigm. Simon argues that it is not necessary to adduce entirely new mechanisms to account for problem formulation in revolutionary science, because, as Kuhn says, the paradigms of any given revolution arise out of the normal science of the previous period. Normal science research leads to the discovery of anomalies, which are new problems that the prospective revolutionaries address. Therefore Simon argues that there is no need for a separate theory of problem formulation. He states that a theory of scientific discovery adequate to explain revolutionary as well as normal science must account not only for the origin of the problems, but also for the origins of representations, the problem spaces or paradigms. Representations arise by modification and development of previous representations as problems arise by modification and development of previous problems. A system that is to explain human problem solving and scientific discovery does not need to incorporate a highly powerful mechanism for inventing completely novel representations. Even in revolutionary science the problems and representations are rooted in the past, and are not cut out of whole cloth.

Later in his "Ramsey Eliminability..." article Simon considers another objection pertaining to the problem space in revolutionary developments. The objection is that in revolutionary science the range of alternative hypotheses that constitute the problem space or representation cannot be delimited in advance. He states that this objection rests on a commonly drawn distinction between well defined problems, which are amenable to orderly analysis such as those in normal science, and ill defined problems, which are thought to be the exclusive domain of creativity, such as those in revolutionary science. Simon argues that the force of the objection depends on the distinctions being qualitative and not just matters of degree. He replies that there is no reason to deny that revolutionary hypotheses can be the result of some kind of recursively applicable rule of generation. He cites as an example of a revolutionary discovery Mendeleev's periodic table, which does not involve a notion of pattern more complex than that required to handle patterned letter sequences. The problem space of possible patterns in which Mendeleev was searching was of modest size, and at least half a dozen of Mendeleev's contemporaries had noticed the pattern independently of him, although they had not exploited it as systematically or as vigorously as he did. Simon concludes that before one accepts the hypothesis that revolutionary science is not subject to laws of effective search, one should await more microscopic studies than have generally been made to date of the histories of revolutionary discoveries. He says that the case of Mendeleev may pose to be not at all exceptional.

Later in "Artificial Intelligence Research Strategies in the Light of AI Models of Scientific Discovery" in *Proceedings of the Sixth International Joint Conference on Artificial Intelligence* (1979) Simon is able to refer to operational discovery systems. He states that discovery systems are distinguished from most other problem-solving systems in the vagueness of the tasks presented to them and of the heuristic criteria that guide the search and account for selectivity, and that because their goals are very general, it is unusual to use means-end analysis commonly used for well structured tasks and to work backward from a desired result. The discovery system solves ill structured tasks and works forward inductively from the givens of the problem and from the new concepts and variables generated from the givens. Simon does not reference Kuhn in this context, but the implication is that discovery systems can routinely produce revolutionary science. Then still later in his *Scientific Discovery* Simon reconsiders his earlier correlation of well structured problems with normal science and ill structured problems with revolutionary science. He notes that normal science is described by Kuhn as involving no development of new laws but simply of applying known laws or developing subsidiary laws that fill in the dominant paradigm. He concludes that a discovery system (no matter what its restrictions to well defined problems) that develops new laws directly from data and not from a dominant paradigm, must be productive of revolutionary science.

Simon's difficulties in relating his ideas to Kuhn's originate with

Kuhn's ideas, not with Simon's. The most frequent criticism of Kuhn's *Structures of Scientific Revolutions* in the philosophy of science literature is that his distinction between normal and revolutionary science is so vague, that with the exception of a few paradigmatic cases his readers could not apply the distinction to specific episodes in the history of science, unless Kuhn had identified a particular episodes himself. The attractiveness of Kuhn's book at the time of its appearance was not its unimpressive conceptual clarity; it was its welcome redirection of the philosophy profession's interest to the history of science at a time when many philosophers of science had come to regard the Logical Positivist philosophy with hardly any less cynicism than Ovid had shown toward the ancient Greek and Roman mythology in his *Metamorpheses*. Simon's application of discovery systems to simulate discovery episodes in the history of science offers analytical clarity and precision that Kuhn could not provide, with or without the Olympian irrelevance of the Russellian symbolic logic.

The Theory of Discovery Systems

Simon's principal work on discovery systems is his *Scientific Discovery: Computational Explorations of the Creative Processes* (1987) co-authored with three colleagues. In the introductory section on the theory scientific discovery he says that the central hypothesis is that the mechanisms of scientific discovery are not peculiar to that activity, but can be subsumed as special cases of the general mechanisms of problem solving. Thus the approach taken is to exhibit a series of computer systems capable of making nontrivial scientific discoveries, which are actually rediscoveries of historic scientific laws and theories including empirical generalizations. And the method of operation of these computer systems is the method of heuristic search. The theory of scientific discovery is also in his view therefore a theory in cognitive psychology. Simon says that he seeks to investigate the psychology of discovery processes, and to provide an empirically tested theory of the information-processing mechanisms that are implicated in that process. He states that an empirical test of the systems as psychological theories of human discovery processes would involve presenting the computer programs and some human subjects with identical problems, and then comparing their behaviors. The computer system can generate a "trace" of its operations, and the human subjects can report a verbal and written protocol of their behavior, while they are solving the problem. Then the system can be tested as a theory of behavior by comparing the trace with the protocol. Simon states that the computer systems described in his book incorporate heuristic search procedures to perform the kinds of selective processes that he believes scientists use to guide them in their search for regularities in data. But he also admits that his book provides little in the way of detailed comparison with human performance. And in discussions of particular applications involving particular discoveries, he says that in some cases the historical discoveries were actually performed differently than the systems performed the rediscoveries.

The principal interest in this book is actually system design rather than psychological testing and reporting, and Simon states that he wishes to provide some foundations for a normative theory of discovery, that is to say, to write a how-to-make-discoveries book. He explains that by this he does not mean a set of rules for deriving theories conclusively from observations. Instead, he wishes to propose a set of criteria for judging the efficacy and efficiency of the processes used to discover scientific theory. Accordingly Simon sets forth a rationality postulate for the scientist: to use the best means he has available - the best heuristics - for narrowing the search down to manageable proportions, even though this effort may result in excluding good solution candidates. If the novelty of the scientific problem requires much search, this large amount of search is rational if it employs all the heuristics that are known to be applicable to the domain of the problem. Thus, his rationality postulate for the scientist is a bounded rationality postulate, not only because of the limits imposed by memory capacity and computational speed, but also because of the limit imposed by the inventory of available heuristics.

BACON and Other Discovery Systems

BACON is a set of successively more sophisticated discovery systems, that make quantitative empirical laws and theories. Given sets of observation measurements for two or more variables, **BACON** searches for functional relations among the variables. The search heuristics in earlier versions of each **BACON** computer program are carried forward into all later ones, and later versions contain heuristics that are more sophisticated than those in earlier versions.* In the literature describing the **BACON**

systems each successive version is identified by a numerical suffix, such as **BACON.1**. The original version, **BACON.1**, was designed and implemented by Pat Langley in 1979 as the thesis for his Ph.D. dissertation written in the Carnegie-Mellon department of psychology under the direction of Simon, and titled *Descriptive Discovery Processes: Experiments in Baconian Science.* He published descriptions of the system in "**BACON.1**: A General Discovery System" in *The Proceedings of the Second National Conference of the Canadian Society for Computational Studies in Intelligence* (1978) and as a co-author with Simon and others in *Scientific Discovery* (1987). **BACON** programs are implemented in a list processing computer language called **LISP**, and its discovery heuristics are implemented in a production-system language called **PRISM.** The system lists the observable measurement data monotonically according to the values of one of the variables, and then determines whether the values of some other variables follow the same (or the inverse) ordering. Picking one of these other variables, it searches for an invariant by considering the ratio (or the product) of these variables with the original one. If the ratio or product is not constant, it is introduced as a new variable, and the process continues the search for invariants. Examples of some of the simpler search heuristics expressed in the conditional form of a production are as follows: (1) If the values of a variable are constant, then infer that the variable always has that value. (2) If the values of two numerical variables increase together, then examine their ratio. (3) If the values of one variable increase as those of another decrease, then examine their product. The general strategy used with these heuristics is to create variables that are ratios or products, and then to treat them as data from which still other terms are created, until a constant is identified by the first heuristic.

BACON.1 has rediscovered several historically significant empirical laws including Boyle's law of gases, Kepler's third planetary law, Galileo's law of motion of objects on inclined planes, and Ohm's law of electrical current. A similar system, **BACON.3** has rediscovered the ideal gas law and Coulomb's law of electrical current. For making these rediscoveries, Simon and his associates used measurement data actually used by the original discoverers, and published by W.F. Magie in *A Source Book in Physics* (1935). **BACON.4** is a significant improvement over earlier versions. It was developed and firstly described by Gary Bradshaw, Pat Langley, and Herbert Simon in "The Discovery of Intrinsic Properties" in *The Proceedings of the Third National Conference of the Canadian Society* *for Computational Studies in Intelligence* (1980), and it is also described in their 1987 book. The improvement is the ability to use nominal or symbolic variables, that take only names or labels as values. For example the nominal variable "material" may take on values such as "lead", "silver", or "water". Values for numerical properties may be associated with the values of the nominal variables, such as the density of lead, which is 13.34 grams per cubic centimeter. **BACON.4** has heuristics for discovering laws involving nominal variables by postulating associated values called "intrinsic properties", by inferring a set of numerical values for the intrinsic properties for each of the postulated nominal values, and then by retrieving the numerical values when applying its numerical heuristics to discover new laws involving these nominal variables. The laws rediscovered by **BACON.4** include: (1) Ohm's law of electrical circuits, where the intrinsic properties associated with the nominal variables are voltage and resistance, (2) Archimedes law of displacement, where the intrinsic properties are density and the volume of an irregular object, (3) Black's law of specific heat, where specific heat is the intrinsic property, (4) Newton's law of gravitation, where gravitational mass is the intrinsic property, and (5) the law of conservation of momentum, where the inertial mass of objects is the intrinsic property. **BACON.4** was further enhanced so that it could rediscover the laws describing chemical reactions formulated by Dalton, Gay-Lussac, and Comizzaro. For example it rediscovered Gay-Lussac's principle that the relative densities of elements in their gaseous form are proportionate to their corresponding molecular weights. Rediscovering these laws in quantitative chemistry involved more than postulating intrinsic properties and noting recurring values. These chemists found that a set of values could be expressed as small integer multiples of one another. This procedure required a new heuristic that finds common divisors. A common divisor is a number which, when divided into a set of values, generates a set of integers. **BACON.4** uses this method of finding common divisors, whenever a new set of dependent values is about to be assigned to an intrinsic property.

BACON.5 is the next noteworthy improvement. It uses analogical reasoning for scientific discovery. **BACON.1** through **BACON.4** are driven by data in search for regularities in the data. Furthermore the heuristics in these previous **BACON** systems are almost entirely free from theoretical presuppositions about domains from which the data are drawn. **BACON.5** incorporates a heuristic for reducing the amount of search for laws in certain special cases, in which the system is given very general theoretical postulates, and then reasons by analogy by postulating symmetries between the unknown law and a theoretical postulate given to the system as an input. The general theoretical postulate that Simon gave to **BACON.5** is the law of conservation. The laws rediscovered by **BACON.5** using analogy with the conservation law include the law of conservation of momentum, Black's law of specific heat, and Joule's law of energy conservation.

The **BACON** discovery system was not the first system developed around Simon's principles of human problem solving with heuristics. In 1976 Douglas B. Lenat published his Ph.D. dissertation written at Stanford University and titled *AM: An Artificial Intelligence Approach to Discovery Mathematics as Heuristic Search.* Allen Newell was one of his dissertation advisors, and Lenat acknowledges that he got his ideas from Herbert Simon. Lenat has since accepted a faculty position in the computer science department of Carnegie-Mellon University. In 1977 he published "The Ubiquity of Discovery" in *The Proceedings of the Fifth International Joint Conference On Artificial Intelligence*, (IJCAI) in which he relates Simon's theory of heuristic problem solving in science and describes the specific heuristics in his **AM** discovery system. While Lenat's article includes discussion of artificial intelligence in empirical science, his **AM** system is not for empirical science, but is a computer system which develops new mathematical concepts and conjectures with these concepts. Also in the 1977 *IJCAI Proceedings* he published "Automated Theory Formation in Mathematics", which offers a more detailed description of the system's two-hundred fifty heuristics, and which also discusses his application of the **AM** system in elementary mathematics. He reports that in one hour of processing time **AM** rediscovered hundreds of common mathematical concepts including singleton sets, natural numbers, arithmetic, and also theorems such as unique factorization. In 1979 Simon published "Artificial Intelligence Research Strategies in the Light of AI Models of Scientific Discovery" in *The Proceedings of the Sixth International Joint Conference on Artificial Intelligence* in which he considers Lenat's **AM** system and Langley's **BACON** systems as useful for illuminating the history of the discovery process in the domain of artificial intelligence itself, and for providing some insight into the ways to proceed in future research and development aimed at new discoveries in that field. He says that AI will proceed as an empirical inquiry rather than as a theoretically deductive one, and that principles for the discipline will be inferred from the computer programs constituting the discovery systems, although he also notes that in the scientific profession the community members' work in parallel, while in the machines the work proceeds serially.

BACON created quantitative empirical laws by examination of measurement data. Simon and his associates also designed and implemented discovery systems, that are capable of creating qualitative laws from empirical data, and three such systems are described in *Scientific Discovery*. They are named **GLAUBER**, **STAHL** and **DALTON**. The **GLAUBER** discovery system is named after the eighteenth century chemist, Johann Rudolph Glauber, who contributed to the development of the acid-base theory. The discovery system was developed by Langley in 1983. For its historical reconstruction of the acid-base theory **GLAUBER** was given facts very similar to those known to eighteenth century chemists, before they formulated the theory of acids and bases. These facts consist of information about the tastes of various substances and the reactions in which they take part. The tastes are "sour", "bitter", and "salty". The substances are acids, alkalis and salts labeled with common names, which for purposes of convenience are the contemporary chemical names of these substances, even though **GLAUBER** makes no use of the analytical information in the modern chemical symbols. Associated with these common names for chemical substances are argument names, such as "input" and "output", that describe the roles of the chemical substances in the chemical reactions in which the substances partake. Finally the system is given names for the three abstract classes: "acid", "alkali", and "salt". When the system is executed with these inputs, it examines the chemical substances and their reactions, and then correlates the tastes to the abstract classes, and also expresses the reactions in a general law that states that acids and alkalis react to produce salts.

The second discovery system is **STAHL**, which creates a type of qualitative law that Simon calls "componential", because it describes the hidden structural components of substances. System **STAHL** is named after the German chemist, Georg Ernst Stahl, who developed the phlogiston theory of combustion. **STAHL** recreates the development of both the phlogiston and the oxygen theories of combustion. Simon states that discovery systems should be able to arrive at laws that have been rejected later in favor of others in the history of science. And he says that since a discovery system's historical reconstruction aims at grasping the main currents of reasoning in a given epoch, then reproducing the errors that were typical of that epoch is diagnostic. Like **GLAUBER**, **STAHL** accepts qualitative facts as inputs, and generates qualitative statements as outputs. The input is a list of chemical reactions, and its initial state consists of a set of chemical substances and their reactions represented by common names and argument names, as they are in **GLAUBER**. When executed, the system generates a list of chemical elements and of the compounds in which the elements are components. The intermediate states of **STAHL**'s computation consist of transformed versions of initial reactions and inferences about the components of some of the substances. When the system begins running, it is driven by data, but after it has made conjectures about the hidden structures, it is also driven by these conjectures, which is to say, by theory. Simon concludes from the rediscovery of the phlogiston and oxygen theories by **STAHL,** that the proponents of the two theories reasoned in essentially the same ways, and that they differed mainly in their assumptions. He also applied **STAHL** to the rediscovery of Black's analysis of *magnesia alba*, and he maintains that the same principles of inference were used by chemists quite widely in their search for componential explanations of chemical substances and their reactions. The principal significance of this diversity to Simon is the demonstration that the reasoning procedures in **STAHL** are not *ad hoc*, and that **STAHL** is a general system.

The third discovery system that creates qualitative laws is **DALTON,** which is named after John Dalton. Like Dalton the chemist, the **DALTON** system does not invent the atomic theory of matter; it employs a representation that embodies the hypothesis, and that incorporates the distinction between atoms and molecules invented by Avogado. **DALTON** is a theory-driven system for reaching the conclusions about atomic weights, that **BACON.4** derived in a data-driven manner. And **DALTON** creates structural laws in contrast to **STAHL,** which creates componential laws. **DALTON** is given information that is similar to what was available to chemists in 1800. The input includes a set of reactions and knowledge of the components of the chemical substances involved in each reaction. This is the type of information outputed by **STAHL**, and **DALTON** uses the same common-name/argument-name scheme of representation used by **STAHL**. **DALTON** is also told which of the substances are elements having no components other than themselves. And it knows that the number of molecules in each chemical substance is important in the simplest form of a reaction, and that the number of atoms of each element in a given molecule is also important. **DALTON**'s goal is to use this input to develop a structural model for each reaction and for each of the substances involved in each reaction, subject to two constraints. The first constraint is that the model of a molecule of a substance must be the same for all reactions in which it is present. The second constraint is that the models of the reactions display the conservation of particles. Simon applied **DALTON** to the reaction involving the combination of hydrogen and oxygen to form water, and the system outputed a model giving a modern account of the water reaction. He also considers applying **DALTON** to elementary particle physics and to classical genetics, but he states that the current version is not adequate to this task.

Since the publication of *Scientific Discovery* Simon and his associates have continued their work on discovery systems and have pursued their work into new directions. While **BACON** and the other systems described in the 1987 book are concerned mainly with the ways in which theories can be generated from empirical data, the question of where the data come from has largely been left unanswered. In "The Process of Scientific Discovery: The Strategy of Experimentation" (1988) in *Models of Thought* Simon and Deepak Kulkarni describe their new **KEKADA** discovery system, which examines not only the process of hypothesis formation, but also the process of designing experiments and programs of observation. The **KEKADA** discovery system is constructed to simulate the sequence of experiments carried out by Hans Krebs and his colleague, Kurt Henseleit, between July 1931 and April 1932, which produced the elucidation of the chemical pathways for synthesis of urea in the liver. This discovery of the ornithine cycle was the first demonstration of the existence of a cycle in the metabolic biochemistry. Simon and Kulkarni's source for this episode is "Hans Krebs and the Discovery of the Ornithine Cycle" in *Federation Proceedings* (1980) by Frederic L. Holmes of Yale University. Holmes also made himself available to Simon and Kulkarni for consultation in 1986 when their study was in progress. The organization of **KEKADA** is based on a two-space model of learning proposed earlier by Simon and Lea in "Problem Solving and Rule Induction: A Unified View" in *Knowledge and Cognition* (1974). The system searches in an "instance space" and a "rule space", each having its own set of heuristics. The instance space is defined by the possible experiments and experimental outcomes, and it is searched by performing experiments. The rule space is defined by the hypotheses and other higher level descriptions coupled with associated measures of confidence. The system proceeds through cycles in which it chooses an experiment from the instance space to carry out on the basis of the current state of the rule space, and the outcome of the experiment modifies the hypotheses and confidences in the rule space. One of the distinctive characteristics of **KEKADA** is its ability to react to surprising experimental outcomes, and to attempt in response to explain the puzzling phenomenon. Prior to carrying out any experiment, expectations are formed by expectations setters, which are a type of heuristic for searching the rule space, and the expectations are associated with the experiment. The expectations consist of expected output substances of a reaction, and expected upper and lower bounds on the quantities or the rates of the outputs. If the result of the experiment violate these bounds, it is noted as a surprise. Comparison of the course of work of Krebs as described by Holmes and of the work of **KEKADA** in its simulation of the discovery reveals only minor differences, which Simon and Kulkarni say can be explanied by focus of attention shifts and small differences in the initial knowledge with which Krebs and **KEKADA** started. The authors also say that a manual simulation of the path that Krebs followed in a second discovery, that of the glutamine synthesis, is wholly consistent with the theory set forth by **KEKADA**. They therefore conclude that the structure and heuristics in **KEKADA** constitute a model of discovery that is of wider applicability than the episode used to develop the system, and that the system is not *ad hoc*.

Simon's Philosophy of Science

Simon's literary corpus is rich enough to contain a philosophy of science that addresses all four of the basic questions addressed by academic professional philosophers. What philosophers of science call the aim of science may be taken as a rationality postulate for basic scientific research. Simon explicitly applies his thesis of bounded rationality developed for economics to scientific research in his autobiography in an "Afterword" titled "The Scientist as Problem Solver", although this explicit statement would not have been necessary for the attentive reader of his literary corpus. Simon describes his theory of discovery as a special case of his theory of human problem solving, because both theories are based on his theory of heuristic search. And he views his theory of heuristic search in turn as a special case of his postulate of bounded rationality. To this meta-theory one need only add that Simon's application of his postulate of bounded rationality to scientific discovery amounts to his thesis of the aim of science. The function of heuristics is to search efficiently a problem space of possible solutions, that is too large to be searched exhaustively. The limited computational ability of the scientist relative to the size of the problem space is the "computational constraint", that is the factor that bounds the scientist's rationality, constraining the scientist from aiming for anything like global rationality. The research scientist is therefore a satisficer, and the aim of the scientist is satisficing within both empirical and computational constraints.

Simon's views on the remaining philosophical topics, explanation and criticism, may also be considered in relation to the discovery systems. Consider firstly his statements on scientific explanation including the topic of theoretical terms. The developers of the **BACON** systems make a pragmatic distinction between observation variables and theoretical variables in their systems. Simon notes that contemporary philosophers of science maintain that observation is theory-laden, and his distinction between observational and theoretical terms does not deny this semantical thesis. He calls his distinction "pragmatic", because he makes it entirely relative to the discovery system, and it is pragmatic in the sense understood in the contemporary Pragmatist philosophy of science. Those variables that receive their associated values by input to the system are considered to be observational variables, while those that receive their values by the operation of the discovery system are considered to be theoretical ones. He states that in any given inquiry we can treat as observable any term whose values are obtained from an instrument that is not itself problematic in the context of that inquiry. Thus Langley considers all the values created by the **BACON** programs by multiplication or division for finding products or ratios to be theoretical terms. And Simon calls the values for nominal variables that are postulated intrinsic properties theoretical terms.

Unfortunately Simon does not follow through with this Pragmatist relativizing to problem-solving discovery systems, but reverts to the Positivist concept of explanation. In his exposition of **DALTON**, which create structural theories, Simon comments that as an area in science matures its researchers progress from "descriptions" to "explanations", and he cites Hempel's *Aspects of Scientific Explanation and Other Essays* (1965). Examples of explanations cited by Simon are the kinetic theory of heat, which provides an explanation of both Black's law and the ideal gas law, and Dalton's atomic theory, which provides explanations for the law of multiple proportions and Gay-Lussac's law of combining volumes. He notes that each of these examples involves a structural model in which macroscopic phenomena are described in terms of their inferred components. Simon contrasts explanation to the purely phenomenological and descriptive analyses carried out by **BACON.4**, when it rediscovered the concepts of molecular and atomic weight, and assigned correct weights to many substances in its wholly data-driven manner. He affirms that **BACON.4**'s analyses involved no appeal to a particulate model of chemical elements and compounds, and that what took the place of the atomic model were the heuristics that searched for small integer ratios among corresponding properties of substances. This concept of explanation is a reversion to the hypothetical-deductive concept of explanation in which theories are said to explain empirical laws by deductive connection, and in which theory and empirical or descriptive generalizations are distinguished by their semantics. This is what Hanson referred to as the almanac view of science. On the Pragmatist view theory and empirical description are not distinguished semantically, but are distinguished pragmatically by their use in the problem-solving activity that is scientific research. Theory is what is proposed for empirical testing, and description is what is presumed for testing. Explanation is language that is theory after it has been empirically tested and not falsified; or one who speaks of "theoretical explanation" is merely speaking of a proposed explanation. This is the functional view of the language of science instead of the Positivist almanac view. Thus given that the discovery systems are problem-solving systems, defining "theory" and "explanation" relative to the discovery system is to define them in a manner consistent with the contemporary Pragmatist philosophy. And on this Pragmatic view the outputed laws generated by **BACON.4** are no less theoretical or explanatory than the outputs of **DALTON**.

In addition to the physical theories that the discovery systems recreated, consideration might also be given to the behavioral and social theories that Simon and his colleagues had not attempted to address with their discovery systems. Why did this Nobel laureate economist never attempt to construct an economic theory with a discovery system? Perhaps one might ask instead: is Simon actually a Romantic in his philosophy of social science? One possible answer is that the economic theory of greatest interest to him, his thesis of bounded rationality, does not lend itself to any discovery system like those he or his colleagues have yet designed. This is an answer in terms of technology rather than philosophy. When Simon found that behaviorism posed a philosophical impediment to his agenda for cognitive psychology, he rejected this variation on the Positivist philosophy, even though he had previously been sympathetic to it. Similarly one might expect that he would not be deterred by any version of Romanticism; he demonstrated the philosophical sophistication to distinguish empirical from ontological criteria for criticism. Simon's view of scientific criticism is based on his theory of heuristics and discovery systems. Philosophers of science such as Hanson, whose interests were focused on the topic of scientific discovery, found that the Positivist separation of the "context of discovery" and the "context of justification" fails to appreciate the interdependent interaction between these two functions in scientific research. Simon also notes this interaction in *Scientific Discovery*, because it is integral to his theory of heuristics and to his discovery system designs. Simon's principal thesis of problem solving is that the availability of evaluative tests during the successive stages of the discovery process carried out with heuristics is a major source of the efficiency of the discovery methods. Each step or group of steps of a search is evaluated in terms of the evidence it has produced, and the continuing search process is modified on the basis of the outcome of these evaluations. The confirmation of partial results accumulate and make the confirmation of the final hypothesis coincide with its generation. Yet Simon does not fail to see the need for predictive testing by observation or experiment of the hypotheses generated by the discovery systems which only find patterns in finite available data.

Muth's Rational Expectations "Hypothesis"

Simon distinguishes three rationality postulates: the prevailing neoclassical postulate of global rationality, his own thesis of bounded rationality, and the rational expectations hypothesis. The reader of Simon's autobiography, however, would never guess that about two decades after its first appearance, the rationality expectations hypothesis had occasioned the development of a distinctive type of discovery system, the Bayesian Vector Autoregression or **BVAR** discovery system. In fact it is doubtful that even its creator, Robert Litterman, or his colleagues recognize the system as a discovery system, even though it does what discovery systems are intended to do: it makes theories. This irony is due to the fact that the prevailing philosophy of science in economics is Romanticism, which has led economists to view **BVAR** models as "atheoretical". But if the term "theory" is understood in the contemporary Pragmatist sense, the equations created by the **BVAR** system are economic theories. Before taking up the **BVAR** system, firstly consider the rational expectations hypothesis.

One of the distinctive aspects of Simon's autobiography is a chapter titled "On Being Argumentative". In this chapter's opening sentence Simon states that he has not avoided controversy, and he adds that he has often been embroiled in it. And on the same page he also says that he has usually announced his revolutionary intentions. But revolutionaries occasionally find others revolting against them. In the preceding chapter of his autobiography he describes a tactical retreat in the arena of faculty politics: his eventual decision to migrate from Carnegie-Mellon's Graduate School of Industrial Administration to its psychology department, which as it happens is not an unsuitable place for his cognitive psychology. This conflict with its disappointing denouement for Simon was occasioned by the emergence of the rational expectations hypothesis, a thesis that was first formulated by a colleague, John F. Muth, and which was part of what Simon calls the ascendency of a coalition of neoclassical economists in the Graduate School of Industrial Administration. Muth's rational expectations hypothesis, which Simon says deserves a Nobel prize even though he maintains that it is unrealistic, was set forth in a paper read to the Econometric Society in 1959, and then published in *Econometrica* (1961) under the title "Rational Expectations and the Theory of Price Movements". Muth explains that he calls his hypothesis about expectations "rational", because it is a descriptive theory of expectations, and is not a pronouncement of what business firms ought to do. The idea of rational expectations is not a pet without pedigree. It is a continuation of an approach in economics known as the Stockholm School, in which expectations play a central role, and which Muth references in his article. A brief consideration of the Stockholm School is in order, to see how the rational expectations advocates depart from it, especially in their empirical modeling.

One of the best known contributors to the Stockholm School is Bertil Ohlin, a Nobel laureate economist, who is best known for his *Interregional and International Trade* (1933), and whose elaboration on the monetary theory of Knut Wicksell anticipated the Keynesian theory in important respects. He called his own theory of underemployment the "Swedish theory of unused resources". In 1949 he published his *Problem of Employment Stabilization*, which contains his own macroeconomic theory and concludes with a critique of Keynes' *General Theory* from the Stockholm School viewpoint. In his critique Ohlin draws upon a distinction between *ex ante* or forward-looking analysis and *ex post* or backward-looking analysis, firstly proposed by Nobel laureate Gunnar Myrdal, his colleague of Stockholm School persuasion and fellow critic of Keynes. Later in life Myrdal evolved his theory of *ex ante* analysis into an institutionalist economic theory, and in his *Against the Stream* (1973) he uses it to explain a phenomenon that is problematic for Keynesian economics: "stagflation", the co-existence of economic stagnation and inflation. However, Myrdal does not address the effect of institutional change on the structural parameters in econometric models, and apparently does not think well of econometrics. In the first chapter, "Development of Economics: Crises, Cycles", he says that when he was still in his "theoretical stage" of thinking, i.e. pre-institutionalist stage, he had something to do with the initiation of the Econometric Society, which he says was planned at the time as a defense organization against the advancing American institutionalists, an advance which was halted in the economics profession by the Keynesian revolution. He says that Keynesian theory is now in crisis because of problems such as stagflation and structural unemployment, and that the future development of economics will be interdisciplinary and institutionalist.

Ohlin, who is not an institutionalist and remains a neoclassical economist, maintains that *ex post* analysis alone cannot provide an explanation in economics, because any explanation must run in terms of factors that govern actions, and actions refer to the future. Any economic explanation must therefore contain *ex ante* analysis, which consists of the expectations or plans of the actors in their economic roles. Ohlin notes that Keynes theory may be said to contain an *ex ante* analysis of investment, because it includes the marginal efficiency of capital, which is similar to Wicksell's "natural" rate of interest: the expected return from newly constructed capital. But Ohlin took exception to Keynes' exclusively *ex post* analysis of saving, in which saving is merely the residual of aggregate income net of aggregate consumption. On the Stockholm School view there must be an *ex ante* analysis of saving, and Ohlin theorizes that *ex ante* saving is determined by the difference between current consumption and the level of income in the prior period. He calls the *ex ante* saving rate the average propensity to save. Ohlin's Stockholm School approach is significant not only because Ohlin offers an explanation of how expectations are formed, but also because it accounts for expectations by explicit variables, the *ex ante* variables, so that their effects need not be incorporated implicitly in the statistically estimated parameters of the econometric models.

Ohlin's explanation notwithstanding, Muth blithely criticizes the Stockholm School for failing to offer an explanation of the way expectations are formed, and he advances his rational expectations hypothesis as the explanation. Muth notes two conclusions from studies of expectations measurements, which he says his rational expectations hypothesis "explains". The principal conclusion is that the averages of expectations made by economic actors in an industry are more accurate than the forecasts made with naive models, and are as accurate as elaborate equation systems, although there are considerable cross-sectional differences of opinion. The rational expectations hypothesis explains this accuracy by the thesis that expectations viewed as informed predictions of future events are essentially the same as the predictions of the relevant economic theory. Muth says that he is not asserting that the scratch work of entrepreneurs resembles a system of equations in any way, although he says that the way expectations are formed depends on the structure of the entire relevant system describing the economy. His more precise statement of his hypothesis is as follows: that expectations of firms (or, more generally, the subjective probability distribution of outcomes) tend to be distributed, for the same information set, about the prediction of the theory (or, the "objective" probability distributions of outcomes). Muth argues that if expectations were not moderately rational, then there would be opportunities for economists to make profits in commodity speculation, running a business firm, or selling information to present owners. In his discussion of price expectations, he offers an equation for determining expected price in a market, and references a paper to be published by him. The equation says that expected price is a geometrically weighted moving average of past prices. He also argues that rationality is an assumption that can be modified to adjust for systematic biases, incomplete or incorrect information, poor memory, etc., and that these deviations can be explained with analytical techniques based on rationality. The second conclusion is that reported expectations generally underestimate the extent of changes that actually take place. Like the Stockholm School, Muth's hypothesis does not assert that there are no expectations errors. He states that in the aggregate a reported expected magnitude such as a market price is an unbiased predictor of the corresponding actual magnitude except where a series of exogenous disturbances are not independent. Muth's explanation of the reported expectations errors of underestimation is his argument that his hypothesis is not inconsistent with the fact that the expectations and actual data have different variances. Muth references Simon's "Theories of Decision-Making in Economics" in *American Economic Review* (1959), and describes Simon as saying that the assumption of rationality in economics leads to theories that are inconsistent with or inadequate to explain observed phenomena, especially as the phenomena change over time. Muth comments that his view is exactly the opposite of Simon's: that dynamic economic models do not assume enough rationality.

Simon's critique of the rational expectations hypothesis is set forth in the second chapter titled "Economic Rationality" in his *Sciences of the Artificial* (1969). In the section titled "Expectations" he notes that expectations formed to deal with uncertainty may not result in a stable equilibrium or even a tendency to stable equilibrium, when the feed forward in the control system has destabilizing consequences, as when each actor is trying to anticipate the actions of others and their expectations. The stock example in economics is the speculative bubble. In the next section titled "Rational Expectations" Simon references Muth's 1961 article. He characterizes Muth's hypothesis as a proposed solution to the problem of mutual outguessing by assuming that actors form their expectations "rationally", by which is meant that the actors know the laws that govern the economic system, and that their predictions of the future position of the system are unbiased estimates of the actual equilibrium. Simon says that the rational expectations hypothesis precludes destabilizing speculative behavior. More fundamentally Simon maintains that there is no empirical evidence supporting the rational expectations hypothesis. And he doubts that business firms have either the knowledge or the computational ability that would be required to carry out the expectations strategy. He concludes that since economists have little empirical knowledge about how people do in fact form expectations about the future, it is difficult to choose at present among the models that are currently proposed by competing economic theories to account for cyclical behavior of the economy.

Ostensibly Muth proposed his rational expectations hypothesis as an explanation of two conclusions about expectations measurements. These empirical measurements should be used to provide the independent semantics and magnitudes needed for empirical testing of the rational expectations hypothesis. What might rationally have been expected of the rational expectations advocates is an attempt to construct conventional structural-equation econometric models using *ex ante* expectations data, to demonstrate and test their explanatory hypothesis. But neither Muth nor the rational expectations advocates took this approach. On the basis of his rational expectations hypothesis Muth shifted from an explanation of empirical measurements of reported expectations to consideration of a forecasting technique that he proposes be used by neoclassical economists. This semantical shift has had three noteworthy effects on subsequent empirical work in the rational expectations school: Firstly there was a disregard of empirical measurements of expectations, measurements that would serve as values for *ex ante* variables; then secondly there was an attack upon the conventional structural type of econometric model and the development of a new type of empirical model as an implementation of the rational expectations hypothesis but with no independently collected expectations measurements; and thirdly there evolved the design and implementation of a computerized procedure for constructing this new type of model, a computerized procedure which is a distinctive type of discovery system. This semantical shift has been consequential for econometric modeling. Haavelmo's structural-equation type of econometric model has been definitive of empirical economics for more than half a century, and it is still the prevailing practice in the economics profession. To the dismay of conventional econometricians the rational expectations advocates' attack upon the conventional structural-equation econometric model is, therefore, hardly less subversive to the *status quo* in the science, than Simon's attack on the neoclassical rationality postulate. And this outcome certainly has an ironic aspect, because the structural-equation econometric model had been advanced as the empirical implementation (at least ostensibly) of the neoclassical economic theory, while the rational expectations hypothesis has been advanced as offering greater fidelity to neoclassical theory by extending rationality to expectations. To understand such a strange turn of events, it is helpful to consider the still prevailing, conventional concept of econometric model, the structural-equation model.

Haavelmo's Structural-Equations Agenda And Its Early Critics

The authoritative statement of conventional econometric modeling is set forth in "The Probability Approach in Econometrics", a supplement to *Econometrica* (July, 1944) written by Nobel laureate econometrician, Trygve Haavelmo. *Econometrica* is the journal of the Econometric Society, which was founded in 1930, and which described itself as "an international society for the advancement of economic theory in its relation to statistics and mathematics" and for "the unification of the theoretical-quantitative and the empirical-quantitative approach" in economics. The society's offices were at the University of Chicago at the time the supplement was published. The supplement essentially advances certain fundamental ideas for the application of the Neyman-Pearson theory of statistical testing of mathematical hypotheses expressing neoclassical economic theory. At the time of its publication most economists believed that probability theory is not applicable to economic time series data, because the data for successive observations are not statistically independent. In her *History of Econometric Ideas* (1990) Mary S. Morgan writes that this introduction of probability theory into economics was a "probabilistic revolution" in econometrics, which shifted the role of econometrics from the measuring of the parameters in a theory to the testing of the theory.

Firstly Haavelmo argued that the time series data are not a set of successive observations, but are one observation with as many dimensions as there are independent variables in the model. This strategy is not mentioned in textbooks today. Haavelmo's more lasting agenda consisted of construing the econometric model as a probabilistic statement of the econometric theory, so that the theory is neither held harmless by falsifying data nor immediately and invariably falsified as soon as it is confronted with the data. He says that the model is an *a priori* hypothesis about real phenomena, that states that every system of values that the economist might observe of the "true" variables, will be one that belongs to the system of numeric values which is admissible within the model. This attempt to construe the model as a third linguistic entity between theory and data leads him to develop an unusual and complicated semantical analysis. The first chapter titled "Abstract Models and Reality", sets forth his theory of the semantics of measurement variables in econometrics. Haavelmo distinguishes three types of "variables", which actually represent three separate meanings associated with each variable symbol that may occur in an empirical economic theory. The first type is the "theoretical variable", which is the meaning a variable symbol has simply by virtue of its occurring in the equations of the model, and its values are subject only to the consistency of the model as a system of one or several equations. The second type is the "true variable", which signifies an ideal experimental design that the economist could at least imagine arranging in order to measure those quantities in real economic life, that he thinks might obey the laws imposed by the model on the corresponding theoretical variable. Haavelmo says that when theoretical variables have ordinary words or names associated with them, these words may merely be vague descriptions that the economist has learned to associate with certain phenomena, or they may signify ideal experimental designs with their descriptions of measurement procedures. But he also says that there are many indications that the economist nearly always has some such ideal experimental design "in the back of his mind", when the economist builds his theoretical models, and that in the verbal description of his model in economic terms the economist suggests explicitly or implicitly some type of measurement design to obtain the measurements for which he thinks his model would hold. Thus the theoretical and true variables are distinguished, but are not separated in the fully interpreted theory proposed for estimation and testing. And associated with the true variables there are true or ideal measurements, which are not only error free, but which are collected in accordance with an ideal experimental design. The third type of variable is the "observational variable", which describes the measurements actually used by the economist for his model construction. Haavelmo says that the economist often must be satisfied with rough and biased measures, and must dig out the measurements he needs from data that were collected for some other purpose. The true variables are those such that if their behavior should contradict a theory, the theory would be rejected as false. On the other hand if the behavior of the observational variables contradict the theory, they leave the possibility that the economist is trying out the theory on facts for which the theory was not meant to hold. This may cause confusion, when the same names are often used for both types of variables. To test a theory against facts or to use it for prediction, either the statistical observations available must be corrected or the theory itself must be adjusted, so as to make the facts the economist considers the true variables relevant to the theory. In Haavelmo's approach to econometrics, probability distributions not only adjust for measurement errors, but also adjust for the deviations between the true and observational values due to their semantical differences.

An experienced economist, Haavelmo is adequately cognizant of the difficulties in the work that makes economics an empirical science. In contrast, most of his contemporaries in the 1940's were ivory-tower theoreticians. Today there is much more adequate data available to economists from government agencies. Nonetheless, economists still sometimes find they must use what they call "proxy" variables, which are recognized as measurements of phenomena other than what the economist is interested in explaining with his models. And sometimes the government statistical agency will use names to identify data that describe phenomena for which the data are a proxy rather than what the data measure. For example in their *Industrial Production* (1986) the Board of Governors of the Federal Reserve System say that when their monthly production index series cannot be based on physical measures of output, such as tons of steel or assemblies of automobiles and trucks, then monthly input measures, such as hours worked or kilowatt hours of electricity consumed adjusted for the observed long-term trend and cyclical relations between input and output, are used to derive the monthly output series. Nonetheless, the Federal Reserve Board calls these proxy series "production". Except in these explicit cases involving proxy variables, however, it is questionable whether the economist has "in the back of his mind", as Haavelmo says, any specific ideal experimental design setting forth ideal measurement procedures. Most often the descriptive words associated with theoretical variable symbols in a mathematical model are vague and are just not given semantical resolution until actual measurements are associated with the model. Then the description of the actual measurement procedures supplies additional information to resolve this vagueness. Only when economists decide that the actual measurements are proxies for what they wish to investigate, such that there is more deviation involved than just errors of measurement, do they find themselves confronting an equivocation like Haavelmo's "true" and "observational" semantics instead of supplying a resolution to the vagueness in the meanings of the terms in the theory.

The second chapter titled "The Degree of Permanence of Economic Laws" contains Haavelmo's theory of scientific law in economics, and specifically his treatment of the degree of constancy or permanence in the relations among economic variables in econometric models. Nonconstancy is manifested by structural breakdown of the traditional structural-equation model, the type that Haavelmo advocates in his monograph. The rational expectations hypothesis is proposed as an explanation for structural breakdown, and the hypothesis is the basis for a new type of model that is an alternative to the structural-equation model. The **BVAR** discovery system constructs a refined version of this new type of model. Haavelmo says that the constancy in a relationship is a property of real phenomena, as the economist looks upon the phenomena from the viewpoint of a particular theory. At the very opening of his monograph he states that theoretical models are necessary to understand and explain events in real life, and that even a simple description and classification of real phenomena would probably not be possible or feasible without viewing reality through the framework of some scheme conceived *a priori.* This statement seems similar to Popper's thesis that there is no observation without theory, and to Hanson's characterization of observation as theory-laden. But the term "theory" in Haavelmo's monograph means specifically the neoclassical economic theory with its rationality postulate, and the basic task of his monograph is to describe his probability approach in econometrics understood as the application of Neyman-Pearson statistical inference theory to neoclassical economic theory for empirical testing.

In the first chapter of the monograph Haavelmo distinguished three types of quantitative economic relations. The first type is the definitional or accounting identity. A common example is the gross national product, which is merely the summation of its component sectors. The second type is the technical relation. The paradigmatic case of the technical relation is the production function, which relates output to inputs such as capital and labor. Technical engineering equations are more properly the tasks of other sciences, but the practice among econometricians has been to estimate production functions with the same statistical techniques that they use for all econometric equations. The third type is the relation describing economic actors. Equations of this type in econometric models are also called behavioral equations or decision functions. The behavioral equations in conventional econometric models are based on economic theory, and are not like the laws and theories developed in the natural sciences such as physics. Neoclassical economic theory purports to describe a decision-making process made by economic actors, notably consuming households and producing business firms. The econometric equation based on neoclassical theory contains independent variables that represent a set of conditions that are considered by the economic actors in relation to their motivating preference schedules or priorities as they make their best or optimized decisions, and the outcome of these optimizing decisions are represented by the dependent variable of the equation. The system of preference schedules is not explicitly contained in the equation, but Haavelmo says that if the system of preference schedules establishes a correspondence between sets of given conditions and optimized decision outcomes, such that for each set of conditions there is only one best decision outcome, then the economist may as it were jump over the middle link in the scheme, and say that the decisions of the individuals or firms are determined by the set of independent variables.

In this traditional neoclassical scheme the econometric model is based on the assumption that individual consumers' and firms' decisions to consume and to produce can be described by certain fundamental behavioral relations, and that there are also certain behavioral and institutional restrictions upon the actor's freedom. A particular system of such relationships with their equations statistically estimated defines one particular theoretical "structure". The problem of finding permanent economic laws becomes the problem of finding permanent structures in this sense; the failure in particular cases to solve this problem is usually manifested by an erroneous forecast with the model, and the failure is called structural breakdown. Haavelmo then considers several reasons for the structural breakdown of an econometric model. In all cases the problem is diagnosed as the absence from the model of a variable representing some operative factor that in reality has a significant effect on the dependent variable of the model, and the solution therefore consists of recognizing the missing factor and then of introducing a variable for it into the model. In the case of a model of a market one of the reasons for structural breakdown is a structural change due to the irreversibility of economic relations. It is a shift in a demand curve, such that price-quantity pairs do not represent movements along the demand curve, because the economic actors are revising their preference schedules as prices change. Haavelmo rejects claims that demand curves cannot be constructed from time series of observed price-quantity pairs, and instead says that the economist should introduce into his model variables representing the factors responsible for the revision of preference schedules. A second explanation for structural breakdown is the simplicity of the model. Economists like simple models, even though the real world is complex. Haavelmo distinguishes potential from factual influences in the real world, and says that models can be simple, because only factual influences need be accounted for in the models. But he says that economists making models may throw away elements of a theory, that would be sufficient to explain apparent structural breakdown that may occur later, because the elements do not exhibit a detectable factual influence over the time series history used to estimate the equation.

Finally a third reason for structural breakdown is the absence of a semantical property that Haavelmo calls "autonomy". Autonomous equations in a multi-equation model have an independence that is not just the formal independence of axioms in a deductive system. The semantical independence or autonomy is due to the success of an equation at identifying the preference schedules of just one social group or social role in the economy. For example the demand equation in a market model represents the decisions of buyers in the market, while the supply equation for the same price-quantity pair represents the decisions of sellers in the same market. If the supply and demand equations for a market model are autonomous, then a structural breakdown in one equation will not also occur in the other. An autonomous equation is one that has successfully identified a fundamental behavioral relation described by economic theory.

In addition to his semantical theory and his theory of scientific law in economics, Haavelmo also gives lengthy consideration to statistical inference. One statistical topic he considers is the meaning of the phrase "to formulate theories by looking at the data". He is concerned with the problem of whether a well fitting statistically estimated model is merely a condensed description of the empirical data, i.e. whether it is *ad hoc*, or whether it is an effective test of a generalization. He maintains that how the economist happens to choose a hypothesis to be tested from within a class of *a priori* admissible theories is irrelevant, and he states that the selection may be made by inspection of the data. But he says that the class of admissible theories must be fixed *a priori* to the testing procedure, so that it is possible to calculate the power of the test and to determine the risk of accepting the hypothesis tested; he rejects the practice of selecting the whole class of admissible theories by the empirical testing process. The class of admissible theories cannot be made a function of the sample data, because then the Neyman-Pearson statistical test no longer controls the two types of errors in testing hypotheses, either the error of accepting a wrong hypothesis or the error of rejecting a true hypothesis. This curious commingling of statistical testing and the investigator's psychology in the Neyman-Pearson statistical inference theory will be ignored by the developers of the new type of model used by rational expectations advocates.

Haavelmo's agenda had its Institutionalist critics long before the Rational Expectations advocates. Morgan notes in her *History of Econometric Ideas* that Haavelmo's paper was very influential both within the Cowles Commission and with others including Herbert Simon, and that acceptance of Haavelmo's approach made econometrics less creative, because data were taken less seriously as a source of ideas and information for econometric models and the theory-development role of applied econometrics was downgraded relative to the theory-testing role. She also notes that Haavelmo's approach was opposed by some economists including the Institutionalist economist, Wesley Clair Mitchell (1874-1948). Mitchell was instrumental in the founding of the prestigious National Bureau of Economic Research, where he was the Research Director for twenty-five years. A biographical memorial volume titled *Wesley Clair Mitchell: The Economic Scientist* edited by Arthur Burns was published by the National Bureau of Economic Research in 1952. Mitchell's principal interest was the business cycle, and in 1913 he published a descriptive analysis titled *Business Cycles*. Haavelmo's proposal to construct models based on existing economic theory may be contrasted with another paper published twenty years earlier by Mitchell in the latter's "Quantitative Analysis in Economic Theory" in *American Economic Review* (1925). Mitchell, who studied philosophy under the Pragmatist John Dewey, predicted that quantitative and statistical analyses in economics will result in a radical change in the content of economic theory from the prevailing type such as may be found in the works of Alfred Marshall. Mitchell said that instead of interpreting the data in terms of subjective motives, which are assumed as constituting an explanation and which are added to the data, quantitative economists may either just disregard motives, or more likely they may regard them as problems for investigation rather than assumed explanations and draw any conclusions about them from the data. He also said that economists will have a special predilection for the study of institutions, because institutions standardize behavior and thus facilitate statistical procedure. While many neoclassical economists view Mitchell's approach as atheoretical, Mitchell had a very erudite knowledge of economic theories as evidenced in his *Types of Economic Theory* (1967).

Mitchell's principal work setting forth the findings from his empirical investigations is his *Measuring Business Cycles*, which was co-authored with Arthur F. Burns and published by the National Bureau in 1946. This five-hundred page over-sized book contains no regression-estimated Marshallian supply or demand equations. Instead it reports on the authors' examination of more than a thousand time series describing the business cycle in four industrialized national economies, namely the U.S., Britain, France and Germany. The authors explicitly reject the idea of testing business cycle theories, of which there were a great many. They state that they have surveyed such theories in an effort to identify which time series may be relevant to their interest, but their stated agenda is to concentrate on a systematic examination of the cyclical movements in different economic activities as measured by historical time series data, and to classify the time series with respect to their phasing and amplitude, in order to trace causal relations exhibited in the sequence that different economic activities represented by the time series reveal in the cycle's critical points. To accomplish this they aggregate the individual time series so that the economic activities represented are not so atomized that the cyclical behavior is obscured by idiosyncrasies of the small individual units.

The merits and deficiencies of the alternative methodologies used by the Cowles Commission group and the National Bureau were argued in the economics literature in the late 1940's. Selections from this literature have been reprinted by the American Economic Association in their *Readings in Business Cycles* (1965). Defense of Haavelmo's structural-equation approach was given by Tjalling C. Koopmans, who wrote a review of Mitchell's *Measuring Business Cycles* in the *Review of Economic Statistics* in 1947 under the title "Measurement without Theory". Koopmans compared Burns and Mitchell's findings to Kepler's laws in astronomy and he compared Haavelmo's approach to Newton's theory of gravitation. He notes that Burns and Mitchell's objective is merely to make generalizing descriptions of the business cycle, while the objective of the structural-equation approach is to develop "genuine explanations" in terms of the behavior of groups of economic agents, such as consumers, workers, entrepreneurs, etc., who with their motives for action are the ultimate determinants of the economic variables. Then he adds that unlike Newton, economists today already have a systematized body of theory of man's behavior and its motives, and that such theory is indispensable for a quantitative empirical economics. He furthermore advocates use of the Neyman-Pearson statistical inference theory, and calls Burns and Mitchell's statistical techniques "pedestrian".

The approach of Burns and Mitchell was defended by Rutledge Vining, who wrote a reply to Koopmans in the *Review of Economics and Statistics* in 1949 under the title "Koopmans on the Choice of Variables to be Studied and the Methods of Measurement". Vining argues that Burns and Mitchell's work is one of discovery, search, and hypothesis-seeking rather than one of hypothesis-testing, and that even admitting that observation is always made with some theoretical framework in mind, such exploratory work cannot be confined to theoretical preconceptions having the prescribed form that is tested by use of the Neyman-Pearson technique. He also argues that the business cycle of a given category of economic activity is a perfectly acceptable unit of analysis, and that many statistical regularities observed in population phenomena involve the behavior of social "organisms" that are distinctively more than simple algebraic aggregates of consciously economizing individuals. He says that the aggregates have an existence over and above the existence of Koopmans' individual units and their behavior characteristics may not be deducible from the behavior characteristics of the component units.

Koopmans wrote "A Reply" in the same issue of the same journal. He admitted that hypothesis-seeking is still an unsolved problem at the very foundations of statistical theory, and that it is doubtful that all hypothesis-seeking activity can be described are formalized as a choice from a pre-assigned range of alternatives. But he stands by his criticism of Burns and Mitchell's statistical measures, because he says that science has historically progressed by restricting the range of alternative hypotheses, and he advocates crucial experiments. He maintains that crucial experiments deciding between the wave and particle theories of light in physics were beneficial to the advancement of physics before the modern quantum theory. He also continues to adhere to his view that it is necessary for economics to seek a basis in theories of individual decisions, and says that he cannot understand what Vining means by saying that the aggregate has an existence apart from its constituent components, and that it has behavior characteristics of its own that are not deducible from the behavior characteristics of the components. He maintains that individual behavior characteristics are logically equivalent to the group's, and that there is no opening wedge for essentially new group characteristics. In the same issue of the same journal Vining wrote "A Rejoinder", in which he said that it is gratuitous for anyone to specify any particular entity as necessarily the ultimate unit for a whole range of inquiry in an unexplored field of study.

The question is not a matter of logic, but of fact; the choice of unit for analyses is an empirical matter.

Contemporary Pragmatist philosophers of science will recognize Vining's appeal to exclusively empirical criteria for deciding the unit of analysis as an application of Quine's principle of ontological relativity. And students of elementary logic will recognize Koopmans' reductionist requirement as an instance if the fallacy of composition, in which one attributes to a whole the properties of its components. Just as the properties of water waves cannot be described exclusively in terms the physical properties of the water molecules, so too for the economic waves of the business cycles. Both types of waves may be described as "real", even if their reality is not easily described as an "entity" as nominalists would require. As it happens in the history of post-World War II economics a reluctant pluralism has prevailed. The U.S. Department of Commerce, Bureau of Economic Analysis, has taken over the National Bureau's business cycle agenda, and now publishes many cyclical time series with charts in the "yellow pages" of their monthly *Survey of Current Business*, which is the Federal agency's principal monthly periodical. They also release their monthly Index of Leading Indicators, which is based on Mitchell's approach, and its forecasts are reported monthly in the national news media. On the other hand the Cowles Commission's structural-equation agenda has effectively conquered the curricula of academic economics. Today fifty years later in the universities empirical economics has become synonymous with "econometrics" in the sense given to it by Haavelmo.

Nevertheless the history of economics has taken its revenge on Koopmans' reductionist agenda. Had the Cowles Commission implemented their structural-equation agenda in Walrasian general equilibrium theory, the reductionist agenda would have appeared to have been realized. But the macroeconomics that was actually used for implementation was not a macroeconomics that is just an extension of Walrasian microeconomics; it was the Keynesian macroeconomics. Even before Smith's *Wealth of Nations* economists were interested in what may be called macroeconomics in the sense of a theory of the overall level of output for a national economy. With the 1871 marginalist revolution economists had developed an economic psychology based on the classical rationality postulate of maximizing behavior, which enabled economists to use the differential calculus to express their theory. And this in turn occasioned the mathematically elegant Walrasian general equilibrium theory, that affirmed that the rational maximizing behavior of individual consumers and entrepreneurs would result in the maximum level of employment and output for the whole national macroeconomy. The Great Depression of the 1930's called this optimism into question, and Keynes' macroeconomic theory offered an alternative thesis of the less-than-full-employment equilibrium. This created a distinctively macroeconomic perspective, because it made the problem of determining the level of total output and employment a different one than the older problem of determining the most efficient interindustry resource allocation in response to consumer preferences.

This new perspective also brought in its train certain other less obvious novelties. Ostensibly the achievement of Keynes' theory was to explain the possibility of the less-than-full-employment equilibrium by the use of the classical economic psychology, the theory of value that explains economic behavior in terms of the maximizing rationality postulate. But supporters as well as critics of Keynes knew that there is a problem in deriving a theory in terms of communities of individuals and groups of commodities from a basic theory set forth in terms of individuals and single commodities. In his *Keynesian Revolution* ([1947], 1966) the Keynesian advocate and Nobel laureate econometrician, Lawrence Klein, called this "the problem of aggregation", and he notes that Keynesians have never adequately addressed this problem. Joseph Schumpeter, a Harvard University economist critical of Keynes, was less charitable. In his review of Keynes' *General Theory* in *Journal of the American Statistical Association* (1936) he described Keynes' "Propensity to Consume" as nothing but a *deus ex machina* that is valueless if we do not understand the "mechanism" of changing situations in which consumers' expenditures fluctuate, and he goes on to say that Keynes' "Inducement to Invest", his "Multiplier", and his "Liquidity Preference", are all an Olympus of such hypotheses which should be replaced by concepts drawn from the economic processes that lie behind the surface phenomena. In other words this brilliant expositor of the Austrian school of marginalist economics regarded Keynes' theory as hardly less atheoretical than if Keynes had used data analysis, and Schumpeter would settle for nothing other than a marginalist macroeconomic theory. For the next quarter of a century economists attempted unsuccessfully to reduce macroeconomics to microeconomics, but econometricians did not wait for the approval of the likes of Schumpeter. Keynesian economics became the principal source of theoretical equation specifications for macroeconometric modeling. In 1955 Klein and Goldberger published their Keynesian macroeconometric model of the U.S. national economy, which later evolved into the elaborate WEFA macroeconometric model of several thousand equations. And this is not the only large Keynesian macroeconometric model; there are now many others, such as the DRI model, and they have spawned an important information-consulting industry marketing to both business and government. But there are considerable differences among these large macroeconometric models, and these differences are not decided by reference to purported derivations from rationality postulates or microeconomic theory, even though econometricians ostensibly subscribe to Haavelmo's structural-equation programme and include relative prices in their equations. The criterion that is effectively operative in the choice among the many alternative business cycle models is unabashedly pragmatic; it is their forecasting performance.

The Modern Critique of Structural-Equation Macromodels

After Muth's papers, interest in the rational expectations hypothesis died, and the rational expectations literary corpus was entombed in the tomes of the profession's periodical literature for almost two decades. Then unstable national macroeconomic conditions including the deep recession of 1974 and the high inflation of the later 1970's created embarrassments for macroeconomic forecasters using the large structural-equation macroeconometric models based on Keynes' theory. These large models had been gratifyingly successful in the 1960's, but their structural breakdown in the 1970's occasioned a more critical attitude toward them and a proliferation of alternative views. One consequence was the disinterrment and revitalization of interest in the rational expectations hypothesis. Most economists today attribute these economic events to the large increases in crude oil prices imposed by the Organization of Petroleum Exporting Countries or "OPEC". These events were also accompanied by large Federal fiscal deficits and by Federal Reserve expansionary monetary policies, and these macroeconomic policy actions became targets of criticism, in which the structural-equation type of models containing such policy variables was attacked using the rational expectations hypothesis.

Robert E. Lucas criticized the traditional structural-equation type of econometric model. He is an economist at Carnegie-Mellon, and came from University of Chicago. Lucas' "Econometric Policy Evaluation: A Critique" in *The Phillips Curve and Labor Markets* (1976) states on the basis of Muth's papers, that any change in policy will systematically alter the structure of econometric models, because it changes the optimal decision rules underlying the statistically estimated structural parameters in the econometric models. Haavelmo had addressed the same type of problem in his discussion of the irreversibility of economic relations, and his prescription for all occasions of structural breakdown is the addition of missing variables. But Lucas does not even mention this remedy. Thomas J. Sargent, economist at the University of Minnesota and also an advisor to the Federal Reserve Bank of Minneapolis joined Lucas in the rational expectations critique of structural models in their jointly written "After Keynesian Macroeconomics" (1979) reprinted in their *Rational Expectations and Econometric Practice* (1981). They state that the verbal statement of Keynes' theory set forth by Keynes himself in his *General Theory* (1936) does not contain reliable prior information that certain variables should be excluded from the right-hand side of the structural equations of the macroeconometric models based on Keynes' theory, and furthermore that neoclassical theory of optimizing behavior almost never implies either the exclusionary restrictions suggested by Keynes or those imposed by modern macroeconometric models. They maintain that the parameters identified as structural by current structural-equation macroeconometric methods are not in fact structural, and that these models have not isolated structures that remain invariant. This criticism of the structural-equation models is perhaps better described as criticisms of the structural models based on Keynesian macroeconomic theory, and they leave open the possibility that structural-equation business-cycle econometric models could nevertheless be constructed, which would not be used for policy analysis, and which are consistent with the authors' rational expectations alternative. While Lucas and Sargent offer the non-Keynesian theory that business fluctuations are due to errors in expectations resulting from unanticipated events, they do not offer another structural-equation type of model. Events took another turn: what happened was the rejection of the use of expectations measurement data by the rational expectations advocates, and the consequent development of a kind of rational expectations macroeconometric model that is different from Haavelmo's structural-equation type of model.

Rejection of Expectations Data and Evolution of VAR Models

The rejection of the use of expectations measurement data antedates Muth's rational expectations hypothesis. In 1957 University of Chicago economist Milton Friedman set forth his permanent income hypothesis in his *Theory of the Consumption Function.* This is the thesis for which he was awarded the Noble prize twenty years later, and in his Nobel Lecture, published in *Journal of Political Economy* (1977) he expressed approval of the rational expectations hypothesis and the contributions of Muth, Lucas and Sargent. In the third chapter of his book, "The Permanent Income Hypothesis", he discusses the semantics of his theory and measurement data. He states that the magnitudes termed "permanent" are *ex ante* theoretical constructs, and that they cannot be observed directly for an individual consumer. He says that the most that can be observed are actual income expenditures and receipts during some definite period, and that these observed measurements are *ex post* empirical data, although they may be supplemented by verbal statements by the consumer about his future expenditures. Friedman explains that the theoretical concept of permanent income is understood to reflect the effect of factors which the income earner regards as determining his capital value, his subjective estimate of a discounted future income stream. Friedman does not explain why the permanent income cannot be directly observed, why *ex ante* empirical reports of expectations cannot function as observations of the permanent income signified by the theoretical concept, or describe the "supplementary" role of empirical *ex ante* measurements. Instead he poses a problem of establishing a correspondence between the theoretical constructs and the observed data. Thus Friedman poses an even more radical semantical dualism between theory and observational description than did Haavelmo. Friedman's strategy for resolving his correspondence problem is to use the statistician's idea of "expected value" of a probability distribution to isolate a permanent income component in the *ex post* measurement data. He calls the concept of permanent income an "analogy" to the statistical concept of expected value. The outcome of his semantical dualism is that empirical *ex ante* or reported expectations data are excluded from any consideration in his empirical analyses based on his theory. Muth does not follow Friedman's neo-Positivist dichotomizing of the semantics of theory and observation. In his rational expectations hypothesis he simply ignores the idea of establishing any correspondence by analogy between the purportedly unobservable theoretical concept and the statistical concept of expected value, and makes the economist's statistical concept of expected value the literal meaning of "expectations". Friedman subdivides total measured income into a permanent part and a transitory part. He says that in a large group the empirical data tend to average out, so that their mean average or expected value is the permanent part, and the residual transitory part has a mean average of zero. In another statement he says that permanent income for the whole community can be regarded as a weighted average of current and past incomes adjusted by a secular trend, with the weights declining as one goes back further in time. When this type of relationship is expressed as an empirical model, it is a type known as an autoregressive model, and it is the type that is very strategic for representation of the rational expectations hypothesis in contrast to the structural-equation type of econometric model.

In 1960 Muth published "Optimal Properties of Exponentially Weighted Forecasts" in *American Statistical Association Journal.* Muth referenced this paper in his "Rational Expectations" paper, but this paper contains no reference to empirically gathered expectations data. Muth says that Friedman's determination of permanent income is vague, and he proposes instead that an exponentially weighted average of past observations of income can be interpreted as the expected value of the time series. He develops such an autoregressive model, and shows that it produces the minimum-variance forecast for the period immediately ahead for any future time period, because it gives an estimate of the permanent part of measured income. The exponentially weighted average type of model had been used instrumentally for forecasting for production and inventory planning, but economists had not thought that such autoregressive models have economic significance. Muth's identification of the statistical concept of expected value with subjective expectations in the minds of the population gave the autoregressive forecasting models a new economic relevance, but the forecasting success or failure of these models does not test the rational expectations hypothesis, because they have no relation to the neoclassical theory and its maximizing postulate with or without expectations.

Nearly two decades later there occurred the development of a more elaborate type of autoregressive model called the "vector autoregression" or "**VAR**" model set forth by Thomas J. Sargent in his "Rational Expectations, Econometric Exogeniety, and Consumption" in *Journal of Political*

Economy (1978). Building on the work of Friedman, Muth and Lucas, Sargent developed a two-equation linear autoregressive model for consumption and income, in which each dependent variable is determined by multiple lagged values of both variables. This is called the "unrestricted vector autoregression" model. It implements Muth's thesis that expectations depend on the structure of the entire economic system; it says that all factors in the model enter into consideration by all economic actors in all their economic roles. The **VAR** model does not have Haavelmo's semantical property of autonomy, because there is no attempt to identify the factors considered in determining the preferences of any particular economic group, since each individual considers everything. In his "Estimating Vector Autoregressions Using Methods Not Based On Explicit Economic Theories" in *Federal Reserve Bank of Minneapolis Quarterly Review* (Summer, 1979), Sargent explains that the **VAR** model is not constructed with the same procedural limitations that must be respected for construction of the structural-equation model. Construction of the structural model requires firstly that the relevant economic theory be referenced as prior information, and assumes that no variables may be included in a particular equation other than those variables for which there is a theoretical justification. This follows from Haavelmo's premise that the probability approach in econometrics is the application of Neyman-Pearson statistical inference technique to equations having their specifications determined *a priori* by economic theory. But when the rational expectations hypothesis is implemented with the **VAR** model, the situation changes, because expectations are viewed as conditioned on past values of all variables in the system and may enter all the decision functions. This makes the opposite assumption more appropriate, namely that in general it is likely that movements of all variables affect behavior of all other variables, and all the econometrician's decisions in constructing the model are guided by the statistical properties and performance characteristics of the model rather than by *a priori* theory. He also notes in this article that **VAR** models are vulnerable to Lucas' critique, and that these models cannot be used for policy analyses. The objective of the **VAR** model is good forecasting with small mean squared errors.

Criticisms of structural-equation models similar to those of Lucas and Sargent were set forth by Christopher A. Sims, a colleague of Sargent at the University of Minnesota, in his "Macroeconomics and Reality" in *Econometrica* (1980), and Sims advocates the rational expectations hypothesis and the development of **VAR** models. He also states that the coefficients of the **VAR** models are not easily interpreted for their economic meaning, and he proposes that economic information be developed from these models by simulating the occurrence of random shocks and then observing the consequences described by the reaction of the model. Sims thus inverts the relation between economic interpretation and model construction advanced by Haavelmo: instead of beginning with the theoretical understanding and then imposing its structural restrictions on data in the process of constructing the equations of the empirical model, Sims firstly constructs the **VAR** model from the data, and then develops an understanding of economic structure from simulation analyses with the model. In the *Federal Reserve Bank of Minneapolis Quarterly Review* (Winter, 1986) Sims states that users of **VAR** models have been using these models for policy analysis in spite of caveats about the practice. Not surprisingly this policy advisor to a Federal Reserve Bank does not dismiss such efforts. He says that use of any models for policy analysis involves making economic interpretations of the models, and that predicting the effects of policy actions thus involves making assumptions for identifying a structure from the **VAR** model. But his technique of using shock simulations admits to more than one structural form for the same **VAR** model, and he offers no procedure for choosing among the alternative structures; his approach is judgmental.

Litterman's BVAR Models and Discovery System

In his "Forecasting with Bayesian Vector Autoregression: Four Years of Experience" in the *1984 Proceedings of the American Statistical Association*, also written as a *Federal Reserve Bank of Minneapolis Working Paper*, Robert Litterman, at the time a staff economist for the Federal Reserve Bank of Minneapolis and who has since moved into the private sector, says that the original idea to use a **VAR** model for macroeconometric forecasting at the Minneapolis Federal Reserve Bank came from Thomas Sargent. Litterman's own involvement, which began as a research assistant at the Bank, was to write a computer program to estimate **VAR** models and to forecast with them. The initial forecasting results with this unrestricted **VAR** model were so disappointing, that a simple univariate autoregressive time series model could have done a better job, and it was evident that the unrestricted **VAR** is not successful. In his

"Are Forecasting Models Usable for Policy Analysis?" Litterman noted that the failure of the unrestricted **VAR** model was the attempt to fit too many variables to too few observations. This failure led to the development of the Bayesian **VAR** model, and the Bayesian technique became the basis for Litterman's doctoral thesis titled *Techniques for Forecasting Using Vector Autoregression* (University of Minnesota, 1980).

In the Bayesian vector autoregression or "**BVAR**" model, there is a prior matrix, that is included in the formula for the ordinary least squares estimation of the coefficients of the model, and the parameters which are the elements in this prior matrix thereby influence the values of the estimated coefficients. This prior matrix is a substitute for the *a priori* imposition of economic theory in the conventional structural-equation econometric model as described by Haavelmo, but it also has the desired effect of restricting the variables in the model. Litterman argues that in the structural models there is rarely an attempt to justify the absence of variables on the basis of economic theory, despite the fact that a zero restriction on the excluded variable implies the existence of very certain prior information. He says that the use of such exclusionary restrictions does not allow a realistic specification of *a priori* knowledge. His Bayesian specification, on the other hand, includes all variables in the system at several time lags, but it also includes the prior matrix indicating uncertainty about the structure of the economy. Like Sargent, Litterman is critical of the adequacy of conventional macroeconomic theory, and he maintains that economists are more likely to find the regularities needed for better forecasts in the data rather than in some economic theory.

The difficult part of constructing **BVAR** models is constructing a realistic prior matrix, and Litterman describes his procedure in his *Specifying Vector Autoregression for Macroeconomic Forecasting*, a Federal Reserve Bank of Minneapolis Staff Report published in 1984. His prior matrix, which he calls the "Minnesota prior", suggests with varying degrees of uncertainty, that all the coefficients in the model except those for the dependent variables' first lagged values are close to zero. The varying degrees of uncertainty are indicated by the standard deviations calculated from benchmark out-of-sample retrodictive forecasts made with simple univariate models, and the variation in the degree of uncertainty is assumed to decrease as the length of the time lags increases. The parameters in the prior matrix are calculated from these standard deviations and from "hyperparameter" factors that vary along a continuum that indicates how likely the coefficients on the lagged values of the variables deviate from a prior mean of zero. One extreme of this continuum is the univariate autoregressive model, and the other is the multivariate unrestricted **VAR** containing all the variables in the model in each equation. By varying such hyperparameters and by making out-of-sample retrodictive forecasts, it is possible to map different prior distributions to a measure of forecasting accuracy according to how much multivariate interaction is allowed. The measure of accuracy that Litterman uses is the determinant of the logarithms of the out-of-sample retrodictive forecast errors for the whole **BVAR** model. Forecast errors measured in this manner are minimized in a search along the dimension between univariate and unrestricted **VAR** models. Litterman calls this procedure a "prior search", and it is unlike anything described by Simon in his theory of heuristic search. The procedure has been made commercially available in a computer system called by a memorable acronym, "**RATS**", which is marketed by VAR Econometrics Inc., Minneapolis, MN. This system also contains the ability to make the shock simulations of the type that Sims proposed for economic interpretation of the **BVAR** models.

Economists including rational expectations advocates typically do not consider the **VAR** or **BVAR** models to be economic theories or "theoretical models". The concept of theory in economics, such as may be found in Haavelmo' paper, originates in the Romanticist philosophy of science, according to which the language of theory must describe the decision-making process in the economic actors' attempts to maximize utility or profits. In other words the semantics of the theory must describe the mental experiences of the economic actors whose behavior the theory explains, and this amounts to the *a priori* requirement for a mentalistic ontology. The opposing view is that of the Positivists, or more specifically the Behaviorists, who reject all theory in this sense. Both views are similar in that they have the same concept of theory. The contemporary Pragmatist on the other hand reject all *a priori* ontological criteria for scientific criticism, even when these criteria are built into such metalinguistic terms as "theory" and "observation". Contemporary Pragmatists instead define theory language on the basis of its use or function in scientific research, and not on the basis of its semantics or ontology: on the Pragmatist view theory language is that which is proposed for testing. Theory is distinguished by the hypothetical attitude of the scientist toward a proposed solution to a problem. Therefore, on the contemporary Pragmatist philosophy of science,

Litterman's system is a discovery system, because it produces economic theories.

Ironically the rejection of the structural-equation type of econometric model by rational expectations advocates is a *de facto* implementation of the contemporary Pragmatist philosophy of science. Sargent described rational expectations with its greater fidelity to the maximizing postulates as a "counterrevolution" against the *ad hoc* aspects of the Keynesian revolution. But from the point of view of the prevailing Romanticist philosophy of science practiced in economics, their accomplishment in creating the **VAR**-type of model is a radical revolution in the philosophy and methodology of economics, because there is no connection between the rational expectations thesis and the **VAR**-type of model. Rational expectations play no role in the specification of the **VAR**-type of model, empirical tests of the model could not test the rational expectations "hypothesis" even if it were an empirical hypothesis, and the exclusion of empirical expectations measurement data justifies the denial that the model even describes any mental expectations experienced by the economic actors. However much the rational expectations advocates might protest, their association of the rational expectations hypothesis with their **VAR** models is merely a rhetoric, a decorative language giving their **VAR** model a Romanticist legitimacy.

The criterion for scientific criticism that is actually operative in the **VAR**-type of model is perfectly empirical; it is the forecasting performance. And it is to this criterion that Litterman appeals. In *Forecasting with Bayesian Vector Autoregressions Four Years of Experience* he describes the performance of a monthly national economic **BVAR** model constructed for the Federal Reserve Bank of Minneapolis and operated over the period 1981 through 1984. He reports that this **BVAR** model demonstrated superior performance in forecasting the unemployment rate and the real GNP during the 1982 recession, which was the worst recession since the Great Depression of the 1930's. The **BVAR** model made more accurate forecasts than the three leading structural models: Data Resources, Chase Econometrics, and Wharton Associates. However, he also reports that the **BVAR** model did not make a superior forecast of the inflation rate measures by the percent change in the GNP deflator. Thereafter Litterman continued to publish forecasts from the **BVAR** model in the Federal Reserve Bank of Minneapolis *Quarterly Review*. In the Fall, 1984, issue he forecasted that the 1984 slowdown was a short pause in the post-1982 recession, and that the national economy would exhibit above-average growth rates in 1985 and 1986. A year later in the Fall, 1985, issue he noted that his **BVAR** model forecast for 1985 was overshooting the actual growth rates for 1985, but he also states that his model was more accurate than the structural-equation models. In the Winter, 1987, issue two of his sympathetic colleagues on the Federal Reserve Bank of Minneapolis research staff, William Roberds and Richard Todd, published a critique reporting that the **BVAR** model forecasts of the real GNP and the unemployment rate were overshooting measurements of actual events, and furthermore that competing structural models had performed better for 1986. The Federal Reserve Bank of Minneapolis continues to publish forecasts from the **BVAR** model in its *Quarterly Review*. Reports in the Minneapolis Bank's *Quarterly Review* also contain descriptions of how the **BVAR** national economic model is revised as part of its continuing development. In the Fall, 1984, issue the model is described as having altogether forty-six descriptive variables and equations, but it has a "core" sector of only eight variables and equations, which receives no feedback from the remainder of the model. This core sector must make accurate forecasts, in order for the rest of the model to function accurately. When the **BVAR** model is revised, changes are made to the selection of variables in this core sector. Reliance on this small number of variables is the principal weakness of this type of model. It is not a vulnerability that is intrinsic to the **VAR**-type of model, but rather is a concession to computational limits of the computer, because construction of the Bayesian prior matrix makes great demands on the computer. In contrast the structural models typically contain hundreds of different descriptive variables interacting either simultaneously or recursively. Eventually improved computer hardware design will enable the **BVAR** models to be larger, but in the meanwhile they must perform heroic feats with very small amounts of descriptive information as they compete with the much larger structural-equation models containing much larger amounts of information.

Unlike Simon's simulations of historically significant scientific discoveries, Litterman cannot separate the merit of his computerized procedures for constructing his **BVAR** models form the scientific merit of the **BVAR** models he makes with his discovery system. Litterman is not recreating what Russell Hanson called "almanac science", but is operating at the frontier of "research science", where there is conflict and confusion. Furthermore, the approach of Litterman and colleagues is much more radical than that of the conventional economist, who needs only to propose a new "theory", and then apply conventional structural-equation econometric modeling techniques. The **BVAR** technique has been made commercially available for microcomputer use, but still the econometrician constructing the **BVAR** model must learn statistical techniques that he had not been taught in his professional education. Many economists will fail to recognize the Pragmatic character of the **BVAR** models, and may reject the technique out of hand, since they reject the rational expectations hypothesis. At this point only time will tell whether or not this new type of modeling will survive and achieve ascendency in the economics profession.

Hickey's Metascience or "Logical Pragmatism"

Thomas J. Hickey, a business economist, received a master's degree in economics from the Department of Economics at the University of Notre Dame, which is located near South Bend, Indiana, and he also studied philosophy of science in the Ph.D. program of their graduate school of philosophy. Hickey found Notre Dame's philosophy faculty to be very paternalistic in contrast to the rapport he enjoyed with their economics department faculty. As a contemporary Pragmatist Hickey maintains a prejudicially favorable view toward the thesis of scientific realism, while the Notre Dame philosophers of science on the other hand demand metaphysical criticism in philosophy of science. After two and a half years in their doctoral program and at the threshold of his dissertation Hickey was told by the philosophy department chairman, a Reverend Professor Ernan McMullin, that because of his prejudicial realism he could not expect to succeed in his studies under their philosophy faculty, that he had a "bad attitude", and that if he preferred to discontinue his enrollment in their philosophy department he might do so. Hickey had the support of a minority of the faculty, but the chairman controlled the selection of faculty to students' boards, and Hickey had no interest in acquiescing to the Reverend's demands for metaphysical reform, especially in philosophy of science. Consequently Hickey wrote a patient farewell letter to the Reverend chairman and the philosophers of science that he had hired, nearly all of whom at this writing are still faculty members in Notre Dame's graduate school of philosophy.

In this letter farewell Hickey stated his intent to pursue his interest independently as an empirical science of science which he called

"Metascience", a term often appearing in the literature of philosophy of science. After leaving Notre Dame he took courses in computer science and numerical analysis programming at San Jose College in California, where he wrote his "**METAMODEL**" computer discovery system. In 1976 he set forth his metatheory and **METAMODEL** discovery system, which was originally intended to have been his doctoral thesis in philosophy at Notre Dame, in a monograph titled *Introduction to Metascience: An Information Science Approach to Methodology of Scientific Research.* Since then Hickey has found that his Metascience agenda seemed philosophically wayward only in comparison to the metaphysical orientation in philosophy of science that confronted him at the University of Notre Dame. Now he also refers to Metascience as "Logical Pragmatism", where the "Pragmatism" involved is the contemporary Pragmatist philosophy of science, and the "Logic" is not the irrelevant Russellian logic applied by the Logical Positivists, but rather the computer logics used in the computerized artificial-intelligence discovery systems such as Herbert Simon's.

This section reports Hickey's current statement of his metatheory, or the "Pragmatist" part of his Logical Pragmatist philosophy, and the following section takes up Hickey's **METAMODEL** discovery system, the "Logical" part of his Logical Pragmatism. His metatheory may be summarized in terms of the four basic topics considered in philosophy of science: explanation, discovery, criticism, and the aim of science. But firstly some preliminary comments are in order, to provide an integrating context. Unlike Simon's cognitive-psychology approach, Hickey follows in the linguistic analysis tradition of the analytic philosophers. Following Carnap, Hickey distinguishes object language and metalanguage, and he views his **METAMODEL** as an object-language processing system written in a metalanguage. Object language consists of the statements used by scientists for expressing their experimental designs and theories, that describe the extralinguistic real world. This language includes both colloquial discourse and the written symbols for which mathematics supplies the grammatical syntax. The metalanguage is used to describe the object language, and it describes not only the syntax, semantics and pragmatics of the object language, but also research practices and procedures including computerized systems procedures used to change the object language. Terms such as "explanation", "falsification", "theory", "test design", and "discovery system" are examples of metalinguistic vocabulary. The Logical Pragmatist philosopher of science, or metascientist, uses the metalanguage for formulating his theory, and it is therefore called a "metatheory"; the computer language used in a discovery system is part of the metalanguage, and the discovery system itself is part of the metatheory.

The state of the object language for a scientific profession at a point in time is exhibited in a "state description". This phrase is also borrowed from Carnap, but it has a different meaning and purpose in Hickey's metatheory. The state description consists of all the object language statements that are relevant to a scientific problem, that have universal logical quantification, and that are believed to be true by at least some members of the cognizant scientific profession. Scientific problems are not like engineering problems; a scientific problem is one that is solved by a new explanation, or what Quine calls an empirically warranted belief, that changes the state description for a profession. A scientific profession is a set of one or several scientists who are directing their problem-solving research efforts to the same problem, and they typically communicate with one another either informally or formally in a shared published literature. Hickey's metatheory is not concerned with the psychology of the individual scientist, but with a whole profession considered as a special language community. One of the functions of the state description is to describe the semantics shared by the cognizant profession at the chosen point in time. On Hickey's metatheory all the universal affirmative statements in a state description having the same descriptive term as a subject term, have predicates that describe the components of the meaning of the common subject term. In other words meanings have parts. For example belief in the statement "All ravens are black" makes the phrase "black raven" semantically redundant, because the concept of blackness is already included as part of the meaning of "raven". A list of all the predicates in such universal affirmations about ravens would describe the parts of the meaning associated with the univocal term "raven". Each statement in the state description may be said to be both a synthetic statement, because it is believed to be empirically true, and an analytic statement, because belief in the statement also enables it to be used for semantical analysis. Hickey thus joins Quine's rejection of the analytic-synthetic distinction in so far as the distinction amounts to a separation of statements into two dichotomous classes, and he accepts Quine's rejection of an analytic theory of truth. But Hickey does not reject analyticity as such, and he says that empirical statements may be used analytically for semantical analysis. Statements need not be in categorical form, to be used for semantical analysis. It is sufficient that they be universal statements in descriptive language, and that they be accepted as empirically true. Mathematical equations containing descriptive variables are universal statements when their numerical values are not assigned; so long as their equality condition is believed to obtain, they exhibit the components of the meanings associated with their measurement variables. The idea that meanings have parts is not commonplace in philosophy. As it happens the piecemeal nature of meanings has recently been proposed in neurology. In "Stroke Patients Yield Clues to Brain's Ability to Create Language" the *Wall Street Journal* (12 Oct 1993) reports Dartmouth College neurologist Alfonso Caramazza as saying that the meaning of a word must be stored "piecemeal" in the brain. For example the meaning of "lemon" is the sum of many attributes that the brain has filed away separately, and a stroke can damage an area of the brain where one of the attributes is stored. The article notes that neurologists are presently using positron emission tomography (PET) to locate such storage areas physically in the brain. However, Hickey's thesis is not a physiological theory, but rather is a thesis in philosophy of language, which is based on such obvious evidence as the fact that lexical entries in a dictionary display the parts of meanings, because the dictionary definitions are believed to be true.

A state description may be called a static or synchronic analysis, since it pertains only to a point in time. A metatheory is a description of the transition from one state description to another, and may be called a dynamic or diachronic analysis. It does not describe institutional change, but rather describes changes within the institution of science. Both empirical testing and theory development produce new state descriptions. A discovery system is a computer system that generates a new state description from the object-language input in the initial state description. The comparative examination of a semantical change resulting from the transition to a new state description may be called a comparative static analysis. Such a semantical comparison between an initial and a later state description for the same scientific problem reveals which parts in the meanings of the descriptive terms in the statements remain unchanged, and which parts have changed due to the change in beliefs. Each descriptive term exhibits semantical continuity and semantical discontinuity. Hickey does not accept either the wholistic doctrine of "paradigm shift" advocated by Kuhn and Feyerabend or the wholistic *gestalt* theory of meaning used by

Kuhn and Hanson.

With these ideas in mind turn now to the four basic topics common to modern philosophies of science, beginning with the aim of science. Hickey's statement of the aim of science could be called a "rationality postulate" for science, if that phrase is not taken to mean that the goal-statement is an incorrigible doctrine used as a criterion for scientific criticism. His thesis of the aim of science is an empirical hypothesis about the regulating institutional value system for empirical science, where science is retrospectively recognized to have performed successfully. The current statement of the aim of science is a statement of the contemporary Pragmatist philosophy of science, the philosophy of science that has itself evolved from the examination of the institutional evolution of twentieth-century physics and that articulates its institutional views and values. At the opening of the twentieth century the prevailing institutional views and values were those of the Positivism, and previously they had been those of Romanticism. Today these anachronistic philosophies of science are still influential in Positivistic behavioral sciences such as Behavioristic psychology and in the Romanticist social sciences such as economics and especially sociology. The lenses of Pragmatism enable the contemporary philosopher to recognize the dysfunctional effects of these earlier philosophies on the current research practices in these sciences, even though the researchers in these sciences are oblivious to their institutional retardation. The contemporary Pragmatist statement of the aim of science may be stated as follows:

> **Scientists aim to construct theories which will satisfy the most critically empirical tests that can be applied to those theories at the current time, and thereby make explanations.**

The meaning of this statement is explained by the other three topics in the Pragmatist philosophy of science: discovery (i.e. theory construction), criticism, and explanation, which includes the distinctively Pragmatist definition of "theory".

Basic research regulated by this institutionalized aim encounters various constraints which scientists view as impediments to be overcome. But there is one constraint that is a voluntary constraint which scientists do not view as an impediment to be overcome, but rather view with an attitude of obligation and respect as integral to the aim of science. That constraint is the empirical constraint; it is like a moral constraint in which scientists are socialized by their professional education, and which is reinforced by social controls that condition professional recognition upon conformity to its regulation of research practices. The operation of this institutional constraint is considered below in the discussion of scientific criticism. All other constraints are impediments to be overcome. Simon's "computation constraint" is an example of such a constraint, as are other more circumstantial conditions such as limited financial resources. But there are two constraining impediments that are more than merely circumstantial, and Hickey calls these the "cognition constraint" and the "communication constraint". These are distinctive in that they are semantical constraints, which are integral to language and therefore to science itself. The production of a new state description will produce a greater or lesser semantical change, depending on how radically or moderately the new state description alters old beliefs. Beliefs are the glue that bonds the structured components of the meaning for each univocal descriptive term. When a scientist develops a new theory that contradicts many previously held beliefs, he dissolves the meanings of the descriptive terms into component parts and must re-associate and restructure those components according to the structure of a new theory. This semantical dissolution and restructuring has a disorienting and confusing effect, which Hanson calls "conceptual resistance", that impedes the scientist's theory-developmental work. This impediment is the cognition constraint. Similarly when the scientist has created a radically new theory, his colleagues to whom he attempts to communicate his new theory also experience the disorientation and confusion as they achieve the dissolution and reconstruction of their semantics for familiar terms. The impediment or conceptual resistance that they encounter in this learning experience due to this semantical change is the communication constraint. A computerized discovery system has no experience of any cognition constraint, but the scientists who read its output may have to overcome a severe communication constraint in their attempt to "communicate" with the machine.

Consider next the topic of scientific explanation. An explanation is the fruitful outcome of successful work in accordance with the regulating institutionalized aim of science: it is a belief that was formerly a theory, which when it was still a theory, had been tested by the most critically empirical test available at the time, and that has not been falsified by the test. A theory in turn is a set of one or several related statements having universal logical quantification, and that is proposed for testing. An explanation is a theory that is no longer a theory, because the successful test has removed its exceptionally provisional status relative to all other empirical statements. It could later be used as a test design statement for defining another problem in another state description. Explanations are not permanent. New test designs that resolve the vagueness in the semantics of current test design statements and thereby improve the accuracy of measurements, may occasion reverting the explanation to the status of a theory for retesting, and the result could be falsification. This definition of "theory" as statements proposed for testing, is a Pragmatic definition in terms of the use or function of theory under the institutional regulation of the aim of science. This contemporary Pragmatist definition of theory language in terms of its function is opposed to the earlier Romanticist and Positivist definitions, which define "theory" in terms of some preferred semantics and ontology. The Romanticist definition still prevails in the social sciences, where "theory" language is that which has a semantics that describes an ontology consisting of subjective mental events. The Positivist view defines "theory" in contrast to "observation" language which also has a specific semantical content, such as sense perceptions, phenomena, or sense data. Both the Romanticist and Positivist views of the aim of science represent anachronistic institutional views and values for empirical science. The contemporary Pragmatist may define "observation" language in contrast to theory in the sense that the vocabulary for observational description is supplied by the test design statements. This vocabulary conceptualizes and articulates the perceptual experiences involved in the test situation including that report of the test outcome. Theory language is defined pragmatically, but theories are individuated semantically. There are two ways: Firstly two theory expressions that address different problems are different theories, but in the sense that they are different theories relative to different scientific professions. What is theory for one profession is not so for some other; theories are proposed solutions for a problem, and each profession is defined in terms of the scientific problem that it is addressing. Secondly two theory expressions that address the same problem but make contrary claims, are different theories, and are different theories for the same scientific profession. In an active science there may be many alternative theories in the same state description, since for each theory there need be only one member of the profession who believes in the theory enough to propose it for testing.

Whereas the topic of explanation pertains to the synchronic perspective, the topics of criticism and discovery pertain to the diachronic perspective. These functions are performed by the research scientist within the regulating institutional matrix defined by the aim of science, and they are the practices that have the effect of changing the state of the object language and its state description. Both of these functions are operative in a discovery system, which produces scientific change. In the dynamic perspective consider firstly scientific criticism by empirical testing. Contemporary Pragmatist philosophy of science admits only empirical criteria for scientific criticism, and it excludes all ontological criteria, such as those required by the Positivist and Romanticist philosophies. Hickey thus joins Quine's rejection of any metaphysics or "first philosophy", which would impose any nonempirical criteria for scientific criticism. And he therefore also joins Quine's doctrines of ontological relativity and scientific realism, which were practiced by Galileo, Einstein and Heisenberg, when these historic physicists affirmed that the real world is as their empirically adequate theories describe it. Hickey maintains an innocuously historical-relativist thesis of the semantics of language implied by the artifactual character of meaning, but he does not maintain a relativist thesis of truth, save for the banal fact that truth is a property of statements and is therefore relative to what is said. Given any set of concepts, it is not possible to say in truth just anything with those concepts. Or more precisely, given a subset of the statements in a state description and the concepts those statements define, only some statements in the remainder set affirmed with these concepts can be true, while others in the remainder will be false. This constraint on the truth of statements in the remainder is not due to formal logical deduction or consistency; it is due to the empirical constraint imposed by nonlinguistic reality that is accessible as available evidence. And since the predicates of these empirically warranted beliefs contribute to the semantics of their common subject terms, the state description's semantics is not arbitrary.

Such is the nature of the situation in the empirical test of a scientific theory. At the time of the test of a theory all the statements in the state description may be viewed a segregated dichotomously into two classes: those that are proposed for testing and those that are presumed for testing. The former are the theory statements. And there may be more than one theory. The statements that are presumed true for testing the theory are the test design statements. Theory statements are included in the state description, because at least one member of the profession, presumably the proponent of the theory, believes that his theory is true. But the test design statements are accepted as true by all the members of the profession, since these statements supply the semantics that characterize the problematic phenomena independently of any theory, define the cognizant profession, and identify the object language that is relevant and thus included in the state description. Execution of the empirical test in accordance with the previously agreed test design changes the state description for the cognizant profession, when it eliminates one or several theories by a falsifying test outcome. By prior agreement the test design statements are those that will be regarded as true in the event of falsification. Regardless of the test outcome, these statements contribute parts of the meanings of the descriptive terms common to both the test design and theory statements. But the parts of the meanings contributed by the theory statements change depending on whether or not the theory was believed true by the particular scientist before the test, and depending on whether or not the theory was falsified by the test. The advocates of the theory believed in it before the test, and therefore believed that its statements supplied a true characterization of the problematic phenomenon in addition to the characterization supplied independently by the test design statements. Both test design and theory statements contribute parts to the meaning of each univocal term common to them, until falsification makes at least one term equivocal. The falsifiying test outcome motivates the proponents and advocates to reconsider, such that the semantics of their theory is no longer thought to supply a characterization of the problematic phenomenon.

However, in the event of falsification the theory's advocates may choose to reconsider the prior agreement about the defining role of the test design statements. Other members of the profession may view this behavior as prejudicial, but even if the response to a falsifying test outcome is merely a strategy to evade falsification, it amounts to redefining the problem into a new one and in effect creating a new state description, in which the falsified theory in the old state description assumes the role of test design statements in the new one. Evidence that this animadversion is not merely an evasion of the empirical constraint is the proposal of a new theory for the new problem. This possibility is permitted by Duhem, when he stated that a falsifying test does not locate the error that caused the falsifying outcome. And it is implied by Quine in his "Two Dogmas", when he stated that it is possible to preserve the truth of any statement come what may by redistributing truth values. Quine was not a Positivist like Duhem, and Quine furthermore admits to no restriction on the redistribution of truth values. The test design statements may be modified for reasons other than a falsifying test outcome. They may be refined by the addition of statements describing new test procedures that offer more accurate measurements or more refined observation techniques, so that the testing may be more critical. These new test design statements have the effect of reducing the vagueness in the semantics of the descriptive terms in the test design statements. All descriptive language is always vague, and vagueness can never be completely eliminated, but it can be reduced. Vagueness occurs to the extent that descriptive terms have not been related to one another in universal affirmations or negations believed to be true. Adding test design statements has the effect of resolving some of the vagueness in the descriptive terms in the test design statements, and the outcome of the consequently more critical test may be the falsification of previously tested and nonfalsified theories.

Finally, turn to the topic of scientific discovery or theory development. The critical elimination of theories from the state description by empirical testing requires consideration of the state description at one point in time. But for the constructional introduction of new theories into the state description, it is necessary to consider the historically accumulated object language from both falsified and nonfalsified theories in many past state descriptions for the same scientific problem. This is because falsified theories have scrap value, since their constituent descriptive vocabulary can be salvaged for new theory construction. In some circumstances the construction of new theories can be predicted and therefore effected by use of the salvaged object language in a cumulative state description. Hickey distinguishes three types of theory construction with the objective of identifying those circumstances: (1) theory extension, (2) theory elaboration, and (3) theory revision. Given a new state description with its statements of test design that identify its scientific problem, the first type of theory construction that the cognizant profession will attempt for a new problem is theory extension. The existing beliefs give phenomena what Hanson called their "intelligibility", and scientists are reluctant to sacrifice intelligibility by disturbing their current beliefs. Theory extension creates minimal disturbance to current beliefs, and it consists of using the statements of an explanation already accepted as a solution for another problem, and then extending that explanation to address this current problem, perhaps because the current problem is viewed as a special case of the solved problem. This extension is something more than just a logical transformation of an existing explanation, because the vocabulary in the test design statements cannot already be in the explanation, for the current problem to be an unsolved problem. The extension must at least consist of relating the explanation to the terms or variables in the test design statements by the addition of new statements, and these new relating statements constitute the new theory, which is tested and may be falsified. Falsification of these new relating statements would not affect the validity of the employed explanation as an explanation of the problem that it had already solved. If successive attempts at theory extension fail to solve the current scientific problem, then some of the members of the cognizant profession will become more willing to depart from the existing stock of accepted explanations.

Theory elaboration is the next most conservative approach. It offers minimal deviance from accepted explanation, and it involves a modification to some previously proposed but since falsified theory for the problem. The modification consists of the introduction of some new descriptive term or variable as a correcting factor that will change the previously falsified theory thereby transforming it into a new theory. Different members will propose different correcting factors as strategic in their theories, but their theories will typically display a recognizable similarity, since they are basically modifications of shared older beliefs. Empirical testing may result in persistent falsification of theories produced in this conservative manner. Some members of the profession will therefore become more willing to deviate more radically, and their theory construction will make new theories that bear increasingly less similarity to past theories produced by theory extension or theory elaboration. As the permutations permitted to theory construction become greater, the only remaining control on the exponentially increasing number of constructional possibilities is the size of the descriptive vocabulary in the state description. But this size approaches a limit, as the persistent failure of theory elaboration provides reason to expect that the solution to the current problem does not consist in the further search for unknown correcting factors, but instead consists in restructuring statements containing a selection from the descriptive vocabulary already in the cumulative state description.

Hickey calls this third type "theory revision", and he maintains that as increasing numbers of researchers abandon theory elaboration in favor of theory revision, the prospects increase for producing an empirically satisfactory explanatory solution by the mechanized theory revision of the object language available in the cumulative state description. The key idea in this strategy for mechanizing theory development is that the descriptive vocabulary that serves as input has been identified, is small, and is available. Hickey notes that the conditions occasioning increased use of the strategy of theory revision suggest something similar to what Kuhn called a "crisis", and also that theory revision produces a much more radically new and different theory, that would readily be called "revolutionary". Ironically revisionary theory development is most often viewed as the most mysteriously muse-inspired type, while on Hickey's metatheory the availability of object-language input from a cumulative state description makes it the type that is most easily mechanized. Mechanization invariably takes the mystery out of musing.

Hickey does not accept Kuhn's thesis that every scientific revolution is a wholistic gestalt switch to a new "paradigm" producing an institutional change. Nor does he accept Feyerabend's radical historicist thesis that there are semantically incommensurable revolutionary developments involving Whorfian covert categories, even when the new theory uses a new patterning mathematics, or that science should be in a state of perpetual revolutionary change. In Hickey's metatheory of semantical description the semantical continuity through theory revision is exhibited in the unchanged semantical contribution to the descriptive vocabulary made by the test design statements, if as Popper says one "sticks to the problem". And the semantical discontinuity is exhibited by the semantical contribution to the descriptive vocabulary by the radically new statements constituting the new theory. Because of the semantical continuity, even the most radical scientific revolution does not create a completely new world view that is semantically incommensurable with the past and that *ipso facto* constitutes an institutional change. Thus there is no semantical basis for maintaining that radical change in theory necessitates institutional change in the science. The extent of semantical restructuring in the new theory produced by theory revision produces a corresponding degree of cognition constraint for the inventor having no discovery system, and a comparable degree of communication constraint for the profession with or without a discovery system.

Furthermore, the contemporary Pragmatist philosophy of science liberates theory from any particular semantics and ontology. When the

Romanticist and Positivist philosophies of science prevailed, they made the predominant scientific theory metaphysically bound to its distinctive semantics and to the ontology its semantics described, thereby giving that theory institutional status. Any revision of theory therefore actually required an institutional change in the views and values in the affected science. The philosophies of science advanced by Kuhn and Feyerabend describe institutional views and values that characterize earlier periods in the history of physics, when science's institutional views, as Hanson noted, defined such concepts of explanation and causality. Pragmatism avoids this outcome by making ontological commitment depend upon empirical adequacy, rather than including any ontology in the criteria for scientific criticism. Its adoption simply means that even the more obdurate physicists and philosophers of science have learned something. Institutional change will continue to occur in sciences in which Pragmatism prevails. It is impossible to predict what the post-Pragmatist philosophy of science will look like. But prediction is possible for those sciences that have not yet matured institutionally by the adoption of the contemporary Pragmatist philosophy of science: in due course psychology will drop Positivist Behaviorism and adopt Pragmatism, and economics and even sociology will outgrow their Romanticism and accept Pragmatism.

Hickey's METAMODEL Discovery System

Hickey's **METAMODEL** discovery system antedates Simon's applications to the problem of scientific discovery of his problem-solving theory of heuristic search by about ten years, but it cannot be said that Hickey's work anticipates the discovery systems of Simon and his colleagues at Carnegie-Mellon. Not only was Hickey unaware of Simon's work in problem solving in 1971-72, when he constructed his **METAMODEL** system, but the **METAMODEL** system is not designed like Simon's discovery systems made in the 1980's. In his autobiography Simon distinguishes three types of discovery systems: expert systems, generate-and-test systems, and heuristic-search systems. Unlike Simon's heuristic-search type, Hickey's is basically a generate-and-test type of system. The generate-and-test procedure in the **METAMODEL** discovery system does not proceed through a lengthy sequence of decision points. Instead its design is basically that of a combinatorial procedure, or what Simon calls a factorial-design system, that generates and tests a very large number of structured nonredundant combinations of language elements. The **METAMODEL** is an exhaustive cognitive exploration of revisionary theory-constructional possibilities that are latent in the input state description. The principal disadvantage of the generate-and-test design is that it is very inefficient, and its utilization of computer resources in comparison to the heuristic-search design is very extensive. On the other hand the principal advantage is that unlike heuristic search, it does not risk overlooking or pre-emptively excluding theories that are worthy of consideration. In other words it is not a satisficing system, but rather is an optimizing system that outputs a small number of output theories. As the computer hardware technology continues to improve, the trade-off between efficiency and thoroughness will continue to move in the direction of thoroughness. Furthermore unlike Simon's cognitive-psychology agenda, Hickey's metascience agenda with its Pragmatist statement of the aim of science places no value or significance in any representation of the computation constraint found in intuitive human problem solving activity in theory development or discovery.

Hickey's *Introduction to Metascience* is divided into two parts. The first part is an exposition of his metatheory, as described above in its essentials. The second part sets forth the design of his **METAMODEL** discovery system together with a description of an application of the system to the trade cycle specialty in economics in 1936, the year in which John M. Keynes published his *General Theory of Employment, Interest and Money*. The **METAMODEL** performs revisionary theory construction to reconstruct the development of Keynes theory, a episode now known as the "Keynesian Revolution" in economics. The applicability of the **METAMODEL**'s revisionary theory construction for the rational reconstruction is already known in retrospect by the fact that, as Lawrence Klein says in his *Keynesian Revolution* (1966, [1947]), all the important parts of Keynes theory can be found in the works of one or another of his predecessors. The **METAMODEL** discovery system has an input and an output state description, and Hickey firstly describes the cumulative input state description containing the object language given to the system. The test design statements are not explicitly displayed in the input state description, since they do not change through the execution of the discovery system. They consist of the statements describing the phenomena symbolized by the descriptive variables occurring in the trade cycle theories, that had been proposed by economists up to 1936, together with the statements describing the measurement procedures for collecting the associated data. The measurement data are those representing the U.S. national economy, which were originally published at the time in annual issues of the U.S. Department of Commerce *Statistical Abstract*, and since reprinted in their *Historical Statistics of the United States* (1958). Hickey searched both the published books and the periodical literature of the economics profession for the interwar years prior to 1937, which pertained to the trade cycle problem. The American Economic Association's *Index of Economic Journals* was a useful bibliographic source, which also revealed that the number of journal articles fluctuated in close correlation with the national average unemployment rate with a lag of two years. This examination of the relevant professional literature yielded ten economic theories of the national trade cycle, which he translated into mathematical form. The ten theories were those of J.A. Hobson, Irving Fisher, Foster and Catchings, J.M. Clark, F.A. von Hayek, R.G. Hawtrey, Gusatv Cassel, Gunnal Myrdal, Johan Akerman, and A.C. Pigou. The descriptive vocabulary occurring in these theories was a highly redundant, and yielded a set consisting of eighteen variables. The data for these variables are annual time series for the period 1921 through 1934, which were available to any economist in 1936. These time series data were converted to index numbers of period-to-period change rates, and together with variable names including one time lag are the input to the **METAMODEL** discovery system for the historical simulation.

The output state description was expected to contain an econometric model of Keynes theory constructed by the discovery system. Therefore Keynes' theory like the other theories was translated into mathematical form. The theory is actually a static theory, but it was made dynamic by including considerations contained in an appendix to the *General Theory* titled "Notes on the trade cycle", in which Keynes explicitly applies his theory of income determination to the phenomenon of the trade cycle. Keynes theory contains ten variables and seven equations with three exogenous variables. All ten variables occur in more than one of the preceding trade cycle theories, and most in several of them. There is no question that all the variables needed for a recognizably Keynesian theory are available in the existing literature in 1936.

The **METAMODEL** contains two initial designations, that must be made prior to execution of the discovery system in the computer. Firstly the user must designate which descriptive variables among the current-valued variables are the problematic variables, which identify the cognizant profession. In the application to the trade cycle problem, the problematic variables are aggregate real employment and aggregate real income for the national economy. Every macroeconometric model printed in the output state description generated by the system will contain these variables and the equations determining their numeric values. Secondly the user must designate which among the current-valued variables are exogenous variables. These variables have their values determined for the model and not by the model; the values are determined by economic policy decisions. The exogenous variables designated in the trade cycle application are Federal real aggregate fiscal expenditures, Federal real aggregate fiscal tax revenues, and the Federal Reserve-determined aggregate nominal money stock. These two types of designations together with other information such as the number of observations in the time series data, are entered into a control record, that is immediately read when the system is executed. The control record is followed by records containing the character names of the input variables with separate identifiers for current values and lagged-valued variables, and then the time series data records follow.

The **METAMODEL** discovery system is a **FORTRAN** system having an architecture consisting of a main program, **SLECTR**, and two called subroutines, **REGRES** and **SOLVER. SLECTR** is a combinatorial procedure that selects nonredundant combinations of language elements. It contains a switch, which is initially set to the open state. When the switch is open, **SLECTR** selects combinations of time series from the input file initially read by the system, and it calls the **REGRES** subroutine. **REGRES** is an ordinary-least-squares-regression procedure, that statistically estimated an intercept and coefficients thereby constructing an equation for the selection of variables passed to it by **SLECTR.** If the estimated equation does not have a satisfactory R-squared or coefficient-of-multiple-determination statistic associated with it, according to a minimum value given to the system and stored in its control record, then control is given back to **SLECTR**. But if the R-squared statistic is satisfactory, the equation is stored as a record in an accumulation file before control is returned to **SLECTR**. After **SELCTR** has made all its selections for nonredundant combinations of as many as six variables, the switch is closed, and **SELCTR** repeats its combinatorial procedure, but this time it makes selections of estimated equations from the accumulation file generated by **REGRES,** and for each selection it calls subroutine

SOLVER. **SOLVER** solves the multi-equation model as a simultaneous-equation model, and then executes the model to generate a reconstruction of the historical data. In order to accomplish this, there are certain criteria that any selection of equations must satisfy, and **SOLVER** checks for these conditions. Firstly the combination of equations constituting the model must contain equations that determine the two designated problematic variables. Secondly the model must be uniquely determined, such that there are as many current-valued endogenous variables as there are equations. Thirdly the model must be recursive, such that there is at least one current-valued variable for each lagged-valued variable describing the same phenomenon. Fourthly the model must be a minimal statement, such that it contains no current-valued variables except the problematic variables, that occurs but once in the model and is not needed to evaluate a lagged-valued variable describing the same phenomenon. When **SOLVER** finds a combination that does not satisfy all these criteria, it returns control to **SLECTR** for another combination of equations. Models that do satisfy all these criteria are capable of being solved, and **SOLVER** solves and iterates the model both to recreate the history with synthetic data for the years 1921 through 1933, and to make a one-period out-of-sample postdictive forecast for the year 1934. The control record for the system also contains a minimum error for the forecasts of the problematic variables, and the final test for the model is for its forecast accuracy. The model that satisfies this criterion also, is outputed to the printer and printed in conventional mathematical form with each equation listed together with its associated R-squared statistic. The output also lists the synthetic data generated by the iteration of the model together with the forecast values for all its endogenous variables.

The computing equipment available at the time the **METAMODEL** discovery system was created did not permit a complete operation of the system, but it was demonstrated with partial runs, that the system would generate a satisfactory Keynesian model. In 1976, four years after designing and testing his **METAMODEL** discovery system, Hickey had occasion and opportunity to use the system to address a contemporary problem in social science. At that time he was a profit analyst in the analysis and statistics bureau of the finance department of a very large steel manufacturer. He had completed a conventionally Keynesian quarterly macroeconometric forecasting model, but found that the model was not performing satisfactorily. This occurred during the years following the large increase in crude oil prices imposed in 1973 by the Organization of Petroleum Exporting Countries (OPEC), and no macroeconometric models available at the time had the consequences of this shock in the sample data used to estimate the models statistically. Many economists reacted to the structural break down of their models with patience, and waited for new data. Others, however, believed that more than oil prices was at fault, and they found more basic reason for dissatisfaction with their models. One such group as mentioned above was the rational expectations economists, and they had their distinctive agenda.

Hickey believed that more was involved than inadequate sample data. And unlike the rational expectations advocates he views the phenomenon of structural breakdown in the same manner as did Haavelmo, who maintained that it is remedied by introducing into the model new variables representing missing factors, the absence of which had caused the breakdown. But unlike Haavelmo, Hickey agrees with the institutionalist economists that neoclassical economics limits economic explanation to an excessively small number of factors, and that it assumes incorrectly that all the other complexities in the real world are irrelevant. Furthermore Hickey is not philosophically sympathetic to the Romanticism in neoclassical economics, and he prefers the explicitly Pragmatic orientation of the institutionalist economists. However, institutionalists did not make econometric models, and furthermore they were usually more interested in the evolution of economic institutions. Hickey ventured beyond conventional institutionalists, and decided to integrate functionalist sociology into his econometric model, even though functionalists do not make econometric models either. Functionalism in sociology is the thesis that all institutions of a national society are interrelated, and so he used his **METAMODEL** discovery system to investigate how variables representing each of the five basic institutions of the American society can be related by statistically estimated equations of the type used in econometric models. Both the sociology in the model generated with the discovery system and the truculent philosophical rejection by the Romantic sociologists to his use of the discovery system, are discussed below. What are noteworthy in the present context are the modifications he made to his **METAMODEL** system, to run it completely.

Hickey had access to much better computer equipment than the machine he used to develop the **METAMODEL**, but some modifications to the design of the system were nevertheless needed. One noteworthy modification made the system store in the accumulation file only one equation of all those constructed for the same dependent variable and having the same number of independent variables. The one saved for further processing is that having the highest R-squared statistic. The effect of this modification is to reduce significantly the number of equations available for selection by **SOLVER**, and therefore to reduce the number of models generated for testing for output. However, this modification has been eliminated from the commercial version of the **METAMODEL**, which now runs without subroutine **SOLVER**. A second modification is the introduction of a new test criterion in subroutine **SOLVER**, that tests for the recreation of an inflection point in the synthetic data. A model assembled by **SOLVER** that cannot recreate the occurrence in the actual data of an inflection point, is rejected by **SOLVER** and is not sent to the printer for display. The modified version of the **METAMODEL** was executed with eleven current-valued input variables, and each was allowed two lagged-valued variables. The total number of equations estimated and stored by **REGRES** for further processing by **SOLVER** was thirteen, and the total number of macrosociological models generated and critically accepted by **SOLVER** for output was three models. As it happens, two of the three models were actually the same model for reasons that **SOLVER** cannot detect, and so the total number of models actually outputed was only two. In response to inquiries during the first year of publication of *Introduction to Metascience*, Hickey released a source-code listing of the **FORTRAN** statements of the **METAMODEL**, and issued it as a supplement to the monograph in 1977. This supplement also contained a listing of the input data for the sociological application and a list of the printed output models generated with the modified version of the discovery system.

The functionalist macrosociometric models generated by the **METAMODEL** discovery system were intended to be used as a guide for integrating sociological, demographic, and human ecological factors into a macroeconometric model of the national society. However, the continuing decline in the domestic steel industry precluded Hickey from accomplishing this objective until several years later. In the 1980's he was Deputy Director and Senior Economist for the Division of Economic Analysis of the Indiana Department of Commerce, and occasion and opportunity were again present for taking up the task of constructing an integrated macrosocio-econometric model of the U.S. national society.

The **METAMODEL** discovery system as originally designed was inadequate to such a project, and it was necessary to revise the design. Subroutine **SOLVER** was the principal limitation; it could only make models with as many as twelve equations, while the integrated model required a number in the range of one hundred equations. Consequently Hickey designed and wrote a new **METAMODEL** discovery system, that performed only the functions of **SLECTR** and **REGRES** in the old system. The new system can accept as many input variables as the installation's computer and its **FORTRAN** compiler can handle, and it can specify, estimate and test equations with as many as twenty-one independent variables, which is more than enough or desirable for a regression model. A description of the resulting integrated macromodel was published in the state agency's *Perspectives on the Indiana Economy* (March, 1985). Later in the September 1985 issue Hickey published "The Pragmatic Turn in the Economics Profession and in the Division of Economic Analysis of the Indiana Department of Commerce", in which he described the new **METAMODEL** and compared it with some **VAR** models and the **BVAR** system constructed by the rational expectations advocates. Hickey has used the new **METAMODEL** system for many other econometric modeling projects for various employers and clients, and the United States Department of Commerce has issued him a registered copyright for this commercial version of the **METAMODEL** discovery system. Hickey submitted a paper describing his macrosociometric model to several sociological journals. Since the prevailing philosophy of science in American academic sociology is still Romantic, and since Hickey is a Pragmatist, the paper was rejected. Before turning to Hickey's model, consider next the philosophies of science that have prevailed in sociology in the twentieth century.

Parsonian Romantic Sociology and its Critics

Twentieth-century sociology and twentieth-century physics offer the philosopher of science a striking contrast. Physics saw the revolutionary developments in the relativity theory and quantum theory, and these in turn occasioned the repudiation of nineteenth-century philosophy of science by both the physicists and the philosophers of science. Sociology on the other hand saw no advancements like the developments in physics, and attempted to rework late eighteenth-century and nineteenth-century philosophies of science. The result has been the intellectual stagnation of the science and the decline of its academic profession. This section examines the reworking of the nineteenth-century philosophies of Romanticism and Positivism by two sociologists, whose names are associated with these efforts in twentieth-century American sociology. The first and most influential of these is the Harvard University Romantic sociologist, Talcott Parsons. This section does not describe any discovery system designed by Parsons. In fact his Romantic philosophy of science is very uncongenial to such modern ideas as computerized discovery systems, but his philosophy is still widely practiced and is enforced by the editors of the periodical literature of American academic sociology. This overview of Parsonian Romanticism is included to explain its inherent long-standing hostility to artificial intelligence.

Talcott Parsons (1902-1979) was a professor at Harvard University from 1927 until his retirement in 1973. He wrote an intellectual autobiography, "On Building Social System Theory", in *The Twentieth-Century Sciences* (1970). He had majored in philosophy at Amherst University, where he was also influenced by the institutionalist economist, Walton Hamilton; he studied under the anthropologist, Bronislaw Malinowski, at the London School of Economics, and he received his doctorate from the University of Heidelberg, where he was influenced by the views of Max Weber of Heidelberg, even though Parsons attended Heidelberg after Weber's death. Parsons' principal work is his *Structure of Social Action: A Study in Social Theory with Special Reference to a Group of Recent European Writers* (1937), an eight-hundred page book in two volumes that examines the social theories of four writers: Alfred Marshall, Vilfredo Pareto, Emile Durkheim, and Max Weber. This *magnum opus* is a historical study in philosophy of social science. Its thesis is that social theory has evolved beyond Positivism by an "immanent" process of development within the body of social theory, and that the outcome has been a convergence to a type of theory that Parsons calls the "voluntaristic theory of action". The voluntaristic theory of social action encompasses its own philosophy of science, which has evolved with it and which in turn describes the evolution of the voluntaristic theory of action set forth in the book.

The principal figure among the four social theorists considered is Weber, whose social theory and *verstehen* philosophy of scientific criticism is represented in Parsons' work as part of an immanent development culminating in Parsons' own voluntaristic theory of action. In the present context what Weber said is of less importance than what Parsons understood and rendered Weber as having said, since it was Parsons who was the principal influence on American sociologists. In summary Weber starts with the concept of action, which he defines as any human attitude or activity, to which the actor or actors associate a subjective meaning. "Social action" in turn is action, which according to its subjective meaning to the actors involves the attitudes and actions of others, and is oriented to them in its course. Finally, sociology is the science which attempts the interpretative understanding, i.e. *verstehen*, of social action, in order to arrive at a causal explanation of its course and effects. The *verstehen* explanation is in terms of a motivation, which he defines as a meaning complex which to the actor or to the observer appears to be an adequate ground for his attitudes or acts. A correct causal interpretation of action is one in which both the outward course and the motive are correctly grasped, and in which their relation to each other is "understandable" to the sociologist in the sense of *verstehen*. The object of *verstehen* in Weber's methodology is to uncover the motivations that cause action.

This philosophy of science is Romantic in two respects: Firstly it requires that the language of explanation contain vocabulary that references an ontology consisting of subjective experiences of the social actors, and it defines the term "theory" in social science specifically as language describing this ontology. Secondly it requires the *verstehen* or introspectively based "understanding" of the motives described by statements referencing this ontology, as a criterion for scientific criticism, and defines "causal explanation" in terms of this *verstehen* imputation of subjective motives for observed behavior. The requirement of *verstehen* may be called a strong version of the Romantic philosophy of social science, since some Romantic social scientists accept a weaker version, in which social science explanation has the subjective ontology but is not required to satisfy the *verstehen* criterion, because the *verstehen* explanations based on the social scientist's empathy have been known to differ widely from one social scientist to another. Some Romantic social scientists who accept the weaker thesis do not believe that the social scientist should have to be able to find an explanation convincing by reference to his own personal or imaginatively vicarious experience. Historically the philosophy of science that evolved in reaction to the Romantic philosophy is Positivism. The Positivist (or Behaviorist) philosophy of science requires the exclusion of the subjective experience required by Romantic philosophy, and either redefines the meaning of "theory" to exclude any mentalist ontology or more typically just forbids all "theory". Finally the contemporary Pragmatist philosophy of science, which has evolved as a criticism of Positivism after the Second World War, rejects the thesis common to both the Romanticist and the Positivist philosophies, that ontological considerations either must or may not function as criteria for scientific criticism, and it defines "theory" by reference to its function in empirical testing rather than to any ontology.

Now consider the Parsonian neo-Weberian Romantic philosophy of science in greater detail. Weber's philosophy of social science is a variation on the distinction between natural science and social science, that originated with the Kantian philosophical separation of the phenomenal and noumenal domains, and that gave rise to the Hegelian historicist view of explanation. Unlike the German Historicists, however, Weber does not reject the use of universal laws in social science. He notes that in practical daily social life people use generalizations to make reasonably reliable predictions of the reactions of other persons to a given situation, and that they succeed by imputing motives to men, by "interpreting" men's actions and words as expressions of motives. He maintains that social scientists similarly use their access to this subjective aspect of human action, and that this access carries an immediate evidence or certainty. The natural and social sciences, therefore, differ in that the former rely on observation of external regularities or *begreifen*, while the latter have the benefit of the internal or subjective knowledge of subjective motives or *verstehen*, which are not present in the sense data of events considered in natural science. Weber postulated different aims for the natural and social sciences. On Weber's view the aim of natural science is the formulation of universally applicable general laws, while the aim of social science is description of the individual uniqueness of an actual or possible historical individual. Weber thus views social science as a historical science, while also admitting its use of general laws. Parsons rejects this correlation of natural and social science to the analytical and the historical respectively; he maintains that both natural and social science are analytical. Also in Weber's view there is a selectivity that every scientist brings to his subject, and he says that this selectivity is determined by the interest of the scientist; the basis for selectivity is the relevance of the subject matter to the values of the scientist. Furthermore, it may be noted that Weber maintains that this value relevance is not the same as value judgments, and that scientific criticism is objective. While recognizing Weber's thesis of value relevance, Parsons says that Weber did not lay sufficient emphasis on the fact that what is experienced is determined by a conceptual scheme, and that conceptual schemes are inherent in the structure of language. It might be said that Parsons thus anticipated in important respects the contemporary Pragmatist theory of observation two decades before the Pragmatist philosophers took it over from the physicists. Parsons says that the principle of value relevance applies to both natural and social sciences making them both analytical instead of historical sciences, and that the difference between the two types is therefore only in their subject matter and not in their logic.

While Parsons may have anticipated the contemporary Pragmatists' philosophy of observation, he had nothing like their metatheory of evidence. He notes that for Weber *verstehen* is not just a matter of immediate intuition; Weber subordinates the immediate evidence from *verstehen* to other considerations: *verstehen* must be "checked" by reference to a logically consistent system of concepts, which Parsons says is equivalent to the situation in the natural sciences, where immediate sense perception of natural events must be incorporated in a system of theoretical knowledge, because what is experienced is always determined by the general conceptual schemes that are already developed. Parsons says that subordination of *verstehen* to a conceptual scheme precludes uncontrolled allegations, and he affirms that Weber had a very deep and strong ethical feeling on this point. Parson's neo-Weberian Romanticism has had a retarding influence in sociology. The editorial practices prevailing today in the American academic sociological journals is that each Romantic sociologist functioning as a referee uses this subjective criterion of "meaningfulness" to advance his own conceptual schemes and to suppress publication of alternative schemes proposed by other sociologists. The result has been an caricature of scientific criticism, that employs a fantisizing wizardry exhibiting such disregard for empirical evidence, that it could have startled even Baum's grand illusionist, the Wizard of Oz. Today the iconoclastic Pragmatist philosopher of science draws back the curtain of self delusion and exposes the Romanticists' "mechanisms".

Weber also takes up the question of how to establish the existence of a validly imputed causal relationship between certain features in the historical individual case and the empirical facts that existed before the historical event. His procedure involves the practice of historical revisionism by means of thought experiments, in which historical events are viewed as cases to which general laws may be applied. Weber calls these cases "ideal types". He lays down as a principal criterion for the correct formulation of an ideal type, that the combination of features used in it should be such that taken together they are meaningful, that they "make sense". Parsons explains this to mean that they must adequately describe a potentially concrete entity, an objectively possible case, in terms of the action frame of reference. Two types of laws are involved in this process, both of which may occur in either the natural or social sciences; they are empirical generalizations and analytical laws. The problem of adequate causal explanation in social science is one of adequate causal imputation to make analytical laws and also involves the relation of empirical generalizations to analytical laws. In social science the elements related by the general laws may be ideal-type units, such as bureaucracy, or they may be more general theoretical categories, such as the rationality of action. The statements of general law which relate these elements may be either empirical generalizations or analytical laws. The former laws are judgments of the probable behavior under certain given circumstances of the type element. The latter are statements of general modes of interaction among the elements and are known by *verstehen.* Interestingly Parsons says that it is perfectly possible for adequate judgments of causal imputation to be arrived at in terms of type units and empirical generalizations alone, i.e. without *verstehen.* But as historical cases become more complex, adequacy of explanation may require resort to more explicit formulations of the cases as ideal types containing ideal-type units related by *verstehen.* But if this approach in turn is not adequate, it may become necessary to resort to more generalized theoretical categories and laws. The less general statements are not dispensed with in this progression from empirical generalizations to analytical laws to more general analytical theory, but the analytical laws serve as an important check on the formulations of the empirical generalizations. Parsons says that the degree to which it is necessary to push forward from empirical generalizations to analytical laws in order to attain adequate explanation, is relative to the given empirical problem at hand. He says that this process may involve probabilistic judgments, when it is necessary to make a very complex judgment of causal imputation, as in the relation of the Protestant ethic to modern capitalism. The historical individual, such as capitalism, must be analyzed into a large number of type-units, each of which is subjected to judgments of probability as to its line of development under the relevant circumstances. In this probabilistic sense Weber speaks of adequacy, when the great majority of the causally relevant type units, such as the Protestant ethic, that might have influenced a given historical individual are favorable to the particular thesis about its development.

Parsons advances his own methodological thesis including an architectonic scheme for the sciences based on his own ontological thesis. Throughout the book he opposes the "reification" of any particular analytical theory, and particularly the reification by Positivists of either classical physics or classical economics. He considers reification to be fallacious and objectionable because it is a "monistic" realism, which requires that all realistic scientific theories be reduced to one if they are not to be regarded as fictional. Parsons proposes his own ontological thesis, which he calls "analytical realism", according to which the general concepts of science are not fictional but adequately grasp aspects of the objective external world. Some earlier philosophers had called this type of realism "perspectivism". This is the realism he affirms for those concepts in analytical laws that are ideal-type units, concepts that he calls analytical elements and that Weber had regarded as fictional. Parsons consequently rejects any reductionist view of the relation between natural and social sciences and explicitly affirms an organicist thesis of emergent properties. This emergentism is the consequence of value relevance, and it is the basis for the frame-of-reference thesis and for Parsons' archetonic for the sciences. Parsons identifies three reference frames that he calls the three great classes of theoretical systems: the systems of nature, the systems of action, and the systems of culture. Parsons says the first two pertain to processes in time and are therefore empirical, while the systems of culture pertain to eternal objects such as art forms and ideas. Examples of sciences of culture are logic, mathematics, and systems of jurisprudence, and Parsons chooses not to consider this type in his book. The empirical analytical sciences are divided into natural sciences and sciences of action. The latter are distinguished negatively by the irrelevance of the spatial frame of reference, and positively by the indispensability of the subjective aspect, i.e. *verstehen*, which is irrelevant to the natural sciences.

The action frame of reference is fundamental to social sciences. It consists in the irreducible framework of relations among analytical elements consisting of ideal-type units and is implied in the conception of these units. Common to all theoretical systems or sciences sharing the action frame of reference are structural elements consisting of ends, means, conditions, and norms. In the relations there is a normative orientation of action and a subjective point of view. These considerations are as basic to the action frame as the space-time aspect is for the framework used for physics. The sciences of action include the social sciences, which Parsons subdivides into economics, politics and sociology, according to three defining emergent properties. The defining emergent property for economics is economic rationality, that for politics is "coercive rationality", and that for sociology is "common-value integration" which Parsons finds in the works of the four authors examined in his *Structure of Social Action*. Thus he defines sociology as the science that attempts to develop an analytical theory of action systems, in so far as these systems can be understood in terms of the property of common-value integration. These defining properties are emergent, because an attempt to analyze the system further results in the disappearance of these properties. Neither economic rationality nor common-value integration is a property of unit acts in an action system apart from their organic relations to other acts in the same action system, and the action system furthermore must be adequately complex so these properties can be observed. Consider further Parsons' ontology: Parsons says that value relevance applies equally to both social and natural science, and he rejects any implication of complete relativism by the thesis of value relevance. Following Weber he limits relativism to specific modes of its application within the action frame of reference and he excludes it from applying to the action frame itself. The reader will note that this exclusion is a completely *ad hoc* limitation. Furthermore Parsons maintains that all different conceptual schemes proceeding from different values or interests must be translatable into one another or into some wider scheme, so that the whole position is not overthrown by skepticism. This too is *ad hoc*; the history of science does not reveal such reductionism, and it is not implied by Parsons' analytical realism. Parsons is unprepared to accept the contemporary Pragmatist ontological relativity and theoretical pluralism, because he thinks that such a view implies skepticism. He says that the development of scientific knowledge is to be regarded as a process of asymptotic approach to a limit, which can never actually be achieved.

In 1951 Parsons published his principal contribution to theoretical sociology, the *Social System*. This work is his implementation at a rather abstract level of the *verstehen* procedure of causal explanation, the vicarious imputation of motivations for social action. In the *Social System* he calls this implementation of *verstehen* "motivational analysis" and "dynamic analysis". Motivated behavior is action that is oriented to the attainment of gratifications or to the avoidance of depredations according to the actor's expectations as defined by the value system in the social culture. Parsons thus sets forth his "fundamental dynamic theorem of sociology": the stability of any social system depends on the integration of a common value pattern into the motivating need dispositions of the personalities of the members of the social system. This integration is achieved by institutionalization. An institution is a cluster of interdependent role patterns, which are integrated into the personalities of the social members by motivational processes or "mechanisms" called socialization. And tendencies to deviance from these role patterns are counteracted by mechanisms called social control. These integrating mechanisms of socialization and social control produce tendencies to social equilibrium. The motivational processes operate to create and maintain social structures such as roles and institutions, and these structures in turn operate to satisfy the functional prerequisites of the social system. Parsons identifies four basic institutional role clusters, which have associated collectivities of social members, and which have their basis in four corresponding functional prerequisites for a social system. They are: (1) the family, which functions to control sex relations and to perform the socialization of new members, (2) the economy, which functions to organize the instrumental achievement roles and the stratification of the society, (3) politics, which functions to organize the roles pertaining to power, force, and territoriality, and finally (4) religion, which functions to integrate value orientations with cognitive orientations and personality. Parsons refers to his sociological theory as structural-functional. The motivational dynamics induces voluntary conformity to prevailing role patterns and thereby produces a tendency to social equilibrium. Changes produced by this tendency are changes within the existing structures of the social system. But there are also changes of the structures of the social system, which is referred to by the phrase "social change". Parsons says that a general theory of the processes of change of social systems is not possible at present, because such a theory would require a complete knowledge of the laws of the motivational processes of the system. He therefore says that the theory of change of the structure of social systems must be a theory of particular subprocesses of change within such systems, and not of the overall processes of change of the system as a system. And in this context he affirms that while it is possible to have knowledge in the form of empirical generalizations that certain changes do in fact occur under certain conditions. But he still maintains that an action theory of social change must include the motivational analyses, and may not merely be a system of empirical generalizations.

Parsonian Romantic sociology has not been without its critics. Not surprisingly the science that was founded by the founder of Positivism, namely Auguste Comte, has offered new Positivist critics to oppose this new variant of Romanticism. The principal protagonist in this critical role, who was contemporary to Parsons, was George Lundberg (1895-1966). As it happens, Lundberg's criticisms did not effectively persuade American sociologists, and post-World War II sociology took the Parsonian path. Nonetheless a brief rendering of Lundberg's criticism will describe the philosophy which for many years American academic sociologists viewed as their principal philosophical alternative to Parsons. Lundberg traces his philosophical heritage to Comte. In his "Contemporary Positivism in Sociology" in *American Sociological Review* (1939) Lundberg gives three quotations from Comte's *Positivist Philosophy*, that he says suggest the principal survivals and modifications of Comte's work that may be regarded as contemporary Positivism in sociology. The first quotation is a statement of the principal aim of science, according to which the business of science is to analyze accurately the circumstances of phenomena, to connect them in invariable natural laws according to the relation of succession and resemblance, and to reduce such laws to the smallest possible number. The second quotation sets forth a secondary aim of science, namely to review existing sciences to show that they have a unity of method and a homogeniety of doctrine. The third quotation affirms the importance of observation and rejects the view that the sciences of human behavior should attempt to study facts of inner experience. Lundberg finds himself at variance with Parsons, and he quotes anti-Positivist comments from a lengthy footnote in Parsons' *Structure of Social Action*, in which Parsons states that all Positivisms are untenable for both empirical and methodological reasons.

Lundberg wrote several methodological works. His principal philosophical work is a monograph of about one-hundred fifty pages titled *Foundations of Sociology* (1939), which includes his views set forth in a previous papers including one titled "Concept of Law in the Social Sciences" published in *Philosophy of Science* (1938). The 1964 edition of the *Foundations* monograph contains an "Epilogue" as a new chapter, in which Lundberg maintains that the Parsonian approach to sociology is converging toward the Positivist view. In 1929 he wrote *Social Research: A Study in Methods of Gathering Data*, which he extensively revised in 1942. In 1947 he wrote *Can Science Save Us?*, and in 1953 he co-authored *Sociology*, a textbook in seven parts with a methodological discussion constituting the first part, and that went through four editions.

Lundberg was very impressed by the successes of natural science especially in comparison to sociology, and he stated that the history of science consists largely of the account of the gradual expansion of the realms of the natural and physical at the expense of the mental and the spiritual. His agenda for sociology therefore is to realize success in sociology by imitating the methods of the natural sciences. The philosophical understanding of natural science during the time of his active career was the Positivist philosophy, which also prevailed in academic philosophy of science at the time. But the classical Machian Positivism implemented in the natural sciences with its phenomenalist ontology is not easily adapted to behavioral and social sciences, and Lundberg therefore developed his own Pickwickian Positivism for sociology. Lundberg's epistemological view has similarities to the classical British empiricists, Locke, Berkeley and Hume, and also to the early Positivists such as Mach. These philosophers started with the thesis that what the human mind knows immediately is its own ideas, sensations, or sense impressions. This is a subjectivist view that occasions the question of how the human mind knows the external or extramental real world. One answer to this problem is the copy theory of knowledge, according to which ideas reveal reality, since they are copies of reality. Another answer is that there is no external world consisting of material substances, such that the ideas themselves become reified, and the result is an idealist and solipsistic thesis, such as Berkeley's *esse est percipi*, "to be is to be perceived". Lundberg also has a subjectivist theory of knowledge, but he has his own solution to the problem of knowledge of reality. Lundberg maintains that the immediate data of all sciences are symbols, by which he means human responses to whatever arouses the responses. And he also calls these responses sensory experience. His subjectivist philosophy of knowledge is nonrealist, because it makes subjective experience instead of extramental reality an object of knowledge rather than making experience constitutive of knowledge. He then goes on to say that the nature of that which evoked these human responses must be "inferred" from these immediate data which are our sensory experience; we infer both the existence and the characteristics of anything from these responses. In his Positivism there are apparently some extramental realities beyond the phenomena. Furthermore this "inference" of the characteristics of reality is not a deductive inference, but consists of operationalist definitions. In his discussion of measurement Lundberg says that since Einstein, physicists have blatantly declared that space is that which is measured by a ruler, that time is that which is measured by a clock, and force is that which is measured by pointers across a dial. A thing is that which evokes a certain type of human response represented by measurement symbols. There is an ironic aspect to Lundberg's epistemological subjectivism, because he uses it to refute the view that the subject matter of social science is subjective, arguing that distinctions between what is subjective and what is objective is not given in the data. Thus objectivity is not given in things, but in those ways of responding that can be corroborated by other persons. He seems unaware that corroboration to establish objectivity or intersubjectivity is itself quite problematic for any subjectivist philosophy of knowledge.

The most distinctive aspect of Lundberg's version of Positivism is his rejection of the naturalistic philosophy of the semantics of language. In discussing quantification he rejects any distinction between natural and artificial units for measurement, and he denies that scientists measure the behavior of some things but not the being, quality or quantity of others. He argues that like physicists, sociologists must recognize that all units are artificial linguistic constructs symbolizing human responses to aspects of the universe relevant to particular problems. Lundberg also implicitly recognizes that his semantical view affirms an ontological relativity, when he says the human knower infers the "nature" of phenomena from his symbolic responses. But his ontological relativity contrasts with that of the contemporary Pragmatists, who are realists and who maintain that the human knower knows reality directly and not indirectly by any kind of inference. Lundberg's rejection of the naturalistic philosophy of the semantics of language absolves him from any need to characterize the observational basis of science. He thus evades a difficult problem for a social or behavioral science attempting to implement the phenomenalist thesis of the Positivist physicist or chemist. Human social behavior is not easily or productively described in terms of phenomenal shapes, colors, sounds, or other purportedly elementary sense data. In contrast the Vienna Circle sociologist, Otto Neurath, was *ad hoc* in his attempt to accomplish the same thing, when he simply announced that a "thing language" as opposed to a phenomenalist language is admissible in Logical Positivism. More importantly Lundberg's artifactual thesis of semantics is strategic to his agenda for rejecting the view that sociology has a distinctive subject matter, i.e. distinctive in its subjective nature, since human knowledge does not immediately apprehend the nature of things. But rejection of the naturalistic semantics undercuts Lundberg's agenda of eliminating vocabulary conventionally referencing subjective experience as opposed to observably objective behavior. His philosophy of the semantics of language does not admit the distinction he tries to enforce as a condition for a scientific sociology.

Lundberg offers several statements of the aim of science. In one statement he says that the primary function of all science is to formulate the sequences that are observable in any phenomena, in order to be able to predict their recurrence. In another he says that the goal of all science is the formulation of valid and verifiable principles as laws comprehending with the greatest parsimony all the phenomena of that aspect of the cosmos which is under consideration. He defines a scientific law in turn as a verifiable generalization within measurable degrees of accuracy of how certain events occur under stated conditions, and he defines a theory as a deductive system of laws. A central thesis in Lundberg's agenda for a natural science approach to sociology is that scientific law in social science means exactly what it means in natural sciences. He therefore rejects any distinctive type of scientific law based on *verstehen*, and he says that understanding in his sense is not a method of research, but rather is the end to which the methods aim. Lundberg's philosophy of scientific criticism is verificationist, and in his textbook he defined a law as a verified hypothesis.

Lundberg offers several statements on the nature of scientific explanation, the topic in which he is most fundamentally at variance with the Romantic sociologists. In one brief statement he says that something is explained or understood, when the situation is reduced to elements and correlations among the elements, which are so familiar that they are accepted as a matter of course, and curiosity is then put to rest. He defines an "element" as any component that is not in need of explanation or of further analysis Another of his statements is given in terms of his thesis of frames of reference. Problematic data are said to be explained when they are incorporated into previously established habit systems of response, which constitute frames of reference. When this is accomplished, the new observations are said to have "meaning" and to be "understood". Consistent with his rejection of naturalistic semantics he says that frames of reference are not inherent in the universe, but are pure constructions for our convenience. He states that the scientist's interest in a problem requiring a response defines the categories in terms of which he reports his experience. When he seeks an explanation, he seeks to associate data reporting the problematic experience with what he already knows, i.e. the familiar, described by his established habit systems of response, which is the relevant frame of reference.

The frame of reference Lundberg considers appropriate for a natural science of social phenomena is behaviorism. In his *Foundations* he references a passage from Robert K. Merton's "Durkheim's Division of Labor" in *American Journal of Sociology* (1934), a relatively early work in Merton's literary corpus, in which Merton states that on the Positivist thesis, which says that science deals only with empirical facts, a science of social phenomena becomes impossible, since it relegates to limbo all ends, i.e. subjective anticipations of future occurrences. Lundberg says that this view fails to recognize that anticipated ends in the sense of conscious prevision exist as words or other symbols to which the organism responds, just as it does to other stimuli to action. In the behavioristic framework words are entities that are just as objective as physical things. No relevant data, even those designated by such words as "mind" or "spiritual" are excluded from science, if these words are manifest in human behavior of any observable kind. Like most Positivists Lundberg is unaware that the meaning of "observable" is philosophically quite problematic. Later in his *Can Science Save Us?* (1947, 1961) he further comments about the word "motives" in relation to frames of reference. He says that it is a word used to designate those circumstances to which it seems reasonable to attribute an occurrence, and that therefore it can have different meanings depending on the frame of reference in which it is used. Lundberg believes that of all reference frames the scientific frame of reference has proved to be the most successful for human adjustment to the environment.

The type of explanation that he explicitly advocates for sociology is what he calls the "field" type, which he also calls relational and situational, and which he opposes to types that refer to unexplained innate traits of social agents. He compares the idea of field to the idea of space as it is used in geography and ecology. The geographer describes behavior in terms of symbolic indices such as birth rates, death rates, and delinquency rates, for a geographical region, and then he correlates these indices. The transition from an ecological map representing delinquency rates as gradients to an organizational or functional representation for sociology involves a transition from a geographical to a social "space" and from a pictorial to a more abstract symbolic representation such as functional equations relating measurements. In "Social Bookkeeping", the concluding chapter of his *Social Research*, Lundberg notes that national demographic statistics have routinely been collected, and that social scientists have made successful objective generalizations on the basis of these data. He maintains that quantitative sociological laws can be just as objective as demographic generalizations.

In the concluding "Epilogue" chapter of the 1964 edition of his *Foundations* Lundberg describes similarities between Parsons' sociology and that of Stuart Dodd. Lundberg takes Dodd's work to be exemplary of the natural science approach in sociology. Dodd was chairman of the Sociology Department at the American University in Beruit, Lebanon. Dodd describes his *Dimensions of Society: A Quantitative Systematics for the Social Sciences* (1942) as a "companion volume" to Lundberg's *Foundations,* which Dodd reports he had sent to Lundberg for prepublication criticism. This distinctive book and its sequel, *Systematic Social Science: A Dimensional Sociology* (1947), set forth a social theory called the *S*-theory, which implements Lundberg's philosophy of science. Dodd's 1942 text contains a distinctive notational system for elaborately describing social "situations" in terms of four "dimensions": the demographic, the cultural, the ecological, and the temporal; and the 1947 text contains representations for eleven social institutions. But the symbols in this notational system serve principally as a kind of shorthand, and seem not to be subject to mathematical computation or transformation, as are theories in natural science. American sociologists did not accept Dodd's *S*-theory or his approach. However, even if the *S*-theory had been mathematical as is, say, Newtonian mechanics or contemporary mathematical economics, the academic sociologists would not have accepted it anyhow, because they are too incompetent in mathematics to assimilate it.

Parsons and Lundberg offer surprising ironies in their attempts at philosophy of science. Each for reasons of his own surpassed the naturalistic thesis of the semantics of language that is common to both the Positivist and the Romanticist traditions in philosophy, and in this respect each had surpassed the academic philosophers of science who were contemporary to them in the 1930's and 1940's. Both of them affirm the artifactual thesis of semantics, the view that the semantics of language is a cultural artifact rather than a product of nature. In this respect these social scientists enjoy the benefit of a professional perspective uncommon at the time to the academic philosophers preoccupied with the philosophy of physics. Ironically, however, neither Parsons nor Lundberg exploited the implications of their philosophically superior view of semantics, because each brought his own agenda to his ersatz philosophizing efforts, which in each case is incompatible with the artifactual-semantics thesis and the realistic epistemology.

Lundberg arrived at his artifactual-semantics thesis at the expense of realism, because he carried forward a subjectivist epistemology from the Positivist philosophy. And his fidelity to Positivism cost him any basis for the objectivity that he thought justifies his natural-science agenda for social science. Historically the Positivist basis for objectivity with the subjectivist epistemology is the naturalistic-semantics thesis of language. The copy theory of knowledge is an old example of a strategy for objectivity with the subjectivist phenomenalist epistemology. Bridgman's operationalist definition is a more contemporary case, which ironically Lundberg calls upon as the basis for his view that the gap between the subjective responses constituting sensory experience and the objective real world is mediated by an inferential process consisting of operationalist definitions. Lundberg may not have realized that both operationalist definitions and the Positivist concept of observation are based on the naturalistic view of semantics.

Parsons arrived at his artifactual-semantics thesis in a more sophisticated manner, when he said that all observation is in terms of a conceptual scheme, and when he said that there is a relativity or selectivity in the conceptual scheme resulting from the value relevance or interest of the scientist. This relativism is consistent with the artifactual-semantics thesis, and is not consistent with the naturalistic-semantics that says the information in concepts is absolutely fixed and predetermined by nature. Furthermore Parsons' approach to the artifactual-semantics thesis is consistent with his realistic epistemology, which he calls "analytical realism". Analytical realism enables scientific observation to describe aspects of the real world with a semantics supplied by the value-relevant conceptual scheme. In these noted respects Parsons' philosophy of science is truly post-Positivist, as he had claimed. But there is a problem, which he attempted to finesse: the artifactual-semantics thesis cannot support his agenda for a voluntaristic theory of social action. This agenda requires a naturalistic-semantics thesis, that would enable Parsons to say that such aspects of reality as ends, norms, or motives are not observable in human behavior, but are causes that must be supplied by imputation by the social scientist by reflection on his own experience, that is by *verstehen*. In order to implement his agenda, Parsons says that the relativism introduced by value relevance obtains within the frames of reference for the natural sciences and for voluntaristic action, but does not obtain between them; and on this basis he distinguishes empirical generalizations about human behavior appropriate for natural sciences from the "analytical laws" appropriate to the action frameworks formed by *verstehen*. This thesis is a completely *ad hoc* addendum that is inconsistent with the artifactual-semantics thesis for language.

The claim made by Parsons that ends, norms, and motives are not observable is erroneous, and it is not erroneous because of behaviorism, as Lundberg maintains. Contrary to Lundberg behaviorism is also dependent on a naturalistic-semantics thesis of language. It is erroneous because, as Parsons says, all observation is in terms of a conceptual scheme, and this means that there is an intellectual component in observation supplied by the linguistic context constituting the conceptual scheme. Contemporary Pragmatists, such as Hanson, have expressed this by saying that observation is "theory-laden". Einstein asserted the same thesis when he told Heisenberg that theory decides what the physicist can observe. The electron was observable by Heisenberg in the Wilson cloud chamber because his quantum theory supplied the conceptual scheme that supplied elementary intelligibility for the phenomenon to be observably recognizable as an electron's track. Similarly the *verstehen* interpretation supplied by the Romantic sociologist is no less contributing to the semantics of the language describing observed human behavior than the quantum theory is to the semantics of the observation report of the vapor tracks in the Wilson cloud chamber. Parsons noted that Weber required that the causal imputation by *verstehen* be checked by reference to a logically consistent system of concepts, which Parsons says in equivalent to the situation in the natural sciences where immediate sense perception must be incorporated into a system of theoretical knowledge. On the Pragmatist view, however, it is the whole theoretical system of beliefs including the *verstehen* analytical laws that is "checked" by empirical testing.

Both Weber and Parsons seem to have failed to see that there can be no requirement for the *verstehen* concept of causality in the sciences of human behavior, just as there is no requirement for the Newtonian or Aristotelian concepts of causality in physics. Weber's and Parsons' attempt to impose such a requirement as a condition for causal explanation in social science, is now recognized to be a fallacy: the fallacy of demanding ontological criteria for scientific criticism. On the contemporary Pragmatist philosophy of science only empirical criteria may operate in scientific criticism. The artifactual-semantics thesis makes all ontologies as dispensable as the empirical theories whose semantics describe those ontologies, and it makes all theories subject only to empirical criticism without regard to how improbable or counterintuitive empirically adequate theories may seem to the individual scientist.

The third and last sociologist considered here is Robert K. Merton (1910-), who studied under Parsons at Harvard University, where he received his doctorate in sociology in 1936, and who later was appointed chairman of the department of sociology at Columbia University. His dissertation, *Science, Technology, and Society in Seventeenth-Century England*, marked the beginning of his career-long interest in sociology of science. His papers in sociology of science written and published between 1935 and 1972 are reprinted in his *Sociology of Science: Theoretical and Empirical Investigations* (1973). While Merton's interest in science is noteworthy, his views in sociology of science are beyond the scope of this history. Here the focus of interest is Merton's *Social Theory and Social Structure* (1949, 1968), where he departs from Parsons' Romanticism with his own rendering of the functionalist type of explanation for sociology, and develops his own concept of scientific theory. He believes that functional analysis is the most promising yet the least codified of contemporary orientations to problems of sociological interpretation. He does not claim to have invented this type of sociological explanation, and he offers several examples of it in the literature of sociology; he says that his major concern in this book is its "codification" by developing a "paradigm" for it. He notes that sociologists often use the term "function" as it is used in mathematics to describe interdependence, but he is not thereby proposing a mathematical type of sociological theory. In fact he explicitly states that his purpose is to codify the procedures of qualitative analysis in sociology.

Merton says that the concept of social function refers to observable objective consequences and not to subjective dispositions such as aims, motives, or purposes, and that the consequences of interest are those for the larger structures in which the functions are contained. The concept of function involves the standpoint of the observer and not necessarily that of the participant. He says that failure to distinguish between the objective sociological consequence and the subjective disposition inevitably leads to confusion. This is because the subjective disposition may but need not coincide with the objective consequence, since the two may vary independently. This concept of functional analysis occasions Merton's distinction between "manifest" function and "latent" function. Manifest functions are those that have objective consequences contributing to the adjustment and adaptation of the social system, and which are intended and recognized by the participants in the social system. Correlatively latent functions are defined as those objective consequences contributing to the adjustment or adaptation of the social system, and which are not intended or recognized by the participants in the social system. As an example Merton says that criminal punishment has manifest consequences for the criminal and latent functions for the community.

Merton describes heuristic purposes for his distinction between manifest and latent functions. The distinction not only precludes confusion between motive and function, which he notes may be unrelated to each other, but it also aids the sociological interpretation of many social practices, that are regarded by observers as merely ignorant "superstitions", yet still persist even though their manifest purposes are clearly not achieved. And it also directs the sociologist's inquiries beyond the manifest or intended aspects of behavior to discover its generally unrecognized consequences. Merton thus affirms that the discovery of latent functions represents significant increments in sociological knowledge, because they represent greater departures from "commonsense" knowledge about social life. This is more philosophically sophisticated than the *verstehen* requirement that hypotheses "make sense". Furthermore he notes that the concept of latent function has significance for social policy or social "engineering". He sets forth a basic theorem, which may be called Merton's theorem of social engineering; it says that any attempt to eliminate an existing social structure without providing adequate alternative structures for fulfilling the functions previously fulfilled by the abolished organization is doomed to failure. More generally Merton's theorem says that to seek social change without due recognition of the latent functions performed by the social organization undergoing change, is to indulge in social ritual rather than social engineering.

Merton's thesis of latent functions reveals the inadequacy of the Parsonian Romantic concept of theory based on motivational analyses, and Merton recognizes that a new concept of theory is needed. But he does not adopt the Positivist's complete rejection of Romanticism. In his discussion of functionalism he says that in preparation for developing a functionalist explanation a fully circumstantial account of meanings, i.e. the cognitive and affective significance attached to a behavior pattern, goes far toward suggesting appropriate lines for a functional analysis. Had he been less sympathetic to the Romantics, he might have followed through to the conclusion that the distinction between manifest and latent functions contributes nothing to the explanatory value of the functionalist explanation, since its explanatory value consists not in a functional factor being either manifest or latent but in its being consequential for other factors to be explained. And this implies that the manifest-latent distinction is informative only for Romantics, who need to be told that motivational analysis is not adequate for explanation in social science, except as one among many possible heuristic devices for developing functionalist hypotheses.

Merton's attack on Parsonian sociology is not a frontal assault on Romanticism, but is part of his own agenda for sociological research. The attack is directed explicitly at the all-inclusive type of system building practiced by many sociologists including notably Parsons. His principal objection to these all-inclusive systems is that they are too vague to be tested empirically, and he refers to them as general orientations toward sociological analysis rather than "theories". The agenda that he advocates for future research in sociology is the development of what he calls "theories of the middle range", theories that he says are somewhere between minor but necessary empirical generalizations or working hypotheses on the one hand and the Parsonian-like all-inclusive systems on the other. Unlike the Romantics, who define theory in terms of the semantics of a vocabulary referring to subjective meanings and motives of social actors, Merton defines theory in terms of its logical structure. He explicitly defines "theory" for both natural and social sciences as a logically interconnected set of propositions from which empirical generalizations can be derived. In another statement he says theory is a set of assumptions from which empirical generalizations are derived. And referencing Lundberg's "Concept of Law in the Social Sciences" he says a scientific law is a statement of invariance that has been derived from a theory. He distinguishes theory from the empirical generalization saying that the latter is an isolated proposition summarizing observed uniformities of relationships between two or more variables. In the history of science there have been significant single-equation theories, such as Newton's theory of gravitation. But Merton does not state explicitly whether or not he intends by his definition to exclude from the domain of theory language the single-equation theories that are found in many sciences.

Referencing Whorf, Merton notes that the empirical researcher's perceptions are fixed by his conceptual apparatus, and that the researcher will draw different consequences for empirical research as his conceptual framework changes. However, Merton does not seem to recognize that this control of language over perception undermines his distinction between theory and empirical generalization, since this semantical control operates by the linguistic context of empirical generalizations, which means that empirical generalizations are never isolated. His distinction is therefore unsustainable. Had he approached this problem by an analysis with an adequate and contemporary philosophy of language, he might have seen that his distinction incurs the same difficulty that both the Romantics and the Positivists encounter, when they purport to distinguish theory from a semantically isolated observation language. The semantics of observational description is not isolated from that of theory; semantics, logical syntax, and belief are interdependent. The only sustainable basis for distinguishing theory from nontheory language is the pragmatics of language, the functions it performs in basic research. As it happens, Merton comments on the functions of theory for empirical research. But his comments presume his distinction between theory and empirical generalizations, and is not definitive of a distinction between theory and nontheory language. Furthermore his list of functions are not applicable to the modern quantum theory, and more generally are not sufficiently universal in the practice of scientific research to serve as defining characteristics of theory language. On the contemporary Pragmatist philosophy of science the only characteristic that distinguishes theory from nontheory language is that the former is proposed for testing, while the latter is presumed for testing.

It may be noted here by way of a postscript to this discussion of Merton, that some economists also recognize what Merton calls "latent functions", even if the economists have no particular name for it. Nobel laureate economist Milton Friedman's "Methodology of Positive

Economics" (1952), reprinted in his *Essays in Positive Economics* (1953), is one of the more popular methodological papers written by an economist for economists in the post-World War II era. A contemporary philosopher of science would likely view this paper as an effort to de-Romanticize neoclassical economics. Although this paper sets forth a somewhat naive semantical thesis, its semantical metatheory is more sophisticated than the neo-Positivist view in his *Theory of the Consumption Function*, and "positive economics" here does not mean Positivist economics. Like the Pragmatists, Friedman says that the only relevant test of the validity of a hypothesis is comparison of its predictions with experience; he thus accepts no ontological criteria in his view of scientific criticism, including the Romantics' mentalistic criteria involving descriptions of motivations. He explicitly rejects objections to the rationality postulates or to any other assumptions employed by economic theory, including the objections of the Institutionalist economists, when they are not based on the predictive performance of the theory. For example he notes that businessmen do not actually calculate marginal cost or marginal revenues and solve a system of simultaneous equations as do economists, and that businessmen seldom do as they report when asked about the factors affecting their decisions. But Friedman says that businessmen must act *as if* they have compared marginal costs and marginal revenues, because they will not remain in business if their behavior is not consistent with the theory of rational and informed maximization of returns. Put in Merton's terms: it is the functional consequences that are relevant, and the motives are latently functional when their unintended consequence is satisfaction of the marginalist conditions.

Hickey and the American Romantic Sociologists

In 1976 Hickey used his **METAMODEL** discovery system to develop a macrosociometric theory of the American national society with historical time series data describing fifty years of American history. In order to display the Romanticist philosophy of science that still prevails in American academic sociology, this section firstly summarizes Hickey's functionalist macrosociometric theory, and secondly examines the responses of the editors and referees of four academic sociological journals to which Hickey had submitted his paper. The editors of all four sociological journals refused to publish Hickey's paper. Hickey has retained all the original correspondences from these editors. In his paper Hickey described his discovery-system generated macrosociometric model as a "quantitative functionalist theory of macrosocial change", and he contrasted it with Parsons' structural-functionalist approach, which Hickey called "classical functionalism". Classical functionalism is a social-psychological theory in the Romantic tradition, and is concerned with the institutionalization of patterns of value orientations. It explains social order and stability by the analysis of motivational processes or "integrative mechanisms" of socialization and social control, which integrate social actors' need dispositions into cultural patterns that always include value orientations. Parsons called this process of integration the fundamental dynamic theorem of sociology and the core phenomenon of the dynamics of the social system. When this stability or equilibrium is extended throughout the macrosociety, the result is called a consensus equilibrium.

Hickey noted that the classical functionalist theory does not explain social change. As Parsons stated, the institutionalization of cultural-value orientations by these integrative mechanisms relate to social change only as forces of resistance except to the degree that the macrosociety is described as malintegrated, in which case these mechanisms legitimate deviant behavior. Hickey therefore maintains that Parsons' paradigm of motivational processes is not an adequate theoretical basis for the analysis of macrosocial change, and he rejects the reductionist dogma that a macrosociological theory must be built up from a social-psychological or microsociological analysis of motivational processes, as had been envisioned by classical functionalists. Hickey also invokes the definition of functionalism in terms of consequences and references the distinction between manifest and latent functions, proposed by Merton, to enable sociological theory to explain the unintended and unforeseen consequences of social actors' behavior that cannot be explained in terms of conscious motivational processes. Furthermore as a Pragmatist, Hickey maintains that sociological theory permits but does not require a Romantic ontology for a valid sociological explanation. Accordingly while his basis for his selection of distinctively sociological variables in his theory is that these variables reference the fact that the behavior of the social actors is voluntary group associational behavior that reveals cultural values distinctive of particular social institutions, he admits to no requirement that he structure his equations on the basis of any postulated motivations, nor does he admit to the more extreme demand of some Romantics that such motivations be empathetically based and imputed to social actors in accordance with the *verstehen* criterion for sociological explanation and criticism.

Hickey's approach is not only an alternative to the reductionist social psychology of classical functionalism that still prevails in sociology, it is also an alternative to attempts to reduce demographic and sociological phenomena to economic motives, as has been proposed by Nobel laureate economist, Gary S. Becker, a University of Chicago professor of economics and sociology. Becker is a Romantic economist, who employs the neoclassical economists' rationality postulate to explain such sociological phenomena as marriage, divorce, education and crime, and also demographic phenomena. The theory of choice approach based on rationality postulates assumes the calculating attitude, which sociologists contrast to the attitude of respect and of voluntary conformity that is institutionalized by socialization and social control. Furthermore, while Hickey's approach is related to the institutionalist tradition in economics, he goes beyond institutionalist economics to include consideration of all of the five basic institutions in the macrosociety, and he models the functional relations among them. Hickey uses the distinctively sociological perspective for the explanation of these sociological phenomena; it is not reducible to neoclassical economic theory, nor is it merely an extension of institutionalist economics. Hickey does not exclude or reject the effects of economic conditions from his explanatory equations; in fact he explicitly includes economic variables. Hickey proposes a distinctively macrosociological perspective that includes macroeconomic conditions.

The variables in Hickey's theory fall into two broad classes: (1) the institutional variables representing types of groups associated with each of the five basic social institutions, and (2) noninstitutional factors. The institutional variables are sociologically relevant, because they are *per capita* rates having numerators describing aggregate voluntary group-associational behavior, thereby making the *per capita* rates measures of voluntary consensus within the macrosociety. They measure degrees of consensus about the institutionalized cultural value systems distinctive of each of the five basic types of institutional groups. Consensus represents the extent of integration of the members of society about the institutional values, integration that is necessary for the type of group to function and to continue in existence. A decline in consensus about the institutional values results in an increase in the incidence of dissolution of the associated type of group. The five basic institutional groups are the family or domestic type of group, the church or religious type of group, the school or educational type of group, the business enterprise or economic type of group, and the law-governed macrosociety itself. The family group is represented by the *per capita* marriage rate, which is denoted in the equations of the theory by the symbol as *MR*, and also by the *per capita* divorce rate, denoted *DR*. These two variables describe new family formation and dissolution respectively. The religious type of group is represented by the *per capita* rate of religious affiliation, denoted *RA*. The educational institutional group is represented by the percent of seventeen-year olds who graduate from high school, denoted *HS*. The economic type of group is represented by the *per capita* new business enterprise formation net of voluntary dissolutions, denoted *BE*. The institution of government is not represented by any political group but by the reciprocal of the *per capita* homicide rate, denoted *LW*; this variable describes the rate of voluntary conformity with the minimum conditions codified into criminal law for membership in good standing in the macrosociety. Except for the divorce rate, increases in the *per capita* rates for all the institutional variables represent increased consensus about their distinctive cultural value systems.

The noninstitutional variables identify factors that have been proposed in the sociological literature as relevant to macrosocial change. Demographic change is represented by the crude birth rate, denoted *BR*. Technological innovation is represented by the *per capita* number of patent applications for inventions, denoted *IA*. Macroeconomic business cycle conditions, are represented by the *per capita* real income measured by the constant-dollar gross national product, denoted *GP*. Military mobilization during wartime is represented by the *per capita* number of armed forces active duty personnel, denoted *AF*. Mass communications media is represented by personal consumption expenditures *per capita* for newspapers, books, periodicals and cinema, plus business income to radio and television broadcasting firms, denoted *CM*. Ecological change is represented both for internal migration due to urbanization represented by the percent of the population living on farms, denoted *UR*, and for international immigration represented by the *per capita* rate of foreign immigration, denoted *IM*.

The data are drawn from the U.S. Commerce Department's *Historical Statistics of the United States* (1976) and annual issues of *Statistical Abstract of the United States*. The historical time series are from 1920 through 1972. Firstly the data are either aggregated into four-year periods

MACROSOCIOMETRIC MODEL OF THE U.S. NATIONAL SOCIETY

$$BR_t = 0.48 + 0.373*LW_t + 1.0279*MR_t + 0.928*CM_t$$
(0.0061) (0.079) (0.0314)
R-Squared 0.8688 Durbin-Watson 2.4321

$$MR_t = 0.82 + 0.638*GP_t + 0.015*AF_{t-1} + 0.495*BR_{t-2}$$
(0.0092) (0.0000) (0.0078)
R-Squared 0.9582 Durbin-Watson 2.0527

$$LW_t = 4.78 + 2.955*RA_t + 1.705*HS_{t-1} + 1.042*BR_{t-1}$$
(0.8509) (0.1263) (0.0333)
R-Squared 0.9156 Durbin-Watson 1.5671

$$HS_t = 1.55 - 0.343*LW_{t-2} - 0.341*BR_{t-2} + 0.396*GP_{t-2}$$
(0.0296) (0.0147) (0.0255)
R-Squared 0.9519 Durbin-Watson 2.2653

$$BE_t = 0.76 - 0.070*HS_{t-1} + 0.450*BE_{t-1} - 0.111*IA_{t-2}$$
(0.0027) (0.0030) (0.0006)
R-Squared 0.9861 Durbin-Watson 1.8646

$$IA_t = 5.05 - 2.519*RA_t + 8.450*UR_t$$
(0.07570) (2.5359)
R-Squared 0.8697 Durbin-Watson 2.5601

$$UR_t = 1.18 - 0.100*HS_t - 0.059*CM_{t-1} + 0.003AF_{t-1}$$
(0.0018) (0.0010) (0.0000)
R-Squared 0.9831 Durbin-Watson 1.3162

$$CM_t = 1.89 - 1.624*RA_t + 0.611*GP_t + 0.2506*GP_{t-1}$$
(0.2610) (0.0110) (0.0105)
R-Squared 0.9555 Durbin-Watson 2.6126

$GP_t = GP$ (exogenous)
$AF_t = AF$ (exogenous)
$IM_t = IM$ (exogenous)

Figure

before *per capita* rates are calculated or are four-year averages of *per capita* rates calculated by the source. Then these four-year *per capita* rates are transformed into period-to-period change rates to enhance sensitivity of the equations and eliminate collinearity, and then the change ratios are transformed into index numbers having the last historical period, which is the first out-of-sample forecast period, as the base period. The quantitative theory is a functionalist theory not only in the sense that it is expressed in mathematical functions, but also in Merton's sense of functionalism, because it describes the interdependence of the types of institutional groups and the consequences of their interaction for the macrosociety as a whole as represented by the whole system of equations. The mathematical model is a recursive, first-degree, higher-order difference equation system. In the *Figure* on the facing page each of the model's equations is displayed with its variance statistic beneath its estimated coefficient and with its coefficient of determination (the R-squared) and its Durbin-Watson statistic. The relative time periods for independent variables are represented in parentheses in the variable symbols; for example *t-2* represents the numeric value of a variable two periods earlier relative to the dependent variable. The dependent variables always have current time periods, which are not indicated explicitly.

This mathematically expressed macrosociological theory is used to examine social change in the U.S. national macrosociety with both a static and a dynamic analysis. Consider firstly the functionalist static analysis. The objective of a static analysis is to determine whether or not there is a stable equilibrium, that is, a solution in which the numeric value of each variable is the same for an indefinite number of successive time periods through which the model may be iterated. Since the numeric values are index numbers of change ratios of *per capita* rates, the equilibrium solution is one of constant change ratios of the *per capita* rates for all of the variables. However these constant change rates may be positive, zero, or negative.

Classical-functionalist-consensus equilibrium extended to macro-social scope is represented by constant *per capita* rates for all the institutional variables, and these *per capita* rates must be high if not actually near their maximum values to represent a high degree of consensus and macrosocial integration. Since constancy of such high *per capita* rates implies zero growth of the change ratios, nonexistence of an equilibrium solution consisting of zero change ratios implies the nonexistence of a classical consensus stable equilibrium. As it happens, examination of the equilibrium solution of the model reveals that a mathematically consistent zero-growth solution for all the institutional variables does not exist, and therefore reveals that consensus equilibrium is not possible. In classical functionalist terms this means that the interinstitutional system of cultural value orientations of the American national society is inconsistent or "malintegrated".

Consider next the dynamic analyses, which consist of iterating the model to examine its response properties. The findings from three simulation experiments were described in the paper. In the first experiment the model is iterated with all its exogenous variables assigned their index-number values set to represent zero growth of their *per capita* rates. The iteration propagates a time path, which oscillates with an increasing amplitude and with a phasing of seven four-year-period iterations, i.e. it generates an intergenerational twenty-eight-year cycle. Examination of the structure of the model reveals that the equations determining the change rates of the *per capita* birth and marriage rates are interacting to create the cycle, and therefore that the cycle reflects changes in the age composition of the population. The explosive instability, however, is likely due to the constancy of the *per capita* real aggregate income variable occurring in the marriage rate equation. The exogenous status of the economic sector means that there can be neither a dampening feedback on *per capita* real income from the exploding birth rates, nor any effect from productivity improvements resulting from the patents for inventions. Hickey maintains that any macrosociological model should be integrated into a model of economic growth, just as the contemporary Institutionalist economist will maintain that conventional neoclassical econometric models of economic growth are inadequate, because they are not integrated into a macrosociological model of institutional change.

In the second experiment and in all succeeding experiments this demographic cycle is eliminated by removing the birth rate equation from the model and by making the birth rate exogenously determined with an assigned constant index number value equivalent to zero growth. When the model thus modified is iterated, it generates a damped oscillating path which is also intergenerational, and which converges into a stable equilibrium of constant growth rates for all its endogenous variables including the institutional variables. The negative feed back producing the dampening effect involves an interaction between the equations determining the homicide rate and the high school graduation rate. A sustained decline in voluntary compliance with the minimum conditions for social order as codified into criminal law, i.e. an increase in the homicide rate, occasions in due course a corrective reaction that involves the socializing function of the educational institution.

However, the resulting growth equilibrium does not necessarily result in a movement toward consensus for all the types of institutional groups. Additional simulations reveal that only when the growth of the exogenous real *per capita* income is made to occur at a rate of no less than four and one-half percent compounded annually, does the equilibrium solution result in positive growth rates for all of the institutional variables excluding the divorce rate. This is the minimum annual rate of *per capita* real economic growth required for the dampening negative feed back to stabilize the national macrosociety in an equilibrium growth toward macrosocial consensus integration. But the static analysis revealed that this state of affairs can only be temporary, since the cultural value system is malintegrated. Furthermore, historically a four and one-half percent annual growth rate for real *per capita* income has not been sustainable by the U.S. macroeconomy, because it results in destabilizing inflation rates.

The third experiment consists of shock simulations in which the model is given an unrealistically large one-period increase after it is iterated sufficiently to settle into its equilibrium growth path. The limits imposed in reality by the *per capita* rates are disregarded in the simulations, to exhibit the dynamic properties of the model. In each simulation the shock consists of a permanent doubling of the *per capita* rate for one selected noninstitutional variable. In all but one case the shock propagates a damped oscillating path, that settles back into a stable equilibrium solution. The exceptional case is the *per capita* urban residence rate, and the outcome of the shock is a nonoscillating explosive destabilization of the macrosociety. In this latter case the equations for both the birth rate and the urban residence rate have been removed from the model, such that the population growth can not be accommodated by internal migration, and the macrosociety is disturbed beyond the stabilizing capacity of its interinstitutional integrative mechanisms.

In summary there are four findings from both the static and dynamic analyses. Firstly the interinstitutional cultural value system of the American national macrosociety is malintegrated, such that a macrosocial consensus equilibrium is not possible. Secondly if demographic cycles are impounded, the national society exhibits a nonsustainable tendency to macrosocial consensus when *per capita* real income grows at a minimum rate of four and one-half percent compounded annually. Thirdly this tendency to consensus equilibrium stabilizes in a growth equilibrium, because the interinstitutional cultural value system contains relationships that create intergenerational negative feed back integrative mechanisms, which involve a corrective reaction to social disorder and operates through the socializing functions of the educational institution. Finally internal migration, an ecological adjustment to population growth, is necessary for the institutional integrative mechanisms to be effective. Such is a summary of Hickey's paper.

The remainder of this section consists of Hickey's criticism of the referee reviews and of the consequent decisions by the editors of four sociological journals to reject the paper. In our society everyone has a right to comment and even criticize severely on matters of public interest and concern. The public that is relevant in this context is firstly the professional philosophers of science and the practicing research scientists in the empirical sciences including sociology; and secondly it is the general public and especially the public officials that they elect to public office, whose public policies may be influenced by papers that are published in these and other sociological journals. Contrary to these editors Hickey regards his paper as worthy of publication. His reports of the sociologists' attempts at scientific criticism are based on the correspondences from the editors, which Hickey has retained in his possession. This sample of seven referee criticisms selected by the editors of three academic sociology journals is not an aberration; it is a consistently representative core sample of academic sociology's institutionalized value system, and it has exposed a pervasive cultural dry rot, as it were, in a dysfunctional scientific profession. It is noteworthy that virtually none of these referee criticisms of Hickey's paper are empirical, but rather are attempted criticisms in philosophy of science. Hickey's basic criticism set forth herein is that the editors and the referees they selected, and indeed the American academic sociological profession at large, are ignorant of the contemporary Pragmatist philosophy of science, and are therefore incapable of distinguishing contrary evidence from contrary opinion. The purpose of Hickey's criticism is firstly to exhibit the institutionalized but philosophically anachronistic Romantic criteria used in attempted scientific criticism by academic sociologists today, secondly to critique these criteria from the perspective of the contemporary Pragmatist philosophy of science, and thirdly by this expose hopefully to educate the American academic sociologists and especially the editors of their journals.

Before constructing the macrosociological theory with his **METAMODEL** discovery system, Hickey undertook an extensive search of the academic sociological literature to determine what factors should govern his selection of apparently sociologically relevant time series data as inputs to the discovery system, and also to find some example of the writing style used in sociology for reporting findings from modeling analyses. In his literature search he could find no precedent for a dynamic macrosociometric model. Empirical work in sociology consisted of various survey research projects using either questionnaire or interview techniques, to examine social-psychological hypotheses. The first and most important consequence of this condition is that any sociologist selected by an editor to be a critic could not reference any previously published equation that is empirically superior to any equations explaining the time series data used by Hickey. From Hickey's contemporary Pragmatist perspective, which accepts only empirical criteria for scientific criticism, this circumstance means that empirical criticism of Hickey's models would have to be made exclusively in terms of the properties of his equations. As it happened none of the referees selected by any of the editors offered any valid criticism of the properties of any of Hickey's equations.

A second consequence of the unprecedented character of Hickey's macrosociometric model is that it reveals that academic sociologists are not educationally prepared to work with higher-order difference equation systems, such as those constituting Hickey's model. Hickey's professional education is in economics, and since the publication in 1939 of "Interactions between the Multiplier Analysis and the Principles of Acceleration" in *Review of Economics and Statistics* by Nobel laureate economist Paul Samuelson, such difference equations have become a staple technique in mathematical economics and econometrics. In fact the technique of shock simulations was introduced into economics in 1933 by the University of Oslo, Norway, economist Ragnar Frisch in his "Propagation Problems and Impulse Problems in Dynamic Economics" in *Economic Essays in Honor of Gustav Cassel.* But these techniques are not common in the curriculum of the sociologists, and even a sociologist who is deemed by an editor to be sufficiently reputable as to be worthy to function as a referee for his journal, is not likely to be professionally adequate to the techniques in a paper such as Hickey's. Thus, even the well respected sociologist may be suspicious of what he cannot cope with, and so may be dismayed by the findings drawn from the simulation and shock analyses.

When people do not know what to do, they do what they know, whatever it may be; and what the critics of Hickey's paper know is a reductionist, social-psychological Romanticism which requires *verstehen* criticism. Not only did these criticizing sociologists omit referencing any previously published contrary empirical findings that are relevant to Hickey's paper, they also used language that makes apparent their *verstehen* practice by the rhetoric and vocabulary in the stated reasons they set forth as criticisms: they criticized Hickey's equations because they do not "make sense", because they are "counterintuitive", "meaningless", "bizarre", "surprising", etc. This rhetoric together with their failure to reference previously published papers describing contrary empirical evidence makes apparent their practice of *verstehen* criticism. The referees selected by the editors to which Hickey sent his paper did not announce explicitly that they practice *verstehen* verification. But just as critics of physics papers do not announce that they practice empirical criticism, so too these sociologists simply go about practicing criticism unreflectively according to the institutionalized value standards that they had learned in their educational experience and that are approved by their colleagues.

Sociologists may use the *verstehen* for *a priori* criticism either before or after testing. It operates before testing when it controls discovery. The sociologist empathetically formulates on the basis of his own personal reality a hypothesis about the mental states that motivate the social actors' behavior that he intends to investigate by survey research, with the result that hypotheses that do not satisfy the *a priori verstehen* criterion are excluded from consideration for empirical testing. And *verstehen* operates after testing when the criticizing sociologist is confronted with a report of findings from another sociologist's empirical work, and when the report leads the criticizing sociologist to reject out of hand an unexpected empirical finding with which he cannot empathize on the basis of his personal or vicarious experience. The perspective in this discussion is philosophical: it examines the academic sociologists' institutionalized criteria for scientific criticism, the concept of evidence exhibited by the referees and enforced by the editors. The editor is the person who decides what will be published and what will not be published. The written correspondences supplied by the three editors show more than enough consistency among their journals, to reveal a widespread institutional dysfunction that pervades contemporary American academic sociology. That institutionalized dysfunction is the outcome of the use of the Romanticist mentalistic criteria for scientific explanation and criticism, of which *verstehen* is a distinctively subjectivist variation, and it consists in the consequent failure by these editors and their selected referees to distinguish contrary evidence from contrary opinion. The outcome of this failure to distinguish contrary opinion from contrary evidence in turn is to retard the advancement of the science by literally censoring its professional literature. In more productive sciences with more empirical standards, such dysfunctional dogmatism is readily recognized by editors of the professional literature and by the literature's readers. For example contemporary information science is interested in professional journals from the perspective of communication theory. A few months before Hickey began submitting his paper to sociological journals for publication, the editor of the *Journal of the American Society of Information Science* (Jan. 1978, Vol 29, No. 1, p. 3.) stated in his "Editor's Notes" that referees sometimes use the peer review process as a means of attacking a point of view, and object to the content of a submitted paper. He said that often rather than rejecting a paper so treated, it would be better to publish the submitted paper with the reviewer's comments.

The first academic sociological journal to which Hickey sent his paper was *Sociological Methods and Research* published by Sage Publications, Inc. Hickey's retained U.S. Postal Service receipt documents that the paper was received on 18 December 1978, the date that Hickey uses to document his priority, although in fact his macrosociological model was actually created in the latter half of 1976. On 22 May 1979 Hickey received a letter rejecting the paper for publication from the editor, a Mr. George W. Bohrnstedt of Indiana University. With the letter were enclosed criticisms by two referees. The principal objection made by the first referee is a philosophical one, that describes Hickey's model as a "reification" of the worst type, and the critic demands that Hickey explain how values are constituted within the population and how they change. This objection could mean anything, but as it happens the term "reification" occurs in Parsons' polemic against Positivism in *Structure of Social Action.* Parsons distinguished two types of realism, one of which he finds to be objectionable. The objectionable type, which he calls "reification", is any monolithic or monistic realism. The unobjectionable type of realism is what he calls "analytical realism", which he uses to justify his voluntarist theory of social action with its subjective semantics for motivational analyses. As a contemporary Pragmatist, Hickey joins Parsons' rejection of monistic realism in favor of ontological pluralism, and as an ontological pluralist he has no objection to a subjectivist semantics or a motivational type of causal analysis even if the hypotheses are developed by *verstehen*, so long as the particular ontology is not required as a criterion for scientific criticism. But he is not willing to comply with a critic's demand that he change the topic of his paper and make his macrosociological theory an extension of any social psychology showing how values are constituted, whether the social psychology is subjectivist, behavioristic, or otherwise.

This critic also ridiculed Hickey's briefly described discovery system, stating that it a stepwise "self-cooking" program that generates a model without researcher intervention, and he said that such atheoretical routinization is "inappropriate" for "structural" modeling. This rhetorical discourse suggests the know-nothingness and Romanticism mentioned by Herbert Simon when describing opposition to artificial-intelligence systems. The Romantic sociologists' attempts to obstruct the mechanization of discovery in science recalls to mind the Luddites' attempts to obstruct the mechanization of production in industry. Just as the Luddites were reacting against the industrial revolution in the nineteenth century, so too these Romantics are reacting against the information revolution in the twentieth century, and both share a role in history as symbols of futile obduracy in the face of progress. Furthermore the critic's statement that Hickey's macro-sociological model is "atheoretical" reveals a characteristically Romanticist concept of theory, and the critic's reference to "structural" modeling suggests that he would have this Romantic sociological theory function to specify equations, as Haavelmo advocated for neoclassical theory in econometrics. As it happens Hickey's macrosociological model might be viewed as a theory from on the philosophical views of either Merton or Lundberg. On Merton's definition a theory is a logical system of equations as opposed to an isolated proposition, and Hickey's model is a system in which each equation from a purely formal viewpoint has axiomatic status. And on Lundberg's definition a theory is not only a deductive system, but is also a system of empirical laws, and Hickey's model is a system of statistically estimated empirical equations. But Hickey rejects these meanings of "theory" as simply different versions of what Russell Hanson called "almanac" science. Hickey uses the contemporary Pragmatist concept of scientific theory, which views science as a research activity productive of new theory language, which when tested yields either empirically falsified expressions or empirically warranted explanations. On the Pragmatist philosophy of science "theory" language is defined as that which is universally quantified logically and proposed for empirical testing; it is not defined by its semantics or its ontology or by the manner in which it is created. On this view Hickey's macrosociological model is a macrosociological theory, since it is the model that is proposed for further empirical testing. And Hickey views the objection of no operator intervention to be irrelevant.

In another objection this critic described Hickey's model as value-based modeling, and said that it is inferior to a demographic accounting framework advocated by a sociologist named Kenneth C. Land. The critic says that the parameters of the models can be interpreted in Land's framework but not in Hickey's. The critic also complains that Hickey did not even cite Land's work. Land set forth his approach in his "A General Framework for Building Dynamic Social Indicator Models: Including an Analysis of Changes in Crime Rates and Police Expenditures" in *American Journal of Sociology* (1976). Hickey had mistakenly overlooked Land's paper in his literature search. In 1971 the Organization for Economic Cooperation and Development published a technical report written by a Professor Richard Stone titled *Demographic Accounting and Model Building,* which describes the concept and illustrates it with data describing the education system for England and Wales, UK. Conceptually the demographic accounting system is analogous to a perpetual inventory accounting system as might be found in a retailing business; it has beginning and ending inventory stocks, and inflows and outflows explaining the changes in stocks over an accounting period. In the demographic system the stock variables represent population head counts, the inflows may be due to births or immigration, and the outflows may be due to deaths and emigration. For describing the educational system the data are disaggregated by age and levels of education. Stone also describes how the accounting system may be expressed as a matrix with the inflows and outflows treated analogously to the economist's input-output models with the rows representing inflows, the columns representing outflows, and the cells representing "transition coefficients" calculated by dividing the aggregates in each row by its row total. Since these transition coefficients will change from period to period, there is an additional problem of projecting these changes if the model is to be used for any kind of policy analysis. Land's paper proposed using the econometric type of models statistically estimated over the time series of transition coefficients, which he furthermore says can be interpreted as measures of opportunities for social benefits. He therefore calls this the "opportunity-structure" approach, which he says is based on ideas originally proposed by the sociologist William F. Ogburn. Land's approach seems interesting and might be fruitful, if and when it is ever carried out. However the equation set forth in his paper is not estimated over vectors of transition coefficients from any demographic model. In any event it is clearly an alternative to Hickey's, and it is one to which Hickey is not indebted. Hickey was therefore unwilling to be conscripted to its support by this critic and this editor as a condition for publication.

The second referee selected by Bohrnstedt started his criticism by saying that he can't quite figure out if the paper is a "put on" or not. Having questioned the author's integrity, the critic then proceeds to dismiss the paper with a few more sentences. The critic objects that the equation specifications have no justification from theory, that statistical inference is totally ignored, that serial correlation is neglected, that the results are not addressed in any depth, that the findings are without meaning or not communicated, that he cannot suggest any improvements, and that he really can't understand what the paper is trying to do. Hickey accepted none of these criticisms including most notably the critic's concept of theory. The only criticism Hickey accepted was the first critic's objection to Hickey's referencing Max Weber. Hickey concluded that *Sociological Methods and Research* is not a suitable publication for his paper, and he did not send the editor any replies. He did not know then, that referees for other sociological journals would offer even more dogmatic criticisms.

The second sociological journal to which Hickey sent his paper is the *American Journal of Sociology*, which is edited by a Mr. Edward O. Laumann at the University of Chicago. The journal acknowledged receipt of Hickey's paper on 19 October 1979, and on 21 November 1979 Hickey received a rejection letter from the editor together with two statements of reasons for rejection written by the referees. The first referee selected by Laumann stated that if knowledge is to progress, sociologists need concrete thinking about specific mechanisms. The words "concrete" and "specific" were underlined in the referee's text. The critic also said that both Land's opportunity-structure and Hickey's reference to functionalist latent control structures are clumsy abstractions that are too vague to illuminate anything about social change that is not obvious. As it happens, latent control structures and latent functions are by definition those that are sufficiently nonobvious, that the social actors do not recognize them. The critic's complaints about obviousness notwithstanding, however, he says in the next paragraph of his criticism, that he is simply not "convinced" about anything reported in the paper, and that the outcome is "bizarre". Hickey observes that criticisms such as describing the equations as unconvincing and bizarre are not empirical criticisms, but rather reveal the use of the Romantic *verstehen* criteria. The critic repeats that what is needed is disciplined and "substantively" informed investigations, that elaborate conjectures about social change, and that the conjectures should rest upon attempts to specify concrete behavioral mechanisms responsible for observed change. He claims that there is a "burgeoning" literature reporting analyses of the various influences of trends in factors such as women's labor force participation, relative economic status, and sex role attitudes, which influence fertility trends, and he says that these analyses rest on attempts to specify concrete behavioral mechanisms. Hickey rejects this vague reference to this purportedly burgeoning literature, and says that if the critic knows of an empirically more adequate equation determining the crude birth rate, the critic must give a specific citation before he calls for suppressing publication of the paper. Hickey says that since the critic offers no specific citation, he is merely suggesting factors that he thinks may be useful for modeling, and that the critic should do his own modeling and should not attempt to make Hickey or any other author his research assistant as a condition for publication.

A reader of this criticism would like to have a better idea of what this critic means when he says that analyses should be based on concrete thinking about specific behavioral "mechanisms". A term such as "mechanism" is a metaphor, and could mean anything since no one expects to find wheels and gears or anything like quantum or celestial mechanics. But there is some hint when the critic says that Hickey avoids "substantive reasoning" and "substantively informed investigations" as a basis for the empirical models, and when the critic objects to Hickey's letting the computer program alone determine equation specifications, and says that Hickey's equations are *post hoc*. The term "substantive" seems to function as a shibboleth in the argot of the Romantic sociologists. The idea of substantive reasoning reveals a distinctively Romanticist concept of explanation. The reasoning this critic has in mind is not the formal syntax supplied by the mathematics of the equations; it is the reasoning processes in the minds of the social actors whose observed behavior is described by the measurement data. Few contemporary Romantic sociologists recognize the influence of Hegel's *Science of Logic* on their methodological thinking. A phrase like "substantive reasoning" echoes Hegel's rejection of the distinction between logical form and content, which is based on his identification of reality and reason or *Geist*. Contemporary Romantic sociologists are not metaphysical Idealists like Hegel, but they may be called methodological Idealists. As a Pragmatist Hickey rejects any such *a priori* ontological requirement. He also objects that the critic has offered no characterization of the concrete thinking and specific behavioral mechanisms, that would inform any reader as to why the factors the critic mentions are behavioral mechanisms and why the factors in Hickey's equations are not, except for the critic's own *verstehen* criticism that makes Hickey's equation specifications seem unconvincing and bizarre to this critic. Furthermore, even if the critic offered such a characterization, the recognition by sociologists of the existence of latent functions reveals the critic's dogmatic Romanticism and renders his criticism untenable. There are unforeseen and unintended consequences, which are latent functions if they operate to mitigate destabilizing forces. They have no place in a *verstehen* explanation acceptable to the Romantic sociologist, but they have an essential explanatory function in the empirically adequate dynamic model. These consequences, which are undetected by the social actors, are detected by the computer discovery system. This critic's objection, that these computer-selected variables do not operate in the social actors' "substantive" reasoning, is irrelevant to the empirical validity of Hickey's quantitative functionalist macrosocial theory.

The second referee selected by Laumann wrote a lengthy commentary, which amounts to three criticisms. The first consists of the critic's rejection of Hickey's interpretation of the institutional variables as measures of social values and his rejection of the model as a description of how social values change. The critic also dismissed Hickey's interpretation of these variables as measures of group associational behavior, and called it "theoretical hocus pocus". This kind of attack suggests that the critic may have his own concept of value in mind, and that he feels threatened by Hickey's construction of *per capita* rates as rates of group associational behavior reflecting consensus about cultural values, although the critic does not characterize any alternative concept. In Hickey's view this critic is merely attempting to suppress publication of Hickey's semantics for the term "value". Hickey defends his interpretation of *per capita* rates as measures of consensus about institutional value systems, because the behavior described by the numerators of the *per capita* rates are measures of voluntary behavior, and because this voluntary behavior furthermore pertains to membership in the various types of institutional groups in the macrosociety. Hickey believes that there may be an analogy between his treatment of cultural values and Nobel laureate economist Paul Samuelson's treatment of utility in the latter's doctrine of revealed preference set forth in "A Note on the Pure Theory of Consumer's Behavior" (1938) and in "Consumption Theory in Terms of Revealed Preference" (1948); both papers are reprinted in *The Collected Papers of Paul A. Samuelson* (1966, Vol. 1). The similarity is in Samuelson's rejection of utility as an introspectively known psychological concept, a view that was advanced by the Austrian school of economics. Samuelson instead describes consumer demand in terms of observed market behavior revealing consumer preference patterns. Just as a commodity's unit price measures economic value in the publicly observable market transaction even though the price does not characterize the economic value, so too the *per capita* rate measures institutional value in the publicly observable group associational behavior even though the *per capita* rate does not characterize the social value. Another analogy that is more familiar is a political election outcome: a landslide election outcome is a measure of a high degree of political consensus, even though the election returns do not characterize the mandate that the winning candidate brings to public office.

In his second criticism this critic attacks Hickey's use of his discovery system, which the critic seems to have confused with a stepwise regression program, and he says that the computer program cannot "replace the complexity of a scientist's intuition". Hickey states that the discovery system is not a stepwise regression. He also says that the manner in which a theory is created is irrelevant to its empirical validity. As the early Pragmatist, William James, tersely said of worthy scientific theories: "By their fruits ye shall know them, and not by their roots". Hickey furthermore rejects the critic's appeal to the "complexity" of intuition. If the critic could actually identify any analytical elements, characterize a complex structure formed from the analytical elements, and describe the functioning of this complex in the process of intuition, then by characterizing the "complexity of intuition" he would have come a long way toward formulating a procedure capable of mechanization. But his criticism gives no hint of any such analytical achievement. Instead this Romantic's apodictic dismissal of mechanization suggests at best that his idea of "the complexity of a scientist's intuition" is as mysterious, wholistic, and obscure as Bergson's "creative intuition" and *elan vital.* In Hickey's view this critic's discourse is nothing but descriptively vacuous rhetoric. It is typically in the context of discussions of creativity in science that the abysmal chasm between the Romantic and the Pragmatist social scientists yawns most widely. It is like the division between the vitalist and molecular biologists, and before them the Aristotelian and mathematical physicists.

The third criticism by this critic, by far the longest one, is an attempt at a technical critique, and is technically inadequate and of little philosophical significance. The attempt shows that the critic does not understand how linear higher-order difference equations can describe nonlinear time paths. He also made a spurious and vague citation to an author whose name never appears in *Review of Economics and Statistics* since it began publication in 1910, and he attacked Hickey's demographic conclusions by comparing them to the models of Alfred Lotka (*Elements of Biology*, 1924). Hickey is flattered by the comparison to Lotka, and he doubts that the critic has the competence to understand the mathematics in Lotka's interspecies equilibrium models. Laumann's second referee concluded his commentary with the strident comment that Hickey's paper is not publishable in *American Journal of Sociology* or anywhere else, and that the paper is thoroughly incompetent in every respect. It is also interesting that this referee cites comments in the criticism written by the other referee. Hickey believes that the interests of the readers of all sociological journals would be better served, if the referees were not informed of the opinions expressed in one another's criticisms of a paper. As a contemporary Pragmatist advocating ontological pluralism, Hickey believes that if the editors had their selected referees write their criticisms independently, then the editors might mitigate the conformism to which sociologists seem to be distinctively susceptible, both because their sociological theory of institutions places a high positive valuation on conformity, and because sociologists cannot distinguish between contrary opinion and contrary evidence. However, the underlying problem in sociology is not just a subculture that places a deviantly high value conformism because of its theory; it is the prior ontological commitments of the Romantic philosophy of science that dictates such conformism.

Hickey submitted his rebuttals to this editor, and on 30 July 1981 he received another rejection letter from Laumann with a brief criticism from a third referee enclosed. The third critic identified as an "internal" reviewer, indicated that he had read the criticisms written by the two other referees and the rebuttals submitted by Hickey. In his own criticism this third critic very briefly summarized the other referees' objections, and then maintained that Hickey does not understand the fundamental objection to the paper: the need for specific mechanisms. Hickey understands quite well, and rejects the critic's reductionist agenda and Romantic criteria for scientific criticism.

The third academic sociological journal to which Hickey sent his paper was *American Sociological Review,* the journal of the American Sociological Association, which was edited by a Mr. William H. Form at the University of Illinois, Urbana. The journal acknowledged receipt of the paper on 13 March 1981. On 10 April Hickey received a rejection letter signed by Form with two referee criticisms enclosed. The first critic selected by Form focused on the idea of theory, which he distinguishes from the idea of model, and he underlines the two terms in his critique. He says that while Hickey's model satisfies statistical criteria, it does not make "substantive sense". He also says that the equations are "puzzling", that demographers will be "amazed", and that criminologists will be "surprised". These objections are dubious; for example examination of the U.S. Federal Bureau of Investigation's *Uniform Crime Statistics* reveals that the factors in Hickey's equation would not be surprising to the criminologist. The critic offers no references to published empirical findings for his reaction. In Hickey's view Form's rejecting empirically adequate equations because a critic finds them to lack "substantive sense" and finds them to be "surprising", "amazing", and "puzzling", is absurd for a scientific journal purportedly publishing original findings. Neither the critic's nor anyone else's psychological reactions are relevant to the scientific value of the empirical findings in a submitted paper. Hickey consigns the use of *verstehen* criticism to the status of an exercise in salesmanship, and he rejects it as irrelevant to scientific criticism.

This critic also purports that Hickey's model needs a "theory" underlying the causal assertions embodied in the equations, and that the causal chains, if they exist at all, are very loose and indirect. At least this critic admits that the relationships may be causal; he does not simply dismiss the invariances in the coefficients of Hickey's empirically adequate equations as describing fifty years of incredible historical coincidence. The critic does not explicitly define his idea of theory, but he says that Hickey's view, which Hickey briefly described and identified as Pragmatist, is "silly". The critic says that Hickey's view amounts to saying that the empirical equations for the motion of a "spring-mass system" constitute a theory that is distinct from the empirical equations for the motion of a rocket leaving the earth. The critic states that in both cases the empirical equations are models of the phenomena in question deriving from the underlying Newtonian mechanics which is a "theory". The critic concludes by saying again that what is missing in the paper is a carefully stated "theory" underlying the author's specification of the quantitative functionalist model of fifty years of social change in American society, and that such a "theory" might help readers to believe, or accept as valid, causal specifications embodied in the author's model that have very little face validity in the absence of such a "theory".

Hickey did not recognize which high school physics laboratory demonstration this critic had in mind when referring to a "spring-mass system", but he found the critic's references to Newtonian mechanics to be ironic, if not also historically ignorant. In his reply Hickey quoted a lengthy passage from Russell Hanson's *Patterns of Discovery* (1958), in which Hanson reports that when Newton's theory was initially published, it was regarded as abstract and merely mathematical, and that critics claimed that it requires supplementary discourse to give it causal significance. Hanson says that with Newton's famous dictum "*hypotheses non fingo*" Newton refused to satisfy this criticism, and then says that when Newtonian mechanics eventually became accepted, it also became definitive of the meaning of causality. Hanson quotes the nineteenth-century physicist, Helmholtz, saying that to understand a phenomenon means nothing else than to reduce it to Newtonian laws, and that then the necessity for explanation has been satisfied in a palpable way. Such was the understanding of the terms "causality", "explanation" and "theory" in physics more than a hundred years ago, and Hanson's point is that the concept of causality is defined by the theories that are prevalent at the time. Later Quine made this point in his more general thesis of ontological relativity, which asserts the realistic claim for all ontological concepts in empirically acceptable theories, and not just for causality. This view is now the common property of the prevailing contemporary Pragmatist philosophy of science, and is known as "scientific realism". On the Pragmatist view

Newton's theory today is a "theory" only in an archival sense or what Hanson calls an "almanac" sense. But in what Hanson calls "research science" it is no longer a theory, since it has already been falsified by empirical testing. In the professional literature of science including the *American Sociological Review* the almanac idea of "theory" is not consistent with the idea of publishing original contributions intended to advance the current state of the science. On Hickey's Pragmatist view, therefore, the requirement imposed by the referee that the statistically estimated and empirically adequate model should have to be related to supplementary language called "sociological theory" in order to be accepted as a "causal" explanation, is just a repetition of an old and dogmatic philosophical blunder in the history of science.

This critic also says that Hickey had distorted the methodological views of the sociologist, Kenneth Land, which are set forth in Land's "Formal Theory" in *Sociological Methodology: 1971*, a book bearing an *imprimatur* identifying it as an "official" publication of the American Sociological Association. In this paper Land distinguishes theories from models, and proposes relating them with a logical schema advanced for the natural sciences by the Logical Positivist philosopher Carl Hempel. This philosophical ecclectism by Land is possible, because both Romantics and Positivists distinguish theory and observation language semantically, even though they reverse the role of semantics. The Positivists such as Hempel believe that theory is meaningless unless it is logically related to observation language, while the Romantics such as this critic believe that empirical models "meaningless" unless they are related to mentalistic sociological theory. The critic says that contrary to Hickey's rendering, Land does not argue that it is imperative to take the equation specifications of sociological models from existing colloquially expressed theory, but merely says realistically that this is the typical level of most current sociological theory. He adds that Land would be the first to indicate that a mathematically stated general theory, such as Newtonian mechanics, would be much better, if such can be constructed in sociology. This critic's unfortunate accusation of distortion reveals his ignorance of the contemporary Pragmatist philosophy of science. According to Hickey's Pragmatist view the issue as to whether theory precedes empirical model or the model precedes the theory is what Heisenberg called a "false question", like the ancient mariner's question about how far he can sail before he falls off the edge of the world. On the Pragmatist view theory is language that is proposed for empirical testing. Hickey therefore says that the empirical equations are theories when they function as such, and that the critic's problem of relating models and theories is a false problem.

The second referee selected by Form wrote a brief commentary consisting of only six sentences. He says that the "metatheoretical" considerations in the paper do not motivate the actual analyses effectively, and that little useful theory is involved, that the particular analyses are similarly little motivated, and that none of them reflect the usually long traditions of research attempting to explain the variables involved. He also calls the paper an "empiricist venture", that is utterly ineffectively related to the empirical traditions involved in explaining certain equations that incur his disapproval. He then attempts a technical criticism about correlations and trends that reveals his inexperience with data analysis, and he concludes that these correlations should not without much more thought be made a basis for causal inference. In his reply to the attempted technical criticisms Hickey referred the critic to an elementary textbook in econometrics. The long empirical traditions referred to by this critic are survey research techniques. For example in an appendix to Clinard's *Anomie and Deviant Behavior* Stephen Cole and Harriet Zuckerman list an annotated bibliography of nearly one hundred empirical sociological studies published since 1897. Many of these papers pertain to crime, but none of them involve quantitative modeling. The principal thrust of this referee, however, is the same Romantic philosophical criticism as had been offered by the previous critics, and in his rebuttals Hickey enclosed a brief annotated bibliography of the relevant philosophy of science literature for the edification of both Form and his selected critics. This bibliography was sent to Form with the unspoken hope that he too would pursue these references, so that he could acquire the understanding of the contemporary Pragmatist philosophy of science, that would enable him to recognize the distinction between contrary evidence and contrary opinion, and thereby recognize that the opinions served up as attempted criticisms by his referees are not valid simply because they have been penned by pontificating referees. It is not necessary that an editor be an expert in the topics published by his scientific journal, in order to distinguish contrary evidence from contrary opinion; but he must understand and respect this distinction in order to perform his editorial function correctly.

The only noteworthy feature of this second critic is his explicit reference to motivation. Like a previous critic who wanted specific mechanisms, this critic demands what Parsons called a motivational analysis of the behavior of the social actors as a basis for the equation specifications. Parsons sociology is a social psychology in which the elementary structure of society consists of an *Ego* and an *Alter*, two or more role players in interaction with each other and conforming to or deviating from one another's complementary expectations because of their common orientation to a system of cultural values. Mechanisms of socialization and social control integrate persons into roles, roles are the mechanisms by which persons are integrated into systems, and a system is and will continue in stable equilibrium so long as *Ego* and *Alter* conform with one another's expectations. Macrosocial system equilibrium in turn is the combined effect of numerous individuals, each of whom is in such a microsocial equilibrium with his associates. This individualistic analysis in terms of specific roles may be contrasted with Durkheim's (and Hickey's) focus on the group as an association of roles that are undifferentiated in the sociological analysis. Hickey's model is a distinctively macrosociological theory, but the referees demand that he develop a macrosociological theory which is a social-psychological theory. In other words an acceptable macrosociology must be extension of microsociology, or conversely any macrosociological theory must be reducible to a microsociological theory.

This reductionist type of criterion for scientific criticism imposed by these referees has not a few precedents in the history of science. It is rejected by the contemporary Pragmatists, who recognize only empirical criteria for scientific criticism and advocate ontological pluralism. Hickey finds it ironic that the Romantics reject the Positivist requirement for the reduction of social science to natural science, while requiring reductionism within the subjective frame of reference for sociology. Hickey submitted rebuttals to Form on 6 May 1981, and received a reply in which Form told Hickey that apparently Hickey does not understand the folkways of his profession, that it is not normative for an article to be resubmitted once it is rejected, and that if this were not the practice, Form would spend his life rereviewing the same manuscript.

Hickey had received seven criticisms from referees of three academic sociological journals, and these criticisms consistently reveal the dogmatic domination of the Romantic philosophy of science in sociology, and the profession's consequent incompetence to distinguish contrary opinion from contrary evidence. Is the prevalence of this philosophy of science in academic sociology consequential for sociology? The practice of an anachronistic philosophy of science yields a retarded science, and American academic sociology is a backward science *in extremis*, if it can even be called a science instead of just a literature. In "Sociology's Long Decade in the Wilderness" the *New York Times* (28 May 1989) reports that universities such as the University of Rochester in New York and Washington University in St. Louis, Missouri, have disbanded their sociology departments, and that the National Science Foundation has drastically cut back funding for sociological research. A graphic display in the article indicates that since the mid-1970's the number of bachelors degrees awarded with majors in sociology has declined by nearly eighty percent, the number of sociology masters degrees by sixty percent, and the number of sociology doctorate degrees by forty percent. Demand for Ph.D. degrees is influenced by many factors not specific to sociology, such as cyclical and secular changes in economic conditions, and changes in population size and demographic profile. The effects of these and other extraneous factors can be filtered by relating the number of sociology doctorates to the number of doctorates in other fields. Data for earned doctorates are available from the United States Department of Education, Office of Educational Research and Improvement. The *Graph* on the facing page is based on these data, and it displays the percent of earned doctorates in sociology relative to the number of earned doctorates in economics and physics. These data corroborate the reported decline of academic sociology, and they also validate the accuracy of Mr. Joseph Berger's reporting for the *New York Times*.

Berger's article also quotes a Mr. Stephen Buff, identified in the article as the assistant executive director of the American Sociological Association, as saying that sociology suffers from not being well defined in the public mind, and that sociology is confused either with social work or with socialism. But contrary to Mr. Buff's explanation public opinion is not operative in these decisions made against academic sociology. Decisions to enroll or not to enroll in graduate schools of sociology are made by students with undergraduate majors in sociology; decisions to support or close sociology graduate schools are made by knowledgeable university administrators; and the funding decisions of the National Science Foundation are made by staff members who are among the best informed in the nation. The cause of these unfavorable decisions originates within academic sociology; it does not lie with an ignorant general public. The article also quotes a more realistic opinion by a professor Egon Mayer, a Brooklyn College sociologist, who said that sociologists are still teaching

EARNED Ph.D. DEGREES IN SOCIOLOGY RELATIVE TO ECONOMICS AND PHYSICS

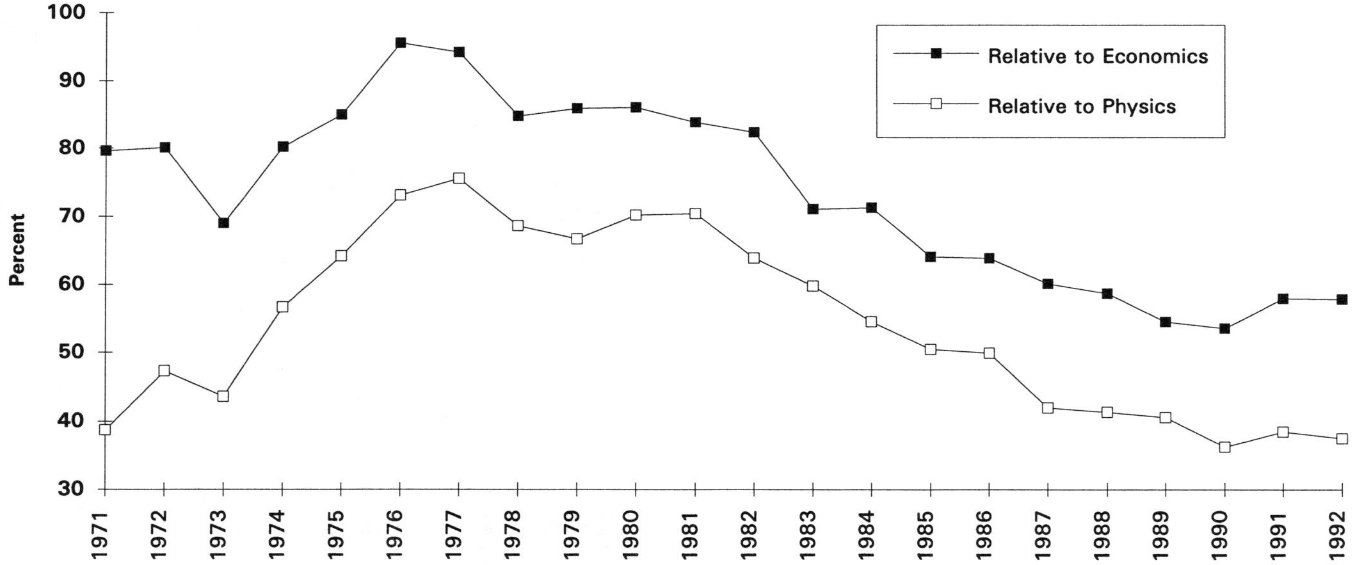

Graph

the same sort of thing that they taught in the 1960's and 1970's, but are not as convinced now that it is worth teaching, and are not quite sure what it should be replaced with. In Hickey's view sociologists will never know what it should be replaced with and what is worth teaching, until they discard their Romantic dogmatism and use the contemporary Pragmatist philosophy of science.

These reports are not encouraging to any young man or woman with the option of graduate-level studies in pursuit of a career in academic sociology. As a Ph.D. graduate in sociology he (or she, of course) would find that there is little demand for what he has to teach, and may expect that he might have to pursue another occupation to earn a living. And were he lucky enough to find employment on the faculty of a university that still has a sociology department, but formulated a view that is contrary to the dominant Romantic dogmas, he would find that he cannot get published in the academic literature. His submitted paper would be rejected for reasons that are a caricature of empirical criticism by an editor who cannot distinguish contrary evidence from the contrary opinions expressed by his selected referees. And were the submitting sociologist so audacious as to presume to submit rebuttals to the comments of the pontificating referees, he may find himself reading a letter from the editor saying that apparently he does not understand the folkways of his profession, that it is not normative for an article to be resubmitted once it is rejected, and that if this were not the practice the editor would spend his life time rereviewing the same manuscript.

The effect of sociology's Romantic dogmatism is not limited to academia. It includes the profession's demonstrated impotence to serve as a guide for the formulation of effective social policy. The same *New York Times* article also cites disillusionment resulting from the failures of the "Great Society" programs of the 1960s', and reports that sociologists today find they have lost Federal funding, must scale down their projects, forsake new inquiries, and disguise their work as anything-but-sociology. Similarly in his *Limits of Social Policy* (1988) Nathan Glazer, Harvard University professor of sociology and formerly an urban sociologist in the Federal Housing and Home Finance Agency during the Administration of John F. Kennedy, writes that the optimistic vision of sociology guiding policy by use of its knowledge of the fine structure of society of how policy impinges on family, neighborhood, and community has faded considerably. Glazer observes that in trying to deal with the breakdown of traditional structures, particularly the family, social policies have weakened them further and have made matters worse. He cites as one noteworthy example the welfare system, which he says undergoes continual expansion and correction with input from social scientists, but which nonetheless damages the family, encourages family breakup, and encourages fathers to abandon children, even though many of the changes in the system were designed to overcome just these untoward effects. He notes that these outcomes have occasioned the rejection of "social engineering", which he describes as the capacity of human foresight using subtly graduated incentives and disincentives and sharply focused programs, to affect human behavior and to improve the human condition. Glazer maintains that the most significant limitation of the effectiveness of the social policies formulated and implemented in the 1960's is lack of knowledge.

However, sociology's failure in the crucible of real-world social policy is not due merely to a lack of knowledge that could be remedied by more research in conformity with the Romantic philosophy of science. Sociologists will never understand these symptoms of failure, until they recognize the pathogen infecting their professional culture: the Romantic dogmatism that operates in their criteria for scientific criticism, and that imposes *a priori* restrictions on their theorizing. As it happens the eighth chapter of Glazer's book "'Superstitions' and Social Policy" could well be taken as an expose of sociologists' failure to recognize latent functions, and it amounts to a vindication of Merton's theorem of social engineering, even though Glazer references the more recent writings of Claude Levi-Strauss and Peter Berger instead of Merton. As long as American academic sociologists accept only theories that reduce to a motivational social psychology, much less to Romantic theories that "make sense" in compliance with the *verstehen* criterion, as long as they reject Romantically inexplicable latent functions and suppress publication of empirically superior theories that seem "surprising" relative to the sociologist's *verstehen*, and most importantly as long as contemporary Pragmatism remains a *terra incognita* to the sociologists - then sociologists will continue to be incapable of contributing to effective social policy, much less establishing their profession as a well functioning and modern empirical science instead of a philosophically retrograde profession.

Twentieth-century physics too had its failures, but when physicists formulated the relativity and quantum theories to remedy these failures, they found that they had to unbundle the ontological commitments which they had conventionally bundled in with the empirical criteria for scientific criticism, and then they had to make some difficult choices about which type of criteria would operate as rules of evidence. Their acceptance of the startlingly counterintuitive but empirically superior relativity and quantum theories led them to discard all ontological criteria including but not limited to those which they recognized as held over from the Newtonian physics, even though the Newtonian ontology had come to define what constitutes causal explanation and what "makes sense" in physics. Had the physicists dogmatically adhered to the old Newtonian ontology, Einstein's relativity theory could not have been accepted, and had they retained the "classical" ontology common to Newtonian and relativity physics as a reduction basis, the quantum theory could not have been accepted. Instead physicists accepted theories on the basis of their test outcomes, even when this meant the rejection of any logical reduction of macrophysical theory to microphysical theory. This is Pragmatism as it initially evolved in the practice of basic research in the Galileo-Einstein-Heisenberg tradition, and then was later articulated firstly by the physicists and then more systematically by the contemporary Pragmatist philosophers of science. Had the physics profession conducted itself as dogmatically as the Romantic sociology profession does today, Newtonian physics with its commitments to the ontology of familiar deterministic causality and spatial continuity would still prevail in physics today.

The optimism of the Great Society social programs to which Glazer referred, has long ago passed into history, even as sociologists continue to bundle their ontological commitments to Romanticism into their criteria for scientific criticism and then demand social-psychological reductionism to this ontology. Glazer's use of the term "optimism" in his *Limits of Social Policy* is an understatement; today only a government of incorrigibly naive Candides would again entrust the philosophically naive sociologists with a guiding role in the formulation of social policy. Before these Panglossian professors of scoiology can restore their credibility with real-world social policy administrators, they must overcome their anachronistic Romanticism and accept the contemporary Pragmatism, which rejects *a priori* commitment to any ontology as a criterion for scientific criticism.

Hickey views the editors' rejections and their referees' criticisms as analogous to the situation confronting an exploring mountaineer in H.G. Wells' "Country of the Blind", who fell into a lost valley inhabited by an isolated community born without eyes for many generations. Since everyone in the community was born without vision, their language had no pictorial semantics for describing visual experience. When these complacent natives heard the descriptive discourse of the intruding mountaineer, they concluded that he was deranged, and they demanded as a condition for his acceptance into their community, that he allow their surgeons to "cure" him by cutting out his eyes. It was not the explorer of course who was in need of remedy, but rather the blind tribe he encountered, and the mountaineer chose to keep his vision and leave the native society. The analogy with sociology should be obvious: American academic sociologists are denizens of the country of the philosophically blind; they have no vision for the contemporary Pragmatism that has revolutionized philosophy of science, and therefore their language has no Pragmatic semantics for describing successful basic science. It is they who are in need of remedy. They are afflicted with a chronic complacency that is symptomatic of their philosophical blindness, and they are in need of extensive remedial education in philosophy of science - in the meanings of such metatheoretical terms as "explanation" and "theory", in the criteria for scientific criticism for the empirical sciences and the distinction between contrary opinion and contrary evidence, in contemporary efforts to proceduralize scientific discovery, and in the aim of science.

There was one more sociological journal to which Hickey had sent his paper, which cannot be treated as the others discussed above, because the editor refused to inform Hickey of the scientific criticisms given as reasons for rejection. This journal is *Social Indicators Research* edited by a Mr. Alex C. Michalos. Michalos is identified on the journal's stationery as Director of the Social Indicators Research Programme at the University of Geulph in Ontario, Canada, and the publisher is identified as D. Reidel Publishing Company, Dordrecht, Holland, and Boston, U.S.A. Michalos acknowledged his receipt of Hickey's manuscript in a letter dated 19 January 1982. In a letter to Hickey dated 4 February 1982 Michalos said that he had received a very unfavorable review of the manuscript and would not be able to publish it. He added that usually he has specific comments from a reviewer to send to authors, but that in this case the reviewer pretty well threw up his hands. In response to a letter from Hickey dated 12 February demanding two written referee comments, Michalos wrote a letter to Hickey dated 22 February 1982 replying that sometimes his reviewers are brutal, and that when the first reviewer is exceptionally critical, as in the case of Hickey's manuscript, he does not go to a second reviewer. He concluded by saying that he had sent Hickey all he had. In Hickey's view no critic is above criticism. The 1995 edition of the *National Faculty Directory* lists Michalos as a faculty member of the Department of Philosophy at the University of Guelph. Having a professional philosopher as editor of a sociological journal might have been a singularly fortunate circumstance both for *Social Indicators Research* and for academic sociology. Instead what Hickey actually encountered is an editorial practice that resembles the judicial practice in Franz Kafka's *The Trial*, in which the accused is arrested, tried, condemned and executed without ever having been informed of the charges brought against him. Hickey has no idea what Michalos actually teaches his philosophy students in the Department of Philosophy at the University of Geulph. But Hickey believes that both Michalos' students and his journal would be better served, were he to accept the decidedly non-Kafkaesque contemporary Pragmatist philosophy of science, and then implement this philosophy in his future editorial decisions. Such is Hickey's contemporary Pragmatist critique of American Romantic sociology. As it happens, however, not all academic sociologists are philosophically and technologically atavistic. Immediately below the reader will find a description of a computerized discovery system developed by an atypically *avant garde* sociologist.

Sonquist on Simulating the Research Analyst with AID

John A. Sonquist (1931-) received a doctorate in sociology from the University of Chicago, and is at this writing a professor of sociology and the Director of the Sociology Computing Facility at the University of California at Santa Barbara, California. He was previously on the faculty at the University of Michigan at Ann Arbor, and was Research Associate and Head of the Computer Services Facility for the University's Institute for Social Research. He is also a past chairman of the Association for Computing Machinery. For his Ph.D. dissertation written in 1963 he developed a computerized discovery system called the **AID** system. "AID" is an acronym for "Automated Interaction Detector" system. His principal publication describing the **AID** system is *Multivariate Model Building* (1970). Today description of the **AID** system can be found in many marketing research textbooks in chapters discussing data analysis techniques for hypothesis development, and his algorithm is commercially available in the **SPSS** computer statistical software package.

In the "Preface" to his 1970 book he says that his interest in such a system started with a conversation with Professor James Morgan, in which the question was asked whether a computer could ever replace the research analyst himself, as well as replacing many of his statistical clerks. He writes that they discarded as irrelevant the issue of whether or not a computer can "think", and instead explored the question of whether or not the computer might simply be programmed to make some of the decisions ordinarily made by the scientist in the course of handling a typical analysis problem, as well as doing the computations. Developing such a computer program required firstly examining the research analyst's decision points, his alternative courses of action, and his logic for choosing one rather than the other course, and then secondly formalizing the decision-making procedure and programming it but with the capacity to handle many variables instead of only a few. An earlier statement of this idea was published in his "Simulating the Research Analyst" in *Social Science Information* (1967). There Sonquist notes that there are three kinds of computer applications in social science: data processing, simulation, and information retrieval. He observes that data processing systems and many information retrieval systems are nothing but an extension of the analyst's pencil and lack really complex logical capabilities. But he adds that there also exist information retrieval systems which are much more sophisticated, because simulating the human being retrieving information is one of the objectives of the system designer. These sophisticated retrieval applications combine both a considerable data processing capability and a logic for problem solving, such that the whole system is oriented toward the solution of a specific class of problems without human intervention in a long chain of decisions.

Sonquist then argues that such a combination of capabilities need not be limited to information retrieval, and that major benefits can be gained from the construction of a new type of simulation program, one in which the phenomenon simulated is the research analyst attempting to "make sense" out of his data. The phrase "make sense", which is a characteristic locution of the *verstehen* Romantics, is placed in quotation marks by Sonquist, and there is no evidence that he is advocating the *verstehen* philosophy of scientific criticism. In fact on the *verstehen* view a computer cannot "make sense" of social data. He says instead that an important function of the research analyst in the social sciences is the construction of models which fit the observed data at least reasonably well, and that this approach to the analysis of data can be likened to curve fitting rather than to the testing of clearly stated hypotheses deduced from precise mathematical formulations. He then offers his own **AID** system as an example of a system that simulates the research analyst.

Sonquist and Morgan initially published their idea in their "Problems in the Analysis of Survey Data, and a Proposal" in *Journal of the American Statistical Association* (June, 1963). The authors examine a number of problems in survey research analysis of the joint effects of explanatory factors on a dependent variable, and they maintain that reasonably adequate techniques have been developed for handling most of them except the problem of interaction. Interaction is the existence of an intercorrelating influence among two or more variables that explain a dependent variable, such that the effects on the dependent variable are not independent and additive. This is contrary to the situation that is assumed by the use of other multivariate techniques, such as multiple classification analysis and multiple linear regression, where each variable associated with an estimated coefficient is assumed to be independent, so that the effects of each variable on the dependent variable can be treated as additive. In "Finding Variables That Work" in *Public Opinion Quarterly* (Spring, 1969) Sonquist notes that it is possible to represent interaction among explanatory variables in a regression, if the interacting variables are combined multiplicatively prior to statistical estimation. But there still remains the prior problem of discovering the interacting variables. The **AID** system is a discovery system which may be used in conjunction with such techniques as regression or multiple classification, in order to detect and identify interaction effects and to assist equation specification for regression.

In *The Detection of Interaction Effects: A Report on a Computer Program for the Optimal Combinations of Explanatory Variables* (1964, 1970) and in *Searching for Structure: An Approach to Analysis of Substantial Bodies of MicroData and Documentation for a Computer Program* (1971, 1973) Sonquist and Morgan describe their algorithm, as it is implemented in the **AID** computer program used at the University of Michigan, Survey Research Center. The program answers the question: what dichotomous split on which single predictor variable will render the maximum improvement in the ability to predict values of the dependent variable. The program divides a sample of at least one thousand observations through a series of binary splits into a mutually exclusive series of subgroups. Each observation is a member of exactly one of these subgroups. The subgroups are chosen so that at each step in the procedure, the arithmetic means of each subgroup account for more of the total sum of squares (i.e. reduce the predictive error more) than the means of any other pair of subgroups. This is achieved by maximizing a statistic called the between-group sum of squares. The procedure is iterative and terminates when further splitting into subgroups is unproductive.

The authors illustrate the algorithm with a tree diagram displaying the binary splits for an analysis of income using data categories representing age, race, education, occupation, and length in present job. When the total sample is examined, the minimum reduction in the unexplained sum of squares is obtained by splitting the sample into two new groups: persons under sixty-five years of age and persons aged sixty-five and over. Both of these groups may contain some nonwhites and varying degrees of education and occupation groups. The group that is sixty-five and over is not further divided, because control parameters in the system detect that the number of members in the group is too small. It is therefore a final grouping. The other group is further subdivided by race into white and nonwhite persons. The nonwhite group is not further subdivided, because it is too small a group, but the system further subdivides the white group into persons with college education and persons without college education. Each of these latter are further subdivided. The college-educated group is split by age into those under forty-five years and those between forty-six and sixty-five. Neither of these subgroups are further subdivided. Those with no college are further subdivided into laborers and nonlaborers, and the latter are still further split by age into those under thirty-five and those between thirty-six and sixty-five. The variable representing length of time in current job is not selected, because at each step there existed another variable which was more useful in explaining the variance remaining in that particular group. The predicted value of an individual's income is the mean value of the final group of which the individual is a member. Such in overview is the **AID** discovery system.

Sonquist offers little by way of philosophical commentary; unlike sociologists such as Parsons and Lundberg, he does not develop a philosophy of science much less a philosophy of language. But there is little imperative that he philosophize, since the application of his **AID** system is less philosophically controversial; there is no necessary conflict between the data his system is designed to process and the mentalistic ontology required by Romantic sociologists, since his system is intended to process data collected by survey research which typically consist in verbal responses revealing respondents' mental states. Such a conflict occurs only with those extreme Romanticists requiring the *verstehen* truth criterion for scientific criticism. In his 1963 paper, "Problems in the Analysis of Survey Data", he considers the problem that occurs when theoretical constructs are not the same as the factors that the sociologist is able to measure, even when the survey questions are attitudinal or expectational questions, and when the measurements that the sociologist actually uses, often called "proxy variables" or "indicators", are not related to the theoretical constructs on a simple one-to-one basis. This is a problem that occurs only in cases in which a theory pre-exists empirical analysis, and in this circumstance Sonquist advocates a role for the **AID** system, in which the system's empirical analyses are used for the resolution of problems involving interaction detection, problems which theory cannot resolve, or which must be solved either arbitrarily or by making untestable assumptions about the connection between theoretical construct and measurement factor. Later he considers the role for the discovery system for the development of theory, and the influence of Robert K. Merton is evident. In *Multivariate Model Building* he states in the first chapter that he is not attempting to deal with the basic scientific problems of conceptualizing causal links or with latent and manifest functions, but only with the apparent relations between measured constructs and their congruence with an underlying causal structure. He defines a "theory" as sets of propositions which describe at the abstract level the functioning of a social system, and proposes that in the inductive phase, *ex post facto* explanations of the relationships found within the data may form a basis for assembling a set of interrelated propositions which he calls a "middle range theory", that describes the functioning of a specific aspect of a social system. The **AID** system facilitates the inductive phase by identifying interacting variables, so that mathematical functions relating sociological variables are correctly specified for statistical estimation.

Sonquist draws upon an introductory text, *An Introduction to Logic and Scientific Method*, written in 1934 by two academic philosophers of science, Morris R. Cohen and Ernest Nagel. Cohen (1880-1947) received his Ph.D. from Harvard in 1906, and Nagel (1901-) studied under Cohen at City College of New York and received his Ph.D. from Columbia University in 1931. The relevant chapter in the book is titled "The Method of Experimental Inquiry", which examines the experimental "methods" for discovering causal relationships advanced by Francis Bacon and later elaborated by John S. Mill. These Baconian experimental methods are anything but Romanticist: the two authors define the search for "causes" to mean the search for some invariable order among various sorts of elements or factors, and the book gives no suggestion that the social sciences should receive any distinctive treatment. Since all discovery systems search for invariant relations, the attractiveness of the Baconian treatment for scientists such as Sonquist is self evident. The propositions which Sonquist views as constituting middle-range sociological theory and which following Cohen and Nagel express a causal relationship, have the linguistic form: *X(1)...X(n)* implies *Y*. The researcher's task in Sonquist's view is to relate the causal proposition to a mathematical functional form, which is statistically estimated, and he concludes that a well specified, statistically estimated mathematical function with a small and random error term, expresses a causal relationship understood as the sufficient condition for an invariant relationship between the dependent or caused variable and the set of independent or causal variables.

In "Computers and the Social Sciences" and "'Retailing' Computer Resources to Social Scientists" in *American Behavioral Scientist* (1977) Sonquist and Francis M. Sim discuss the inadequate social organization in universities for the effective consumption of computer resources, especially by social scientists, whom they report are described derisively by other academicians as "the marginal computer users". The authors present some arguments for changing the professional roles and social organization of computing in social science departments, and they propose some organizational forms. While the authors' sociological analysis of computing in social science is interesting, and while their reorganization proposals may offer benefits, the underutilization of computer resources and techniques by social scientists cannot be remedied by such measures as academic reorganization, so long as the prevailing philosophy of science is still Romanticist. Reorganizing roles could do no more for Romantic sociology than rearranging deck chairs did for the sinking *H.M.S. Titanic*.

Examination of Sonquist's writings in their chronological order suggests that, as he had attempted to expand the discovery function of his system, he discovered that he had to move progressively further away from the Romanticism prevailing in contemporary academic sociology. He would have been better served by the contemporary Pragmatist philosophy of science, than he had been by invoking the 1930's Positivist views of Cohen and Nagel. Both Positivism and Romanticism give a semantically based definition of "theory" and ontologically based criteria for scientific criticism. On the Pragmatist view "theory" is defined by the pragmatics of language, its function is what Hanson called research science as opposed to almanac science. And according to the Pragmatist realism practiced by Galileo, Einstein and Heisenberg and formulated as "ontological relativity" by Quine, every causal claim is based exclusively on the empirical adequacy of a tested theory. Discovery systems therefore are not only able to discover causality, they can also make causal theories.

Hickey's correspondence with the editors of the academic sociological journals described above reveals that these editors and their selected referees are committed to the perpetuation of archaic criteria for scientific criticism, criteria based on an institutionalized Romantic philosophy of science. In this context Sonquist's **AID** discovery system was not merely an anomaly in academic sociology. Developed in 1963 at the apex of Parsonian Romanticism, the **AID** system was a portent. It signaled the widening cultural lag in academic sociology between technological modernity represented by the innovative computerized discovery systems on the one hand and anachronistic philosophy represented by an atavistic paleo-Romanticism on the other hand. This conflict is not merely between alternative theories; a multiplicity of empirically acceptable conflicting theories is consistent with Pragmatism. This conflict is institutional. And it is irreconcilable. The new technology is an implementation of ontological relativity, which subordinates ontology to empirical criticism, while the old philosophies subordinate empirical criticism to their peculiar ontological dogmas. Only one or the other type of criterion can prevail, and the prevailing one will at least banalize if not completely exclude the other.

Comment and Conclusion

Artificial intelligence started as an agenda in psychology as cognitive psychology. Will the development of discovery systems continue as a psychology agenda of artificial intelligence? In "A Split in Thinking Among Keepers of Artificial Intelligence" the *New York Times* (18 July 1993) reported that scientists attending the 1993 meeting of the American Association of Artificial Intelligence expressed disagreement about the goals of artificial intelligence. Some maintained the traditional view that artificial-intelligence systems should be designed to simulate intuitive human intelligence, while others maintained that the phrase "artificial intelligence" is merely a metaphor that has become an impediment, and that AI systems should be designed to exceed the limitations of intuitive human intelligence The article also notes that the division has fallen along occupational lines with the academic community preferring the traditional goal and the business community expressing the alternative goal. This alignment is not inherent, since the academic community need not view artificial intelligence exclusively as an agenda for psychology, but the alignment is understandable, since the business community financially justifies artificial-intelligence systems pragmatically like any computer systems, and not as faithful replicas of human limitations. The article mentions that large AI systems have been installed in such U.S. companies as American Express and Northwest Air. This is the same pragmatic justification that applies generally in basic scientific research: scientists do not use discovery systems to replicate the scientist's limitations but to transcend them and enhance their research capabilities. These limitations include the computational constraint described in Simon's thesis of bounded rationality, and the semantical impediment described by Hanson and referred to by Hickey as the cognition constraint. Artificial intelligence may have outgrown its original home in academic psychology.

As it happens, one of the things shown in the above survey of discovery systems is that each of the systems except Simon's **LOGIC THEORIST** had been developed by a graduate student for his doctoral dissertation (except Hickey's, for reasons noted above). Another is that none of these dissertations were written in academic philosophy departments. Could such a doctoral dissertation be written in a graduate school of philosophy? That would depend in part on the particular philosophy department's concept of philosophy of science. There will always be philosophers for whom "philosophy of science" is a comparison of some recent view with the metaphysics of Aristotle, Kant or Hegel. This mentality may originate in some kind of religious agenda or it may reflect a preference for a nontechnical or even recreational interest euphemistically called "discursive" philosophy. But if a metaphysical approach to methodology of science has any place in philosophy, it is in history of philosophy rather than philosophy of science. The contemporary philosophy of science is Pragmatism, which is not only the philosophy of science that is adequate to the revolutionary achievements of twentieth-century physics, but is also the philosophy of science that is compatible for the mechanization of theory development. It is the philosophy that has discarded the ontological dogmas that inhibit mechanization, and has redefined the fundamental concept of "theory" in the pragmatic sense, as Hanson described it and Simon used it.

Could such a dissertation be written in a contemporary academic philosophy department? It would be unusual to find a philosophy faculty professionally prepared even minimally in the requisite techniques of computer logic, even though traditionally philosophy has been the academic home for logic. The computer systems skills needed for a contemporary methodology of science are already well established with their own academic curriculum, computer science. Herbert Simon managed the interdisciplinary approach that he believed was required, by his appointment to both the psychology and computer science department faculties at Carnegie-Mellon University. But it is not necessary to be a faculty member of a computer science department, much less a Nobel-laureate calibre academician, to learn computer languages and systems techniques for the design of new discovery systems.

Could a dissertation setting forth a new discovery system be written in a philosophy department with a faculty competent enough in computer science to practice Logical Pragmatism? The Logical Pragmatist philosophy of science alone is not sufficient for developing a discovery system. Knowledge of contemporary Pragmatism is very facilitating, because it removes the impediments created by the older and less adequate philosophies of science. But the above overview reveals that most successes to date have involved a working knowledge of the empirical science in which the discovery system is developed and implemented. Sonquist is a sociologist who developed his **AID** system, although his inadequacy in contemporary philosophy of science led him to turn to Positivism in order to evade the obstructionist Romanticism in academic sociology. Pragmatism would have served him better. For Litterman evasion of Romanticism was easier. He is an economist who developed his **BVAR** system under teachers who were rational expectations advocates. His teachers were economists, who had rejected Haavelmo's structural-equation agenda, and for whom Romanticism is inoperative for determining equation specifications for model construction. And Hickey was even more fortunate, since he is a Pragmatist philosopher of science and also an

Institutionalist economist. He developed his **META-MODEL** system independently for both economics and sociology - even though the academic sociologists have yet to forgive him for exploiting their data by using modeling and simulation techniques in which they are incompetent and by violating the sacrosanct Romanticism in which they are incorrigible.

Could an interdisciplinary dissertation be written setting forth a new discovery system in an academic philosophy department practicing Logical Pragmatism? In fact a doctoral dissertation setting forth a worthy and original contribution consisting of a new discovery system design and implementation, could be written in just about any academic department of just about any graduate school, whether it is a department of mathematics, computer science, philosophy, psychology, natural science or social science. And the reason is, that writing a worthy and original contribution ultimately depends upon the faculty's self-confidence and level of aspiration. All faculties have limitations, but limitations are not invincible impediments. When the needed skills or background knowledge are not available in-house, the faculty will draw upon the services of professionals who are outside its own department or even its university. Several of the graduate students who developed the dissertations setting forth the discovery systems described above, were supported by their department faculties in this manner. And it might be added that these students did not have to earn multiple doctorates to write their interdisciplinary dissertations.

BIBLIOGRAPHY

Alston, William P. *Philosophy of Language*. Prentice-Hall, Englewood Cliffs, NJ, 1964.

American Economic Association. *Readings in Business Cycles*. (edited by Robert A. Gordon and Lawrence Klein). Richard D. Irwin, Homewood, IL, 1965.

Bar-Hillel, Yehoshua and Rudolf Carnap. "Semantic Information", *British Journal for the Philosophy of Science,* IV (1953), Pp. 147-157.

Bar-Hillel, Yehoshua. *Language and Information.* Addison-Wesley, Reading, MA, 1964.

Belkin, Nicholas J. "Some Soviet Concepts of Information for Information Science," *American Society for Information Science Journal*. Vol. 26 (January, 1975), Pp. 56-60.

Bell, J.S. *Speakable and Unspeakable in Quantum Mechanics:* Collected Papers on Quantum Philosophy. Cambridge University Press, Cambridge, England, 1987.

Black, Max. *Models And Metaphors*. Cornell University Press, Ithica, NY, 1962.

Blackmore, John T. *Ernst Mach: His Work, Life and Influence*. University of California Press, Los Angeles, CA, 1972.

Bohm, David. *Causality and Chance*. D. Van Nostrand, New York, NY, 1957.

Bohm, David. "A Proposed Explanation of Quantum Theory in Terms of Hidden Variables at a Sub-quantum Mechanical Level" in *Observation and Interpretation*. Edited by S. Korner. Butterworth Scientific Publications, London, 1957.

Bohm, David. *Wholeness and the Implicate Order*. Routledge & Kegan Paul, London, England, 1980.

Bohm, David . "Hidden Variables and the Implicate Order" in *Quantum Implications: Essays in Honor of David Bohm*. Edited by B.J. Hiley and F. David Peat. Routledge & Kegan Paul, New York, NY, 1987. Pp. 33-45.

Bohm, David and F. David Peat. *Science, Order and Creativity.* Bantam Books, New York, NY, 1987.

Bohr, Niels. *Atomic Theory and the Description of Nature*. Cambridge University Press, Cambridge, England, 1961 [1934].

Bohr, Niels. "Can Quantum Mechanical Description of Physical Reality Be Considered Complete?," *Physical Review,* Vol. 48 (1935), pp. 696-702.

Bohr, Niels. "Discussion with Einstein", in *Albert Einstein: Philosopher-Scientist.* Edited by Paul A. Schilpp. Open Court, LaSalle, IL, 1949. Pp. 201-41.

Bohr, Niels. *Atomic Physics and Human Knowledge.* John Wiley & Sons, New York, NY, 1958.

Bohr, Niels. *Essays 1958-1962 on Atomic Physics and Human Knowledge.* Interscience Publishers, New York, NY, 1963.

Born, Max. "Physical Reality", *The Philosophical Quarterly.* Vol. 3 (1953), pp. 139-149.

Born, Max. *My Life and My Views.* Charles Scribner's Sons, New York, NY, 1968.

Braithwaite, Richard B. *Scientific Explanation: A Study of the Function of Theory, Probability, and Law in Science.* Cambridge University Press, Cambridge, England, 1968 [1953].

Brookes, Bertram C. "The Foundations of Information Science. Part I. Philosophical Aspects," *Journal of Information Science*, Vol. 2 (October, 1980). Pp. 125-133.

Burns, Arthur F. *Wesley Clair Mitchell: The Economic Scientist.* National Bureau of Economic Research, New York, 1952.

Carnap, Rudolf. *The Logical Construction of the World.* Translated by R.A. George. The University of California Press, Berkeley, CA, 1967 [1928].

Carnap, Rudolf. "On the Character of Philosophical Problems" (1934) in *The Linguistic Turn. Edited by Richard Rorty.* The University of Chicago Press, Chicago, IL, 1967. Pp. 54-62.

Carnap, Rudolf. "Logical Foundations of the Unity of Science", in *International Encyclopedia of Unified Science*, Vol. I. Edited by Otto Neurath, Rudolf Carnap and Charles Morris. The University of Chicago Press, Chicago, IL, 1938. Pp. 42-62.

Carnap, Rudolf. "Foundations of Logic and Mathematics", in *International Encyclopedia of Unified Science*, Vol. I. Edited by Otto Neurath, Rudolf Carnap, and Charles Morris. The University of Chicago Press, Chicago, IL, 1938. Pp. 140-213.

Carnap, Rudolf. *Introduction to Semantics and Formalization of Logic.* Harvard University Press Cambridge, MA, 1958 [1942,1943].

Carnap, Rudolf. Meaning and Necessity: *A Study in Semantics and Modal Logic.* The University of Chicago Press, Chicago, IL, 1964 [1947].

Carnap, Rudolf. *Logical Foundations of Probability*. The University of Chicago Press, Chicago, IL, 1950.

Carnap, Rudolf. *The Continuum of Inductive Methods*. The University of Chicago Press, Chicago, IL, 1952.

Carnap, Rudolf and Yesohua Bar-Hillel. "Semantic Information," *British Journal for the Philosophy of Science*, IV (1953). Pp. 147-157.

Carnap, Rudolf. "Meaning and Synonymy in Natural Languages" (1955) in *Meaning and Necessity: A Study in Semantics and Modal Logic*. The University of Chicago Press, Chicago,IL, 1964 [1947]. Pp. 233-247.

Carnap, Rudolf. "On Some Concepts of Pragmatics" (1956) in *Meaning and Necessity*: A Study in Semantics and Modal Logic. The University of Chicago Press, Chicago, IL, 1964 [1947]. Pp. 248-253.

Carnap, Rudolf. "The Methodological Character of Theoretical Concepts" in *The Foundations of Science And The Concepts of Psychology and Psychoanalysis*. Edited by Herbert Feigl and Michael Scriven. University of Minneapolis Press, Minneapolis, MN, 1956.

Carnap, Rudolf. "Intellectual Autobiography" in *The Philosophy of Rudolf Carnap*. Edited by Paul Arthur Schilpp. Open Court, LaSalle, IL, 1963, Pp. 3-84.

Carnap, Rudolf. *Philosophical Foundations of Physics: An Introduction to the Philosophy of Science*. Edited by Martin Gardner. Basic Books, New York, NY, 1966.

Carnap, Rudolf. *Studies in Inductive Logic and Probability*. Edited by Richard C. Jeffery. Vol. 1 (1971) and Vol. 2 (1980). University of California Press, Berkeley, CA.

Cohen, Morris R. and Ernest Nagel. *An Introduction to Logic And Scientific Method*. Harcourt, Brace, and World, New York, NY, 1934.

Commons, John R. *Institutional Economics: Its Place in Political Economy*. Two Volumes. University of Wisconsin Press, Madison, WI, 1959 [1934].

Commons, John R. *Economics of Collective Action*. University of Wisconsin Press, Madison, WI, 1970 [1950].

Conant, James B. *On Understanding Science: An Historical Approach*. Yale University Press, New Haven, CT, 1947.

Conant, James B. *Science and Common Sense*. Yale University Press, New Haven, CT, 1951.

Conant, James B. *Modern Science and Modern Man.* Columbia University Press, New York, NY, 1952.

Conant, James B. *My Several Lives: Memoirs of a Social Inventor*. Harper & Row, New York, NY, 1970.

Conant, James B. *Two Modes of Thought*. Trident Press, New York, NY, 1964.

Duhem, Pierre. *The Aim and Structure of Physical Theory.* Translated by Philip P. Wiener. Princeton University Press, Princeton, NJ, 1962 [1906].

Duhem, Pierre. *To Save the Phenomena: An Essay on the Idea of Physical Theory from Plato to Galileo.* Translated by Edmund Doland and Chaninah Maschler. The University of Chicago Press, Chicago, IL, 1969 [1908].

Einstein, Albert, B. Podolsky, and N. Rosen. "Can QuantumMechanical Description of Physical Reality Be Considered Complete?" *Physical Review*, Vol 47 May, 1935), Pp. 777-780.

Einstein, Albert. "Physics and Reality" *The Journal of the Franklin Institute*, Vol 221 March, 1936), Pp. 349-82. Translated by Jean Piccard.

Einstein, Albert. "Autobiographical Notes", in *Albert Einstein: Philosopher-Scientist.* Edited by Paul A. Schilpp. Open Court, LaSalle, IL, 1949. Pp. 2-96.

Einstien, Albert. "Remarks Concerning the Essays Brought Together in This Cooperative Volume", in *Albert Einstien: Philosopher-Scientist.* Edited by Paul A. Schilpp. Open Court, LaSalle, IL, 1949. Pp. 665-88.

Feyerabend, Paul K. "On the Quantum-Theory of Measurement", in *Observation and Interpretation.* Edited by S. Korner. Butterworths Scientific Publications, London, England, 1957. Pp. 121-130.

Feyerabend, Paul K. "An Attempt at a Realistic Interpretation of Experience" in *Proceedings of the Aristotelian Society.* Vol LVIII. Harrison & Sons, Ltd., London, England, 1958. Pp. 143-170.

Feyerabend, Paul K. "Complementarity", in *Aristotelian Society* (Supplementary Volume XXXII, 1958). Harrison and Sons, Ltd., London, England, 1958. Pp. 75-104.

Feyerabend, Paul K. "Problems of Microphysics", in *Frontiers of Science and Philosophy.* Edited by Robert G. Colodny. University of Pittsburgh Press, Pittsburgh, PA, 1962. Pp 189-283.

Feyerabend, Paul K. "Explanation, Reduction and Empiricism", in *Minnesota Studies in the Philosophy of Science*. Edited by Herbert Feigl and Grover Maxwell. University of Minnesota Press, Minneapolis, MN, 1962. Pp. 28-97.

Feyerabend, Paul K. "Problems of Empiricism", in *Beyond the Edge of Certainty*. Edited by Robert G. Colodny. Prentice-Hall, Englewood Cliffs, NJ, 1965. Pp. 145260.

Feyerabend, Paul K. "Problems of Microphysics", in *Philosophy of Science Today*. Edited by Sidney Morgenbesser. Basic Books, New York, NY, 1967. Pp. 136-147.

Feyerabend, Paul K. *Against Method: Outline of an Anarchistic Theory of Knowledge*. Verso, London, England, 1975.

Feyerabend, Paul K. *Science In A Free Society*. NLB, London, England, 1978.

Feyerabend, Paul K. *Realism, Rationalism, and Scientific Method: Philosophical Papers*. Vol. 1. Cambridge University Press, Cambridge, England, 1981.

Feyerabend, Paul K. *Problems of Empiricism: Philosophical Papers.* Vol. 2. Cambridge University Press, Cambridge, England, 1981.

Feyerabend, Paul K. *Farewell to Reason*. Verso, London, England, 1987.

Fox, James A. *Forecasting Crime Data.* Lexington Books, Lexington, MA, 1978.

Friedman, Milton. "The Methodology of Positive Economics" in *Essays in Positive Economics.* The University of Chicago Press, Chicago, IL, 1953. Pp. 3-43.

Friedman, Milton. *A Theory of the Consumption Function*. Princeton University Press, Princeton, NJ, 1957.

Glazer, Nathan. *The Limits of Social Policy*. Harvard University Press, Cambridge, MA, 1988.

Haavelmo, Trygve. "The Probability Approach in Econometrics" *Econometrica*. Vol. 12 (July, 1944), Supplement.

Hagstrom, Warren O. *The Scientific Community*. Basic Books, New York, NY, 1965.

Hanson, Norwood Russell. "On Elemantary Particle Theory", *Philosophy of Science*, Vol 23 (1956). Pp. 142-148.

Hanson, Norwood R. *Patterns of Discovery*. Cambridge University Press, Cambridge, England, 1958.

Hanson, Norwood R. "Copenhagen Interpretation of Quantum Theory" in *The American Journal of Physics*, Vol 27 (1959). Pp. 1-15.

Hanson, Norwood R. "Niels Bohr's Interpretation of the Quantum Theory" in *Current Issues in the Philosophy of Science.* Edited by Herbert Feigl and Grover Maxwell. Holt, Rinehart and Winston, New York, NY, 1961. Pp. 371-390.

Hanson, Norwood R. "Postscript" in *Quanta and Reality.* American Research Council, New York, NY, 1962. Pp. 85-93.

Hanson, Norwood R. *The Concept of the Positron.* Cambridge University Press, Cambridge, England, 1963.

Hanson, Norwood R. "Newton's First Law: A Philosopher's Door into Natural Philosophy", in *Beyond the Edge of Certainty.* Edited by Robert G. Colodny. Prentice-Hall, Englewood Cliffs, NJ, 1965. Pp. 6-28.

Hanson, Norwood R. "Notes Toward A Logic Of Scientific Discovery" in *Perspectives on Peirce.* Edited by Richard J. Bernstein. Yale University Press, New Haven, CT, 1965. Pp. 43-65.

Hanson, Norwood R. "The Genetic Fallacy Revisited", in *American Philosophical Quarterly.* Vol 4 (April, 1967), Pp. 101-113.

Hanson, Norwood R. "Quantum Mechanics, Philosophical Implications of", in *The Encyclopedia of Philosophy.* Vol. VII. Edited by Paul Edwards. The Macmillan Company & The Free Press, New York, NY, 1967. Pp. 41-49.

Hanson, Norwood R. "Observation and Interpretation" in *Philosophy of Science Today.* Edited by Sidney Morgenbesser. Basic Books, New York, NY, 1967. Pp. 89-99.

Hanson, Norwood R. *Perception and Discovery: An Introduction to Scientific Inquiry.* Edited by Willard C. Humphreys. Freeman, Cooper & Company, San Francisco, CA, 1969.

Heelan, Patrick A., S.J. *Quantum Mechanics and Objectivity: A Study of the Physical Philosophy of Werner Heisenberg.* Martinus Nijhoff, The Hague, Netherlands, 1965.

Heisenberg, Werner. *The Physical Principles of the Quantum Theory.* Translated by Carl Eckart and F. C. Hoyt. Dover Publications, Inc., New York, NY, 1950 [1930].

Heisenberg, Werner. *The Physicist's Conception of Nature.* Translated bt Arnold J. Pomerans. Harcourt, Brace and Company, New York, NY 1955.

Heisenberg, Werner. "The Development of the Interpretation of the Quantum Theory" in *Niels Bohr and The Development of The Quantum Physics*. Edited by Wolfgang Pauli. McGraw-Hill, New York, NY, 1955.

Heisenberg, Werner. Physics and Philosophy: *The Revolution in Modern Science.* Harper and Row, New York, NY, 1958.

Heisenberg, Werner. "Quantum Theory and Its Interpretation", in *Niels Bohr*. Edited by S. Rosental. Interscience Press, New York, NY, 1964. Pp. 94-108.

Heisenberg, Werner. *Physics and Beyond: Encounters and Conversations.* Translated by Arnold J. Pomerans. Harper and Row, New York, NY, 1971.

Heisenberg, Werner. *Across the Frontiers.* Translated by Peter Heath. Harper and Row, New York, NY, 1974.

Heisenberg, Werner. *Philosophical Problems of Quantum Physics.* Translated by F. C. Hayes. Ox Bow Press, Woodbridge, CT, 1979. Formerly titled *Philosophic Problems of Nuclear Science.* Pantheon, New York, NY, 1952.

Heisenberg, Werner. "Remarks on the Origin of the Relations of Uncertainty" in *The Uncertainty Principle and Foundations of Quantum Mechanics.* Edited by William C. Price and Seymour S. Chissick. John Wiley & Sons, New York, NY, 1977.

Hempel, Carl G. and Paul Oppenheim. "Logic of Explanation", *Philosophy of Science.* Vol. 16 (1948). Pp. 135-175.

Hempel, Carl G. *Philosophy of Natural Science.* PrenticeHall, Englewood Cliffs, NJ, 1966.

Henry, D.P. *Medieval Logic & Metaphysics.* Hutchinson, London, England, 1972.

Hesse, Mary B. *Models and Analogies in Science.* Sheed and Ward, London, England, 1953.

Hesse, Mary B. "Models and Matter" in *Quanta and Reality.* American Research Council, New York, NY, 1962. Pp. 49-60.

Hesse, Mary B. "The Explanatory Function of Metaphor" in *Logic, Methodology, and Philosophy of Science.* Edited by Y. Bar-Hillel. Amsterdam, Holland, 1965.

Hickey, Thomas J. *Introduction to Metascience: An Information Science Approach to Methodology of Scientific Research.* Hickey, Oak Park, IL 1976.

Hickey, Thomas J. "The Indiana Economic Growth Model", *Perspectives on the Indiana Economy.* Vol. 2, (March, 1985). Pp. 1-11.

Hickey, Thomas J. "The Pragmatic Turn in the Economics Profession and in the Division of Economic Analysis of the Indiana Department of Commerce", *Perspectives on the Indiana Economy.* Vol 2, (September, 1985).Pp. 9-16.

Jammer, Max. *The Conceptual Development of Quantum Mechanics*. McGraw-Hill, New York, NY, 1966.

Jammer, Max. *The Philosophy of Quantum Mechanics*. John Wiley & Sons, New York, NY 1974.

Keynes, John Maynard. *The General Theory of Employment, Interest, and Money*. Harcourt, Brace & World, New York, 1964 [1936].

Klein, Lawrence R. *The Keynesian Revolution*. MacMillan, New York, 1966, [1947].

Kneale, William and Martha Kneale. *The Development of Logic*. Clarendon Press, Oxford, England, 1962.Kuhn, Thomas S. *The Copernican Revolution: Planetary Astronomy in the Development of Western Thought*. Harvard University Press, Cambridge, MA, 1957.

Kuhn, Thomas S. *The Structure of Scientific Revolutions*. The University of Chicago Press, Chicago, IL, 1970 [1962].

Kuhn, Thomas S. "The Function of Dogma in Scientific Research" in *Scientific Change*. Edited by A.C. Crombie. Basic Books, New York, NY, 1963.

Kuhn, Thomas S. *The Essential Tension*. The University of Chicago Press, Chicago, IL, 1977.

Kuhn, Thomas S. *Black-Body Theory and the Quantum Discontinuity* 1894-1912. The Clarendon Press, Oxford, England, 1978.

Land, Kenneth C. "Social Indicators" in *Social Science Methods*. Edited by R.B. Smith. The Free Press, New York, 1971.

Land, Kenneth C. "Formal Theory" in *Sociological Methodology* 1971. Jossey-Bass, San Francisco, CA, 1971.

Land, Kenneth C. and Marcus Felson. "A General Framework for Building Dymanic Macro Social Indicator Models", *American Journal of Sociology*. Vol. 82, No. 3 (Nov. 1976), Pp. 565-604.

Lande, Alfred. "Probability in Classical and Quantum Physics" in *Scientific Papers Presented to Max Born*. Hafner Publishing Company, New York, NY, 1954.

Lande, Alfred. *Foundations of Quantum Theory: A Study in Continuity and Symmetry*. Yale University Press, New Haven, CT, 1955.

Lande, Alfred. *New Foundations of Quantum Physics*. Cambridge University Press, Cambridge, England, 1965.

Litterman, Robert B. *Techniques for Forecasting Using Vector Autoregressions.* Federal Reserve Bank of Minneapolis, Minneapolis, MN, 1979.

Litterman, Robert B. *Specifying Vector Autoregressions for Macroeconomic Forecasting.* Federal Reserve Bank of Minneapolis, Minneapolis, MN, 1984.

Litterman, Robert B. "Above-Average National Growth in 1985 and 1986" *Federal Reserve Bank of Minneapolis Quarterly Review.* Vol. 8 (Fall, 1984), Pp. 3-7.

Litterman, Robert B. *Forecasting with Bayesian Vector Autoregressions - Four Years of Experience.* Federal Reserve Bank of Minneapolis, Minneapolis, MN 1985.

Litterman, Robert B. "How Monetary Policy in 1985 Affects the Outlook" *Federal Reserve Bank of Minneapolis Quarterly Review.* Vol. 9 (Fall, 1985), Pp. 2-13.

Lucas, Robert E. "Econometric Policy Evaluation: A Critique" in *Phillips Curve and Labor Markets.* Edited by K. Brunner and A. H. Meltzer. North Holland, Amsterdam, The Netherlands, 1976.

Lundberg, George A. "Contemporary Positivism in Sociology", *American Sociological Review.* Vol. 4 (February, 1939). Pp. 42-55.

Lundberg, George A. *Foundations of Sociology.* Greenwood Press, Westport, CT, 1964 [1939].

Lundberg, George A. *Social Research.* Longmans, Green & Co., New York, NY, 1942.

Lundberg, George A. *Can Science Save Us?* David McKay, New York, NY, 1961 [1947].

Mach, Ernst. *The Analysis of Sensations and the Relation of the Physical to the Psychical.* Translated by C. M. Williams and Sydney Waterlow. Dover Publications, New York, NY, 1959 [1885].

Mach, Ernst. *The Science of Mechanics: A Critical and Historical Account of Its Development.* Translated by Thomas J. McCormack. Open Court, LaSalle, IL, 1960 [1893].

Mach, Ernst. *Popular Science Lectures.* Translated by Thomas J. McCormack. Open Court, LaSalle, IL, 1943 [1898].

Mach, Ernst. *The Principles of Physical Optics: An Historical and Philosophical Treatment.* Translated by John S. Anderson and A.F.A. Young. Dover, New York, NY, 1953 [1921].

Merton, Robert K. *Social Theory and Social Structure.* The Free Press, New York, NY, 1968 [1947].

Merton, Robert K. *Sociology of Science: Theoretical and Empirical Investigations*. University of Chicago Press, Chicago, IL, 1973.

Mitchell, Wesley C. *Business Cycles*. Uniuversity of California Press. Los Angeles, CA, 1913. Part III published as *Business Cycles and Their Causes*. 1941.

Mitchell, Wesley C. "Quantitative Analysis in Economic Theory", *American Economic Review.* Vol. 15 (1925), Pp. 1-12.

Mitchell, Wesley C. *Measuring Business Cycles*. National Bureau of Economic Research, New York, 1946.

Mitchell, Wesley C. *Types of Economic Theory*. Vols. I and II. Edited by Joseph Dorfman. Augustus M. Kelley, New York, NY 1967.

Morgan,Mary S. *The History of Econometric Ideas*. Cambridge University Press, Cambridge, England. 1990.

Muth, John F. "Optimal Properties of Exponentially Weighted Forecasts", *Journal of the American Statistical Association,* Vol. 55 (1960), Pp. 299-306.

Muth, John F. "Rational Expectationa and the Theory of Price Movements" *Econometrica*. Vol. 29 (1961)Pp. 315-335.

Myrdal, Gunnar. *Against the Stream: Critical Essays on Economics.* Pantheon Books, New York, NY, 1973.

Nagel, Ernest. *The Structure of Science*. Harcourt, Brace, World, New York, NY, 1961.

National Science Foundation. *Science and Engineering Doctorates: 1960-1988*. Washington, DC, 1989.

Neurath, Otto. "Foundations of the Social Sciences", in I*nternational Encyclopedia of Unified Science,* Vol. II. Edited by Otto Neurath, Rudolf Carnap and Charles Morris. University of Chicago Press, Chicago, IL, 1944.

Neurath, Otto. *Empiricism and Sociology*. Edited by Marie Neurath and Robert S. Cohen. D. Reidel, Dordrecht, Holland, 1973.

Ohlin, Bertil G. *The Problem of Employment Stabilization Greenwood Press,* Westport, CT, 1977 [1949].

Orenstein, Alex. *Willard Van Orman Quine*. Twayne, Boston, MA, 1977.

Parsons, Talcott. *The Structure of Social Action*. The Free Press, New York, IL, 1968, [1937]. Vols. I and II.

Parsons, Talcott. *The Social System*. The Free Press, New York, IL, 1964, [1951].

Parsons, Talcott. "On Building Social Systems Theory" in *The Twentieth-Century Sciences: Studies in the Biography of Ideas.* Edited by Gerald Holton. W.W. Norton, New York, NY, 1972 [1970]. Pp. 99-154.

Peirce, Charles S. *Philosophical Writings of Peirce*. Edited and Selected by Justice Buchler. Dover, NY, 1955 [1940].

Peirce, Charles S. *Essays in the Philosophy of Science*. Edited with introduction by Vincent Thomas. Bobbs-Merrill, New York, NY, 1957.

Popper, Karl R. *The Poverty of Historicism.* Harper and Row, New York, NY, 1961 [1957].

Popper, Karl R. *Conjectures and Refutations*: The Growth of Scientific Knowledge. Basic Books, New York, NY, 1963.

Popper, Karl R. *The Logic of Scientific Discovery.* Harper and Row, New York, NY, 1968 [1959,1934].

Popper, Karl R. "Normal Science and its Dangers", in *Criticism and the Growth of Knowledge*. Edited by Imre Lakatos and Alan Musgrave. Cambridge University Press, Cambridge, England, 1970. Pp. 51-58.

Popper, Karl R. "Autobiography of Karl Popper" in *The Philosophy of Karl Popper*. Edited by Paul A. Schilpp. Two Vols. Open Court, LaSalle, IL, 1972. Pp. 3-181.

Popper, Karl R. *Objective Knowledge: An Evolutionary Approach.* Clarendon Press, Oxford, England, 1974.

Popper, Karl R. "The Rationality of Scientific Revolutions" in *Problems of Scientific Revolution:* Progress and Obstacles to Progress in the Sciences. Edited by Rom Harre. Clarendon Press, Oxford, England, 1975.

Popper, Karl R. and John C. Eccles. *The Self and Its Brain.* Springer International, New York, NY, 1977.

Popper, Karl R. *Realism and the Aim of Science.* Edited by W.W. Bartley, III. Rowman and Littlefield, Totowa, NJ, 1983.

Popper, Karl R. *The Open Universe.* Edited by W.W. Bartley, III. Rowman and Littlefield, Totowa, NJ, 1983.

Popper, Karl R. *Quantum Theory And The Schism in Physics*. Edited by W.W. Bartley, III. Rowman and Littlefield, Totowa, NJ, 1983.

Quine, W.V.O. *Methods of Logic.* Harvard University Press, Cambridge, MA, 1982 [1950].

Quine, W.V.O. *From A Logical Point of View.* Harvard University Press, Canbridge, MA, 1980 [1953].

Quine, W.V.O. *Word and Object.* M.I.T. Press, Cambridge, MA, 1960.

Quine, W.V.O. *Selected Logic Papers.* Random House, NY, 1966.

Quine, W.V.O. *Ontological Relativity.* Columbia University Press, NY, 1969.

Quine, W.V.O. *Philosophy of Logic.* Harvard University Press, Cambridge, MA, 1970.

Quine, W.V.O., and J.S. Ullian. *The Web of Belief.* Random House, NY, 1970.

Quine, W.V.O. *The Roots of Reference.* Open Court, LaSalle, IL, 1974.

Quine, W.V.O. *Theories and Things.* Belknap Press, Cambridge, MA, 1981.

Quine, W.V.O. "Autobiography of W. V. Quine" in T*he Philosophy of W.V. Quine*. Edited by Lewis Edwin Hahn and Paul Arthur Schilpp. Open Court, LaSalle, IL, 1986, Pp. 3-46.

Quine, W.V.O. *The Time of My Life*: An Autobiography. MIT Press, Cambridge, MA, 1985.

Quine, W.V.O. *Quiddities: An Intermittently Philosophical Dictionary.* Belknap Press, Cambridge, MA, 1987.

Quine, W.V.O. and Rudolf Carnap. *Dear Carnap, Dear Van: The Quine-Carnap Correspondence and Related Works.* Edited with introduction by Richard Creath. University Of California Press, Berkeley, CA, 1990.

Radnitzky, Gerald. *Contemporary Schools of Metascience.* Vols. I and II. Akademiforlagst, Goteborg, Sweden, 1968.

Redman, Deborah A. *Economic Methodology: A Bibliography with References to Works in the Philosophy of Science.* Complied by Deborah A. Redman. Greenwood Press, Westport, CT, 1989.

Rogers, Rolf E. *Max Weber's Ideal Type Theory.* Philosophical Library, NY, 1969.

Sargent, Thomas J. "Rational Expectations, Econometric Exogeniety, and Consumption" J*ournal of Political Economy.* Vol. 86 (August, 1978), Pp. 673-700.

Sargent, Thomas J. "Estimating Vector Autoregressions Using Methods Not Based on Explicit Economic Theories" *Federal Reserve Bank of Minneapolis Quarterly Review,* Vol. 3. (Summer, 1979). Pp. 8-15.

Sargent, Thomas J. "After Keynesian Macroeconomics" in *Rational Expectations and Econometric Prectice.* Edited by Robert E. Lucas and Thomas J. Sargent. University of Minnestoa Press, Minneapolis, MN, 1981.

Schroedinger, Erwin. *Science and the Human Temperament.* Translated by James Murphy and W.H. Johnston. W.W. Norton & Company, NY, 1935 [German, 1932].

Schroedinger, Erwin. *My View of the World.* Translated by Cecily Hastings. Ox Bow Press, Woodbridge, CT, 1983 (1961).

Schumpeter, Joseph A. *Essays on Economic Topics of Joseph A. Schumpeter.* Edited by Richard V. College. Kennikat Press, Port Washington, NY, 1951.

Shannon, Claude E. and Warren Weaver. *The Mathematical Theory of Communication.* The University of Illinois Press, Urbana, IL, 1949.

Shreider, Yu A. "On the Semantic Characteristics of Information," *Information Storage and Retrieval.* Vol. II (August, 1965).

Shreider, Yu A. "Semantic Aspects of Information Theory" in *On Theoretical Problems On Informatics.* All-Union Institute for Scientific and Technical Information, Moscow, USSR, 1969.

Shreider, Yu A. "Basic Trends in the Field of Semantics" in *Statistical Methods in Linguistics.* Sprjakfhorlaset Skriptor, Stockholm, Sweden, 1971.

Simon, Herbert A. T*he Sciences of the Artificial.* MIT Press, Cambridge, MA, 1981 [1969].

Simon, Herbert A. and Laurent Silkossy. *Representation and Meaning with Information Processing Systems.* PrenticsHall, Englewood Cliffs, NJ, 1972.

Simon, Herbert A. *Models of Discovery And Other Topics in the Methods of Science.* D. Reidel, Dordrecht, Holland, 1977.

Simon, Herbert A., et al. *Scientific Discovery: Computational Explorations of the Creative Process.* The MIT Press, Cambridge, MA, 1987.

Simon, Herbert A. *Models of My Life.* Basic Books, Dunmore, PA, 1991.

Sims, Christopher A. "Macroeconomics and Reality" *Econometrica.* Vol. 48. (Jan, 1980), Pp. 1-47.

Sims, Christopher A. "Are Forecasting Models Usable for Policy Analysis?" *Federal Reserve Bank of Minneapolis Quarterly Review.* Vol 10 (Winter, 1986), Pp. 2-16.

Sonquist, John A. "Problems in the Analysis of Survey Data, and A Proposal" *Journal of the American Statistical Association.* Vol 58, (June, 1963), Pp. 415-35.

Sonquist, John A. and James N. Morgan. *The Detection of Interaction Effects: A Report on a Computer Program for the Selection of Optimal Combinations of Explanatory Variables.* Survey Research Center, Institute for Social Research, University of Michigan, Ann Arbor, MI, 1964.

Sonquist, John A. "Simulating the Research Analyst" in *Social Science Information.* Vol VI, No 4 (1967), Pp. 207-215.

Sonquist, John A. *Multivariate Model Building: Validation of a Search Strategy.* Survey Research Center, Institute for Social Research, University of Michigan, Ann Arbor, MI, 1970.

Sonquist, John A., Elizabeth L. Baker, and James N. Morgan. *Searching for Structure: An Approach to Analysis of Substantial Bodies of Micro-Data and Documentation for a Computer Program.* Survey Research Center, Institute for Social Research, University of Michigan, Ann Arbor, MI, 1973.

Sonquist, John A. "Computers and the Social Sciences" *American Behavioral Scientist.* Vol 20, No. 3 (1977), Pp. 295-318.

Sonquist, John A. and Francis M. Sim. "'Retailing' Computers to Social Scientists" *American Behavioral Scientist.* Vol 20, No. 4 (1977), Pp. 319-45.

Stone, Richard. *Demographic Accounting and Model-Building.* Organization for Economic Co-operation and Development, Paris, 1971.

Tarski, Alfred. *Logic, Semantics, Metamathematics.* Trans. by J.H. Woodger. Clarendon Press, Oxford, England, 1956.

Todd, Richard M. and William Roberds. "Forecasting and Modeling the U.S. Economy" F*ederal Reserve Bank of Minneapolis Quarterly Review.* Vol. 11 (Winter, 1987), Pp. 7-20.

Von Weizsacker, Carl F. "The Copenhagen Interpretation" in *Quantum Theory and Beyond.* Edited by T. Bastin. Cambridge University Press, Cambridge, England, 1971.

Weber, Max. *The Methodology of the Social Sciences.* Translated and edited by Edward A. Shils and Henry A. Finch. The Free Press, Glencoe, IL, 1949.

Weber, Max. *Critique of Stammler*. Translated by Guy Oakes. The Free Press, NY, 1977.

Whorf, Benjamin Lee. *Language, Thought and Reality*. Edited by John B. Carroll. The M.I.T. Press, Cambridge, MA, 1956.

Wittgenstein, Ludwig. *Tractatus Logicus-Philosophicus*. Trans. by D.F. Pears and B.F. McGuinness. Macmillan, New York, NY 1961.

Wittgenstein, Ludwig. *Philosophical Investigations*. Trans. by G.E.M. Anscombe. Macmillan, New York, NY, 1953.

INDEX OF NAMES